The Theory of Finslerian Laplacians and Applications

Mathematics and Its Applications

Managing Editor:

M. HAZEWINKEL
Centre for Mathematics and Computer Science, Amsterdam, The Netherlands

Volume 459

The Theory of Finslerian Laplacians and Applications

edited by

Peter L. Antonelli

Department of Mathematical Sciences,
University of Alberta,
Edmonton, Alberta, Canada

and

Bradley C. Lackey

Department of Mathematical Sciences,
University of Alberta,
Edmonton, Alberta, Canada

SPRINGER SCIENCE+BUSINESS MEDIA, B.V.

A C.I.P. Catalogue record for this book is available from the Library of Congress.

DOI 10.1007/978-94-011-5282-2

Printed on acid-free paper

TABLE OF CONTENTS

SECTION III. Stochastic Lagrange Geometry

SECTION IV. Mean-Value Properties of Harmonic Functions

SECTION V. Analytical Constructions

PROLOGUE

1 Early History of the Laplacian

Laplace's equation was formed within one of natural philosophy's most noble pursuits: understanding the nature of gravity. Two of the main investigations of eighteenth century science were to verify Newton's inverse square law of gravitational attraction, and with that determine the shape of the Earth. Newton, himself, claimed that the Earth was flattened about the poles; the ratio of the length of the equator to a that of a typical longitude he predicted to be $\frac{231}{230}$. The French Academy of Sciences was eager to put this prediction to the test, and in the 1730's sent expeditions to Lapland and Peru to measure the gravitational attractions. The former expedition was led by Maupertuis and Clairaut, who confirmed Newton's predicted shape, but claimed the ratio was $\frac{179}{178}$ (which is far less accurate than Newton's).

In 1740, Maclaurin published a proof that a fluid of constant density with constant angular velocity would have an oblate spheriod as an equilibruim state. He later gave an expression for the gravitational attraction of such a body on an external particle lying on the axis of revolution. Both these results relied on geometric methods which proved of limited use as they could only be applied to such special configurations. Similar results were obtained by Clairaut about the same time, but using analytical means. These are compiled in his famous book, *Théorie de la figure de la terre* (1743).

The key idea previous authors failed to apply, was the use of a potential to derive the force. The concept of using a potential function was already popular in fluid dynamics. In fact, the term "potential function" appears in Bernoulli's text, *Hydrodynamica* (1738). The first explicit appearance of Laplace's equation occured in Euler's paper, "Principles of the Motion of Fluids" (1752). Euler was unable to analyze the equation, but did provide some simple polynomial solutions.

Little progress was made in determining the force of gravitational attraction until 1782, when Legendre wrote the paper, "Recherches sur l'attraction des sphéroïdes." Oddly enough, this paper did not appear in print until 1785, the same year of Laplace's famous paper on the subject. Legendre proved the following statement: if the attraction of a solid of revolution in known along the axis of revolution, then it is known for every external point. By introducing what are now called the Legendre polynomials, he could compute the radial component of force based on the given data. With this, he was able to derive the potential, and compute the remaining components of the force.

Inspired by Legendre's paper, Laplace considered the general problem of gravitational attraction by a body in his paper, "Théorie des attractions des sphéroïdes et de la figure des planètes" (1785). Without any justification, he begins by stating that the potential outside the body must satisfy Laplace's equation (in spherical coordinates). He then assumes the potential is of the form $V = \frac{1}{r}U_0(\theta, \phi) + \frac{1}{r^2}U_1(\theta, \phi) + \cdots$, and in effect separates the variables. The U_n he wrote in terms of the Legendre polynomials; much later, Lord Kelvin dubbed these functions the *spherical harmonics*. Finally, he expands the function defining the boundary of his "spheroid" in terms of these harmonics, solving for the potential, entirely.

Legendre and Laplace continued their study of the Legendre polynomials and the spherical harmonics through the 1780's. Laplace only made one significant error in his gravitational theory: he assumed that Laplace's equation holds for the potential on the interior of the body, as well. This was corrected by Poisson in 1813, introducing what is now called Poisson's equation. Yet, Poisson even criticized his own derivation as not being sufficiently rigorous.

Little effort was made to study the general properties of harmonic functions until Green's privately published booklet in 1828. In this he derived what we now call Green's identity. Later, Green also provided the first existence and uniqueness proof for the Dirichlet problem. His idea was to minimize the value

$$\int_{\Omega} \left[\left(\frac{\partial f}{\partial x^1} \right)^2 + \cdots + \left(\frac{\partial f}{\partial x^n} \right)^2 \right] dx^1 \cdots dx^n .$$

Riemann would later call this the *Dirichlet principle*, although Green was the first to introduce it. As a note, Green worked in n-dimensions, radical thinking at the time, because credit for the formation of n-dimensional geometry is given to Grassmann, with *Ausdehnungslehre* (1844).

In 1833, one of the most significant conceptual steps in potential theory was taken by Lamé. He pointed out in the context of the heat equation, that the mathematicians of his time were only analyzing the Laplacian in rectangular coordinates. He considered this of limited value and proposed that one build a coordinate system appropriate to the problem at hand, and then express the Laplacian in that system. He constructed various types of orthogonal curvilinear coordinates in $\mathbb{R}^3$, and presented the form of the Laplacian in each. In a later work, he gave the general technique for transforming the Laplacian into any such coordinate system; this procedure is the same algorithm taught in engineering science, today.

Beltrami, who was quite familiar with Riemann's differential geometry, began the study of differential invariants in the 1860's. Lamé, as a side result, had shown that the two expressions

$$\left(\frac{\partial f}{\partial x}\right)^2 + \left(\frac{\partial f}{\partial y}\right)^2 + \left(\frac{\partial f}{\partial z}\right)^2$$

and

$$\frac{\partial^2 f}{\partial x^2} + \frac{\partial^2 f}{\partial y^2} + \frac{\partial^2 f}{\partial z^2}$$

retain their form under orthogonal rotation of the coordinates. Using a method he learned from Jacobi, Beltrami found the invariant form of both these objects in the context of Riemannian manifolds. The former is just $g^{jk}\frac{\partial f}{\partial x^j}\frac{\partial f}{\partial x^k}$, and the later is what is now called the Laplace-Beltrami operator

$$\Delta f = \frac{1}{\sqrt{g}}\frac{\partial}{\partial x^j}\left(\sqrt{g}\, g^{jk}\frac{\partial f}{\partial x^k}\right),$$

where we have written $g = \det(g_{jk})$.

Note that the Dirichlet principle now just reads: the functions which minimize the integral of the first invariant form are those for which the second invariant form vanish. The first invariant has a canonical generalization to Finsler spaces; this is the approach taken by Shen, Section IV. Yet, as this form is not quadratic, the ensuing Shen Laplacian is nonlinear.

Euler's work in hydrodynamics spawned another avenue of investigation. Starting with the question of when one can interchange the order of integration, Cauchy developed the theory of complex integration in his seminal 1814 paper, "Mémoire sur la théorie des intégrales définies". His approach was to introduce the Cauchy-Riemann equations as integrability criteria for finding an antiderivative of a complex valued function. He showed that functions which satisfy these equations, (Briot and Bouquet would later call these *holomorphic*), have harmonic real and imaginary parts. Riemann provided the converse relationship between harmonic and holomorphic functions, showing that a harmonic function naturally induces a holomorphic function. Actually, his interest in potential theory centered around Riemann surfaces and Abelian integrals. He introduced the notion of connectivity of a Riemann surface, which Betti would later generalize and which in turn Poincaré would build into *homology theory*. This connectivity he related to the genus of the surface, and used these ideas to give the values of Abelian integrals as linear combinations of "periods". This was just a very early precursor of what would become Hodge-deRham theory.

The generalization of potential theory from functions to differential forms is usually credited to Poincaré, deRham, and Hodge; yet, there is no doubt the essential ideas in the theory were known to Volterra, but versed in his version of the calculus of variations. Volterra viewed Cauchy's integral theorem in this context as the statement: the "function of a line" associated to a holomorphic function vanishes on any closed curve. In 1889, he even took to extending this theorem to higher dimensions. In his paper, [V1], he constructs differential forms in three dimensions, expressing them as alternating tensors, and provides the now ubiquitous translation of the vector calculus operations, into this setting. He provides a proof a Stokes' theorem, which he states is the natural generalization of Cauchy's theorem. His next paper, [V2], extends these results to n-dimensions. Volterra's third paper on the subject, [V3], contains the deepest results. The first theorem of this paper is precisely Poincaré's lemma! He later defines the Laplacian of a form as $\Delta\theta = (d^*d + dd^*)\theta$, and considers the structure of harmonic forms, again in terms of his functionals.

2 Mean-value properties of harmonic functions

One of the most important properties a harmonic function satisfies is the mean value property: the value of the function at a point is equal to the average of the function over a "sufficiently small" sphere about the point. We say "sufficiently small" as the sphere and its interior must lie in the domain of the function. Undoubtedly, this property was suspected to hold for physical reasons, from the earliest formulation of potential theory. But it was Gauss, [Ga], who explicitly exhibited it.

In fact, this property characterizes harmonic functions as was shown in 1906 by Koebe, [Koe]. On Riemannian manifolds, harmonic functions do not satisfy the mean value property, in general. Yet, they almost have this property: if $S_\epsilon(p)$ is the geodesic sphere about p with radius ϵ, then for any harmonic function, f, the difference of $f(p)$ and the average of f on

$S_\epsilon(p)$ is of order ϵ^4. In 1982, Gray and Willmore provided a proof that this "approximate" mean value property characterized harmonic functions on Riemannian manifolds, [GW], as well as computing some of the coefficients of the ϵ^k, for $k \geq 4$. These terms are complicated functions of the curvature and its covariant derivatives. Spaces where all these coefficients vanish, and hence the mean value property is satisfied, are called *harmonic spaces* and have been well-studied, see for instance [RWW] and citations therein.

The use of the mean value property to analyze diffusions in inhomogeneous and anisotropic media is immediate. Let u measure the quantity which is diffusing, and p be a point in our medium. Then in every direction emanating from p, we use the physics of the situation to determine the flow in that direction induced by a unit difference in u. With some regularity assumptions, the equiflow surfaces near p will be concentric spheroids, and with perhaps more regularity assumptions, we can take the limit and get a Finsler function which measures the "infinitesimal flow" in all directions from p. This function will be induced from a Riemannian metric when the concentric spheroids are asymptotically ellipses centered at p. If S_ϵ represents the Finsler geodesic ϵ-sphere about p, then the diffusion equation is

$$\frac{\partial u}{\partial t} = -\lim_{\epsilon \to 0} \frac{1}{\epsilon^2} \left[u(p) - \frac{1}{\operatorname{Vol}(S_\epsilon(p))} \int_{S_\epsilon(p)} u(x)dx \right],$$

where some dimensionality constants may need to be inserted. The analysis of the right side of this equation is precisely the concern of Centore's paper, Section III. Considering the relationship of isotropic transport processes and diffusions, it not surprising that the Laplacian obtained by Antonelli and Zastawniak, Section III, is very similar to that of Centore. It should be pointed out, however, that the use of the mean value property is not restricted to physical theories; famous biologist Goodwin, [Go], has used this notion with great success in modeling development of biological organisms.

3 Brownian motion and stochastic analysis

The erratic motion of tiny particles suspended in liquid is known as *Brownian motion*, after the botanist who discovered the phenomenon in, 1828. In 1905, A. Einstein and M. Smoluchowski realized this behaviour could be described by parabolic 2nd order p.d.e.'s whose fundamental solutions could be interpreted as *transition densities* for Brownian motion, modelled as a *Markov process* in the state (i.e. position) space. From this perspective the covariance of position changes could themselves be dependent on the position, and according to Kolmogorov (1937), [Kol], the diffusion equation could be cast in *geometrically invariant form*, using the inverse of the *covariance tensor* as the *fundamental metric tensor* of a *Riemannian geometry*. Thus, the *Laplace-Beltrami operator*, from Riemannian geometry, acting on the transition density for Brownian motion, could yield the time rate of change of that density. In particular, as time went to infinity that density would become a *harmonic function* in this intrinsic Riemannian geometry.

This established a link between ergodic theory and the theory of harmonic functions, foreshadowing the theory of A. Milgram and P. Rosenbloom (1950), who studied harmonic p-forms on a compact, oriented, Riemannian manifold without boundary, using heat kernel methods and proved the Hodge Theorem, [MR], to the effect that each real cohomology class in dimension p has a representative p-form realized as an ergodic heat equilibrium and which is annihilated by the Laplace-Beltrami operator. Could this theory be of interest to engineers or physicists?

In many applications after World War II, one was faced with a system of deterministic ode's which had to be augmented with "noise" in order to model random perturbations from external influences, e.g. telecommunications, finance, cybernetics, etc. Led by French physicist, Paul Langevin, the subject of these so-called *stochastic differential equations* (Sde's) started to develop. Spurred on by the rigorous work of K. Îto and H. McKean [IMc], [Mc], N. Ikeda and S. Watenabe [IWa], and K.D. Elworthy [El], the subject grew over the next thirty years into the modern edifice that is today, *Stochastic Riemannian Geometry*, (see [Pin1], [Pin2] for recent reviews). At this stage, such topics as the role of Riemann scalar curvature in Feynman-Kac solutions to parabolic PDE's and in Onsager-Machlup formulas, in nonequilibrium statistical thermodynamics [Gra1], [Gra2] could be rigorously discussed and the techniques could be used on applied problems like filtering [AES], large deviations [Va], and quantum mechanics via the stochastic mechanics of E. Nelson [Nel].

In another direction, however, the work of Antonelli and his associates in the 80's on coral reef ecology and evolution, chemical warfare in plants and marine invertebrates [Ant1], [Ant2], [Ant3], [Ant4] and in epidemiology, marked a return to the biological world from which Brownian motion had originally come. It is in this applied context that Finslerian ideas would suddenly emerge, [ABL1], [ABL2], [ABKS]. To explain how this occurred we will use the terminology of Volterra-Hamilton systems [AIM], [AB], [AZ]. Letting (X^i, N^i) denote the natural phase space coordinates in a local coordinate system of the tangent bundle TM^n, consider the 2nd

order system

$$\sum : \begin{cases} \frac{dX^i}{dt} = k_{(i)}, N^i \quad \text{(not summed)} \\ \frac{dN^i}{dt} = -\Gamma^i_{jk} N^j N^k + r^i_j N^j + e^i, \end{cases}$$

where all coefficients (possibly) depend on X^i, N^i, t and the n^3 functions Γ^i_{jk} are homogeneous of degree zero in the variables N^i, and with smooth initial conditions X^i_0, N^i_0, t_0. The coordinates X^i are *Volterra's production variables* whose constant per capita rate of increase is k_i, while the second part of the system, often called the *population equations*, is a description of how different species or subpopulations of a colonial organism (i.e. castes) $N^i \geq 0$, *grow* (r^i_j), *interact* (Γ^i_{jk}) and *react* (e^i) to external influences. The condition that Γ^i_{jk} are functions of *ratios* of N^i, (i.e. degree zero) signals the presence of so-called *social interactions*, which are higher-order, density-dependent effects. Whereas, classical ecological theory would have the Γ^i_{jk} merely constants, the theory of density-dependent social interactions, introduced by the great ecologist G.E. Hutchinson in 1946, [Hutch1], [Hutch2], found experimental verification in subsequent work in the 70's of Wilbur, Hairston and others [Wilb1], [Wilb2], [Hars]. In 1991, working with marine biologist Roger Bradbury, Antonelli found that Hutchinson's theory should be recast using zero degree homogeneous interactions (i.e. Γ^i_{jk}) in order to be consistent with the data. Unfortunately, the approach of Hutchinson was mathematically intractable and his theory of social interactions lay fallow for more than 40 years. *Thus,* 1991, *marks the birth of the modern theory of Hutchinsonian social interactions, and the realization that this theory must be Finslerian in order to account for these social interactions.* Moreover, such interactions could be a consequence, in models of cost-effective growth and physiology as occur in superorganisms like siphonophores and ants, but could also appear in other areas, such as disease outbreak models. One of the best examples of this is given in the paper on myxomatosis, the European Wild Rabbit disease, in Section I of the present collection.

It should be noted that the Volterra-Hamilton systems $\sum$, have other interpretations of importance. For example, Gabriel Kron (1934) used them to describe the unified theory of rotating electrical machines, [Kr1], [Kr2]. Also, there is the interpretation of $\sum$ given by K. Kondo on a problem of the flight stability of an airplane, [Kondo]. Our general theory of noise naturally applies to these and other possible interpretations.

In order to model noise in a Volterra-Hamilton system $\sum$, it may not be enough to merely add white noise to the population equations. Using an ecological example, the reason can be simply illustrated. A coral grows mainly due to the sun's energy which induces photosynthesis in a polyp's endosymbiont algae, allowing the entire colony to produce a calcium carbonate exoskeleton. Since the coral colony is very sensitive to sunlight variations caused by erratic cloud cover variations, the effect on the deposition of calcium carbonate (X^i) is also highly irregular, or noisy. Yet, noise added to the production equations (the first system in $\sum$) must reflect the interactions Γ^i_{jk} through which the sunlight fluctuations are propagated, for these Γ^i_{jk} express the physiology which ultimately produces the exoskeleton.

The *noise ansatz* used for Volterra-Hamilton systems in which Γ^i_{jk} are Levi-Civitá coefficients of a *Riemannian metric* is just the usual ansatz for the *population equations* (i.e. the white noise ansatz). On the other hand, for the production process, *a point in state space* (X^i) *will be displaced by a Riemannian distance proportional to the magnitude of the perturbation.* In the case where the geometry is *Finsler*, the fundamental metric tensor and interaction coefficients Γ^i_{jk} both depend on ratios of N^i. The noise added to the population equation is now *Minkowskian* and the distance concept of the noise ansatz is *Finslerian* for the *production equations*.

It can now be understood why the very first theory of Finslerian Laplacians was formulated on the slit tangent bundle $\widetilde{TM}^n$, (i.e. with zero-section removed) rather than on the base manifold itself, as was done later by David Bao and Brad Lackey, motivated by finding Finslerian Hodge Theory. The tangent bundle with the Sasaki lift Riemannian metric is the natural setting for Finsler Geometry as developed by Radu Miron and his school [Mir1], [Mir2], and which is called *Lagrange geometry*. In the context of the Lagrange spaces we have the three papers in Section III on generalizations of the Finslerian h- and hv-diffusion theory.

A main result in the Stochastic Finslerian theory is the *Stochastic Imbedding Theorem* which expresses the Finslerian Laplacian in terms of the Laplace-Beltrami operator plus drift, on $\widetilde{TM}^n$ with Sasaki lift Riemannian metric. The drifts are induced by generally nonzero torsions C and P of the Cartan connection. This result holds for diffusions restricted to the Indicatrix bundle (the Finslerian sphere bundle), as well. The results for $\widetilde{TM}^n$ are generalized to a class of Lagrange manifolds called *φ-Lagrange spaces*. Also, the hv-diffusion is generalized to vector bundles, while the h-diffusion is generalized to a different class of Lagrange spaces. Results on stochastic imbedding are of importance for *short time asymptotics* of Finslerian or Lagrangian diffusions, because they reduce the problem to a Riemannian one, albeit, with torsion-induced drifts, which is well-understood. The remaining problem would be one of evaluation of line integrals of these drifts which enter the asymptotic kernel [Mol].

In Section II we obtain Feynman-Kac solutions to hv-diffusions and in Section I prove an Onsager-Machlup formula for a class of Finsler spaces which arise in applications in biology, the so-called *positive definite Berwald spaces*. These formulas involve the Berwald scalar curvature $\mathbb{R}$, as in the asymptotic sojourn time formula for a Riemannian diffusion, [TW], or the nonequilibrium statistical thermodynamics [Gra1], [Gra2]. We prove an Onsager-Machlup formula for h-diffusions only, by introducing a *stochastic version of kinetic energy*, called *quadratic dispersion*, which in normal coordinates has the form

$$\frac{1}{2}\, g_{ij}(x_s, y_s) E_s\{x^i_{s+h} x^j_{s+h}\}$$

where E_s denotes expectation conditional on h-diffusion (x_s, y_s) on $\widetilde{TM}^n$. Expanding in powers of h, one obtains the famous, $R/12$, as coefficient of the quadratic term.

In Section IV, #1, there is constructed a Finslerian Laplacian Δ_{AZ} on the based manifold M^n using a generalization of Mark Pinsky's isotropic transport on the tangent bundle [Pin1]. Using arguments of David Bao and Brad Lackey, [BL], it is shown that each qth de Rham cohomology class of M^n has a q-form representative which is annihilated by Δ_{AZ}.

4 Hodge-deRham theory

Although Volterra's work in the late 1880's hinted at the work we attribute to Hodge, two major constructions needed to occur before Hodge could recognize the significance of the Laplacian: the rigorization of the exterior calculus and the formulation of deRham's cohomology.

Differential forms arose in the early 1800's under the name of total differential equations. The pioneer into their study was Pfaff, [Pf]; even today, differential one-forms are sometimes referred to as Pfaffian forms. The exterior derivative was introduced by Frobenius, [Fr], under the name "bilinear covariant" in his study of Pfaff's problem of the equivalence of forms. It was Cartan, however, who recognized the value of the exterior calculus, converting it from a peculiar topic in partial differential equations to a powerful algebraic tool. All the essential elements of the exterior differential calculus appear in modern form (except for some minor changes in notation), in his paper, [Ca1]. The global issues, such as integration of forms, appeared in Cartan's text, [Ca2].

Poincaré provided the other major step in the harmonic theory of forms in his paper, "Analysis Situs", [Po]. Following Riemann and Betti, he considered a collection submanifolds (of a fixed dimension) trivial if their union formed the boundary of a submanifold of one higher dimension. Unlike his precessors, Poincaré allowed for arbitrary sums and (using the notion of orientation invented by Klein and von Dyck) differences of submanifolds, constructing what we now call the homology groups with integer coefficients.

In section seven of "Analysis Situs", Poincaré makes the first clear progress into the theory of harmonic forms. Just as Volterra had, Poincaré considers skew-tensors which satisfy an integrability condition (that is, what we now call closed forms). With Stokes' theorem, he proves that the integral of such a form over an arbitrary submanifold of the proper dimension yields a linear combination of the "periods".

Based on the results of Poincaré's "Analysis Situs", Cartan predicted the results which we now term deRham's theorems, in the context of Lie groups, [Ca3]. DeRham, aware of Cartan's claims, proved these theorems in chapter three of his thesis, [dR1]. The main result is in essence: given any basis for the homology of a given dimension, and a period for each chain therein, there is a closed form whose integral over each of the chains yields precisely the desired periods.

Hodge, who was familiar with the work of Cartan, Volterra, and deRham, formulates the final link in his paper, [Ho1]. Working with a Euclidean metric, he constructs what we now call the Hodge star, and defines how the Laplacian operates on closed differential forms. He then shows there is a unique harmonic form which has the properties indicated by deRham. In later paper, [Ho2], he allows for arbitrary Riemannian metrics, and also changes the definition of a harmonic form to one which is closed and coclosed. In section three of this paper, he presents what we call the Hodge decompositions theorem. In the two previous papers, Hodge worked exclusively in the analytic case, as his existence proof required this. But under the suggestion of H. Kneser, he uses Hilbert's parametrix method to provide the existence proof, [Ho3]; many modern treatments of Hodge's decomposition theorem follow this line of argument.

Kodaira independently repeated much of Hodge's work, [Kod1]. In a later paper, [Kod2], he unified his constructions with those of Hodge and deRham, and provided another existence proof of harmonic forms using Weyl's method of orthogonal projection. Since Hodge's work, the theory of harmonic forms has undergone many generalizations and modernizations, see [dR2], [dRK], or [Ho4] for some early such ones.

Hodge's work on harmonic integrals indicated the topological side of harmonic forms; Weitzenböck showed the relationship between the Laplacian on forms and curvature, [We] or [dR2]. Bochner extended this relationship, proving that the Ricci curvature has a great effect on harmonic one-forms (he considered harmonic vector fields), showing that if the curvature is nonnegative, then harmonic one-forms are parallel, and if the curvature is positive, then the only harmonic one-form is zero, [Bo]. Bochner's technique has been used repeatedly by many authors, for instance Lichnerowicz in the case of spinors, [Li], and Lackey, in Section V, for general elliptic complexes.

5 Modern topics

Perhaps one of the most significant construction associated to the theory of harmonic forms was proposed by Atiyah and Bott, [AtBo]. They formulated the elliptic complex, which is the abstract version of the exterior algebra. This object has just enough of the features of the exterior algebra for one to formulate the Hodge decomposition theorem. The importance of this construction is realized in that many new ideas arising from theoretical physics fall naturally into the domain of elliptic complexes. We will consider two here: the signature complex and the spin complex.

Spinors are not a particularly new idea; Cartan discovered them as a peculiar representation of the orthogonal group, [Ca4]. While studying relativistic quantum mechanics, Dirac rediscovered spinors and their natural differential operator (now called the Dirac operator). Dirac's formulation came from the attempt to "square-root" the Laplacian. He found that one may write $\Delta = D^2$, D being the Dirac operator, but this equation needs to be applied to spinors (or another representation of the Clifford algebra) to hold.

Atiyah and Singer recognized spinors and the Dirac operator as an example of an elliptic complex, and analyzed it with their index theorem. Later, Witten used spinors to give an elegant proof of the positive energy theorem, [Wi]; the original proof given by Shoen and Yau, [SY] used complicated estimates. Very recently, Seiberg and Witten introduced a pair of differential equations involving spinors and a curvature from within the framework of quantum field theory, [SW1], [SW2]. It has been recognized that these equations, and the moduli space of their solutions, are a powerful tool in the study of four-manifolds, see [Mo] and reference therein.

The practical use of spinors in Finsler geometry is not in its final form. Flaherty, Section V, provides the construction of horizontal spinors and their Dirac operator. The Weitzenböck formula for the square of the Dirac operator involves the torsion of Cartan's connection, so appears quite intractable. Nonetheless, the corresponding equation for Riemannian manifolds with only a $Spin^c$ structure involves a complicated curvature term; this is one way to use the Seiberg-Witten equations: if the Ricci curvature is positive, then the moduli space only contains trivial solutions.

The signature complex was introduced and analyzed by Hirzebruch, [Hi]. This complex is built from the spaces of self-dual and anti-self-dual (inhomogeneous) forms being mapped to each other by the differential operator $d + d^*$ (one may note the close relationship between this and the spin complex of a Kähler manifold). The index of this operator is the signature of the manifold, hence the name.

Self-dual and anti-self-dual forms are more than a mathematical curiosity. Minima of the Yang-Mills functional (called instatons or pseudoparticles) are realized by connections whose curvatures are self-dual or anti-self-dual. Mathematically, a practical use of such forms is to avoid constructions with spinors. To this end, Lackey, Section V, proves a Finslerian analogue of Lichnowicz's vanishing theorem using the signature complex rather than the spin complex.

P.L. Antonelli and B.C. Lackey
Edmonton, Canada

February 1998

References

[AB] Antonelli, P.L. and Bradbury, R. (1996) Volterra-Hamilton Models in the Ecology and Evolution of Colonial Organisms, *Series in Mathematical Biology and Medicine*, World Scientific Press, Singapore.

[AtBo] Atiyah, M.F. and Bott, R. (1967) A Lefschetz Fixed Point Formula For Elliptic Complexes I.II., *Ann. Math.*, **86**, 347-407; *Ann. Math.*, **88** (1968), 451–491.

[ABKS] Antonelli, P.L., Bradbury, R., Krivan, V. and Shimada, H. (1993) A Dynamical Theory of Heterochrony: Time-Sequencing Changes in Ecology, Development and Evolution, *J. Biol. Systems*, **1**, 451–487.

[ABL1] Antonelli, P.L., Bradbury, R. and Lin, X. (1991) A Higher-Order Predator-Prey Interaction with Application to Observed Starfish Waves and Cycles, *Ecol. Modelling*, **58**, 323–332.

[ABL2] Antonelli, P.L. and Bradbury, R. and Lin, X. (1987) On Hutchinson's Competition Equations and Their Homogenization: A Higher-Order

Principle of Competitive Exclusion, *Ecol. Modelling*, **60**, 309–320.

[AES] Antonelli, P.L. and Elliott, R.J. and Seymour, R.M. (1987) Nonlinear Filtering and Riemannian Scalar Curvature IR, *Adv. in Appl. Math.*, **8**, 237–253

[AIM] Antonelli, P.L. and Ingarden, R.S. and Matsumoto, M., (1993) *The Theory of Sprays and Finsler Spaces with Applications in Physics and Biology*, Kluwer Academic Publishers, Dordrecht-Boston-London.

[AZ] Antonelli, P. L. and Zastawniak, T. J. (1994) Density-Dependent Host/Parasite Systems of Rothschild Type and Finslerian Diffusion, *Math. Comput. Modelling*, **20**, no. 4-5, 117–129.

[BL] Bao, D. and Lackey, B. (1996) A Hodge Decomposition Theorem for Finsler Spaces, *C.R. Acad. Sci., Paris, Ser.* 1, **323**, 51–56.

[Bo] Bochner, S. (1946), Vector Fields and Ricci Curvature, *Bull. Amer. Math. Soc.*, **52** 776–797.

[Ca1] Cartan,É. (1899) Sur Certaines Expressions Différentielles et le Problème de Pfaff.

[Ca2] Cartan, É. (1922) *Leçons Sur les Invariants Intégraux*, Hermann.

[Ca3] Cartan, É. (1928) Sur les Nombres de Betti des Espaces de Groupes Clos, *C.R. Acad. Sci. Paris*, **187**, 196–8.

[Ca4] Cartan, É. (1967) *Theory of Spinors*, Cambridge.

[El] Elworthy, K.D. (1982) Stochastic Differential Equations on Manifolds, *London Mathematical Society Lecture Notes Series*, **70**, Cambridge University Press, Cambridge.

[Fr] Frobenius, F.G. (1877) Ueber das Pfaffsche Problem, *J. de Crelle*, **82**, 230–315.

[Ga] Gauss, K.F. (1840) Allgemeine Lehrsatze in Bezeihung auf die im Verkehrten Verhältnisse des Quardrets der Enternungwirkenden Anziehungs und Abstossungskrafte, *Werke*, **5**, 197–242.

[Go] Goodwin, B.C. (1976) *Analytical Physiology of Cells and Developing Organisms*, Academic Press.

[Gra1] Graham, R. (1977) Covariant Formulation of Non-Equilibrium Statistical Thermodynamics, *Z. Phys. B*, **26**, 397–405.

[Gra2] Graham, R. (1977) Path Integral Formulation of General Diffusion Processes, *Z. Phys. B*, **26**, 281–290.

[GW] Gray, A. and Willmore, T.J. (1982) Mean-Value Theorems for Riemannian Manifolds, *Proc. Royal Soc. Edinburgh*, **92A**, 343–364.

[HARS] Hairston, N., Allan, J., Colwell, R., Futuyma, D., Howell, J., Lubin, M., Mathias, J. and Vandermeer, J. (1969) The Relationship Between Species Diversity and Stability: and Experimental Approach with Protozoa and Bacteria, *Ecology*, **49**, 1091–1101.

[Hi] Hirzebruch, F. (1966) *Topological Methods in Algebraic Geometry*, Springer.

[Ho1] Hodge, W.V.D. (1932) A Dirichlet Problem for Harmonic Functionals, with Applications to Analytic Varieties, *Proc. London Math. Soc.* (2) **36**, 257–303.

[Ho2] Hodge, W.V.D. (1933) Harmonic Functionals in a Riemannian Space, *Proc. London Math. Soc.* (2) **38**, 72–95.

[Ho3] Hodge, W.V.D. (1936) The Existance Theorem for Harmonic Integrals, *Proc. London Math. Soc.* (2) **41**, 483–496.

[Ho4] Hodge, W.V.D. (1941) *Harmonic Integrals*, Cambridge.

[Hutch1] Hutchinson, G.E. (1947) A Note on the Theory of Competition Beteween Social Species, *Ecology*, **28**, 319–321.

[Hutch2] Hutchinson, G.E., (1978) *An Introduction to Population Biology*, Yale University Press, New Haven, Conn.

[IM] Itô, K. and McKean, Jr., H.P. (1964) *Diffusion Processes and Their Sample Paths*, Academic Press, New York.

[IW] Ikeda, N. and Watanabe, S. (1989) *Stochastic Differential Equations and Diffusion Processes*, (Second Ed.), North-Holland, Amsterdam-Oxford-New York. (1st edition: Kodansha, Tokyo, 1981).

[Kod1] Kodaira, K. (1944) Über die Harmonischen Tensorfelder in Riemannschen Mannigfaltigkeiten, *Proc. Imp. Acad. Tokyo*, **20**, 186–198, 257–261, 353–358.

[Kod2] Kodaira, K. (1949) Harmonic Fields in Riemannian Manifolds (Generalized Potential Theory), *Ann. Math.*, **50**, 587–665.

[Koe] Koebe, P. (1906) Herleitung der Partielle Differentialgleichung der Potentialfunktion aus Integraleigenschaft, *Sber. Berl. Math. Ges.*, **5**, 39–42.

[Kol] Kolmogorov, N. (1937) Zur Umkehrbarkeit der Statistischen Naturgesetze, *Math. Ann.*, **113**, 766–772.

[Kondo] Kondo, K. (1954) On the Dynamics of the Aeroplane and Non-Riemannian Geometry, Parts I and II, *Jour. Japan Soc. of Aeronautical Engineering*, 161–166, 193–196.

[Kr1] Kron, G. (1934) Non-Riemannian Dynamics of Rotating Electrial Machinery, *J. Math. Phys.*, **13**, 103–194.

[Kr2] Kron, G. (1955) A Physical Interpretation of the Riemann-Christoffel Curvature Tensor (The Distribution of Damping and Synchronizing Torques in Oscillatory Transmission Systems), *Tensor*, **4**, 150–172.

[Li] Lichnerowicz, A. (1963) Spineurs Harmonique, *C.R. Acad. Sci. Paris*, Ser. A-B, **257**, 7–9.

[McK] McKean, Jr., H.P. (1969) *Stochastic Integrals*, Academic Press, New York-London.

[MIR1] Miron, R. and Anastasiei, M. (1994) *The Geometry of Lagrange Spaces: Theory and Applications*, Kluwer Academic Press, Dordrecht.

[MIR2] Antonelli, P.L. and Miron, R. (eds.) (1996) *Lagrange and Finsler Geometry. Applications to Physics and Biology*, Kluwer Academic Presss, Dordrecht.

[Mol] Molchanov, S.A. (1975) Diffusion Processes and Riemannian Geometry, *Russian Math. Surveys*, **30:1**, 1–63.

[Mo] Morgan, J.W. (1996) *The Seiberg-Witten Equations and Applications to the Topology of Four-manifolds*, Princeton.

[MR] Milgram, A.N. and Rosenblum, P.C. (1951) Harmonic Forms and Heat Conduction. I. Closed Riemannian Manifolds, *Proc. N.A.S.*, **37**, 180–184.

[Nel] Nelson, E. (1985) *Quantum Fluctuations*, Princeton University Press, Princeton, New Jersey.

[Pf] Pfaff, J.F. (1814-1815) Methodus Generalis Æquationes Differentiarum Partialium Necnon Æquationes Differentiales Vulgares, Utrasque Primi Ordinis, Inter Quotcumque Variabiles Complete Integrandi, *Abh. d. K.-P. Akad. d. Wiss. zu Berlin*, 76–136.

[Pin1] Pinsky, M. (1978) Stochastic Riemannian Geometry, *Probabilistic Analysis and Related Fields*, (Bharucha-Reid, A.T., ed.), Academic Press, New York,199–235.

[Pin2] Pinsky, M. (1992) Inverse Questions in Stochastic Differential Geometry, *Probability Theory*, (Chen, L., Choi, K., Hu., K. and Lou, J.H., eds.), Walter de Gruyter & Co., Berlin-New York, 3–28.

[Po] Poincaré, H. (1895) Analysis Situs, *J. École Poly.*, **1**, 1–121.

[dR1] deRham, G. (1931) Sur l'Analysis Situs des Variétés a n Dimensions, *J. Math. Pur. Appl.*, (9) **10**, 115–200.

[dR2] deRham, G. (1984) *Differentiable Manifolds* (English edition), Springer.

[dRK] deRham, G. and Kodaira, K. (1950) *Harmonic Integrals*, Princeton.

[RWW] Ruse, H.S., Walker, A.G. and Willmore, T.J. (1961) *Harmonic Spaces*, Edizioni Cremonese.

[SY] Shoen, R. and Yau, S.T. (1981) On the Proof of the Positive Mass Conjecture in General Relativity I.II., *Comm. Math. Phys.*, **65**, 45–76; *Comm. Math. Phys.*, **79**, 231–260.

[SW1] Seiberg, N. and Witten, E. (1994) Electromagnetic Duality Monopole Condensation and Confinement in $N = 2$ Supersymmetric Yang-Mills Theory, *Nucl. Phys.*, **B426**, 19–52.

[SW2] Seiberg, N. and Witten, E. (1994) Monopoles, Duality, and Chiral Symmetry Breaking in $N = 2$ Supersymmetric QCD, *Nucl. Phys.* **B431**, 581–640.

[TW] Takahashi, Y. and Watanabe, S. (1980) The Probability Functionals (Onsager-Machlup Functions) of Diffusion Processes, Stochastic Integrals, Proceedings of LMS Durham Symposium, *Lecture Notes in Math.*, **851**, Springer-Verlag, New York-Berlin, 433–463.

[Var] Varadhan, S.R.S. (1984) Large Deviations and Applications, *SIAM Publications*, Philadelphia.

[V1] Volterra, V. (1889) Sur une Generalisation de la Théorie des Fonctions d'une Variable Imaginaire, *Acta Math.*, **12**, 233–286.

[V2] Volterra, V. (1889) Delle Variabili Complesse Negli Iperspazi, *Rend. Accad. dei Lincei* (4) $\mathbf{5}_1$, 158–165.

[V3] Volterra, V. (1889) Sulle Funzioni Coniugate, *Rend. Accad. dei Lincei* (4) $\mathbf{5}_1$, 599–611.

[We] Weitzenbock, R. (1923) *Invarientがあtheorie*, Groningen, Noordhoff.

[Wi] Witten, E. (1981) A New Proof of the Positive Energy Theorem, *Comm. Math. Phys.*, **80**, 381–402.

[Wilb1] Wilbur, H. (1971) Competition, Predation and the Structure of the Abystoma-Rana Sylvatica Community, *Ecology*, **53**, 3–21,

[Wilb2] Wilbur, H. (1971) The Ecological Relationship of the Salamander Abystoma Laterale to Its All-Female Gynogenetic Associate, *Evolution*, **25**, 168–179,

PREFACE

Section I

The first section is concerned with the application of stochastic analysis to model the dynamics of interacting populations. The central theme is that if the dynamics is to include density-dependent social interactions, then a Finslerian theory of stochastic development is required.

In "Introduction to Diffusions on Finsler Manifolds", diffusion theory on Finsler manifolds is very briefly introduced via the notions of stochastic parallel transport and stochastic development (Rolling) in the context of arbitrary h- and v-*metrical*, deflection-free, v-symmetric Finsler connections including the Wagner connection. The Stochastic Imbedding Theorem is stated in full generality. It is used in the paper "Density Dependent Host/Parasite Systems of Rothschild Type and Finslerian Diffusion", a famous model of host/parasite interactions in which the host has hormonal control over reproduction of its parasite is presented. Social interactions arising from evolution for smaller size and simpler morphology for parasites and higher reproductivity for hosts (i.e. progenesis) are incorporated in building the deterministic equations. Since progenesis necessarily occurs in a noisy environment typical of any r-selective evolutionary regime, the emergent density-dependent effects require Finslerian Diffusion Theory to obtain the stochastically perturbed model equations. Application is made to the myxomatosis epizootic of the European wild rabbit as it occurred in the 1950's in Great Britain. Important qualitative information about this devastating disease is obtained from a Feynman-Kac formula for the forward Cauchy problem in the physiological space which is a positive definite Finsler space of Wagner class.

In "Stochastic Finsler Geometry in the Theory of Evolution by Symbiosis", a dynamical model is presented of the Margulis Theory of Evolution by Endo-symbiosis in which modern cells of plants and animals arise from separately living bacterial species. The incorporation of chemical exchanges between two bacterial colonies necessarily implies a Finslerian basis for the dynamics, while the noise ansatz entails use of *hv*-Diffusion Theory. The Feynman-Kac solution to the forward Cauchy problem is computed and shows explicit dependence on the scalar curvature of the Finslerian metric, which is positive. This proves that a stable relationship can exist, even in the presence of noise, for the subpopulations of the symbiont organism. The Finsler space involved in this model is of Berwald type, so one can apply the general theorem on the *Onsager-Machlup formula for Berwald spaces.* The notion of *quadratic dispersion*, as a sort of *kinetic energy,* is introduced and a Taylor expansion gives *relative vigor, V,* for *h-diffusions.* This in turn, is shown to be, $-\mathbb{R}/12$, where $\mathbb{R}$ is the Berwald-Gauss scalar curvature. V is small for the model of evolution by symbiosis.

Section II

The second section provides the rigorous mathematical development of diffusions on Finsler spaces. As prompted by the applications of the previous section, the diffusions are constructed on the slit tangent bundle and the indicatrix bundle of the Finsler space.

In "Diffusions on Finsler Manifolds", the concept of stochastic parallel transport along a sample path of Finslerian Brownian motion is developed by geodesic approximation, for the processes of *h*-diffusion and *hv*-diffusion. Lifts are made to stochastic processes in the orthonormal Finsler frame bundle which then enable computation of the generators for these two types of Finslerian diffusion. Diffusions of Finslerian tensor fields are also defined and their generators are obtained for the *h*- and *hv*-stochastic developments. Minkowski space and Berwald space examples are given.

In "Stochatic Calculus on Finsler Manifolds and an Application in Biology", the theory of Brownian motion and stochastic development is extended from the Riemannian case to Finslerian manifolds. The construction is based on the notions of Finslerian stochastic parallel transport and rolling. Elementary examples are presented and the radial behaviour of a Finslerian Brownian motion is studied. An application is given to a Volterra-Hamilton system of Berwald type occurring in the biology of colonial animal growth in the presence of environmental or developmental noise.

In "Diffusion on the Tangent and Indicatrix Bundles of a Finsler Manifold", Finslerian *hv*-diffusion is realized as processes in the slit tangent bundle and on the indicatrix bundle of the base manifold. There are two versions of *the Stochastic Imbedding Theorem*, one for each bundle. The result is that Finslerian *hv*-Brownian motion can be viewed as Riemannian Brownian motion with drift on the tangent bundle with diagonal lift

Riemmanian metric. The generally non-zero drifts are induced by the torsions of the Finslerian connection. Likewise, for the indicatrix bundle case, where the Riemannian metric is induced from the tangent planes to their Finslerian indicatrices.

Section III

In Section III, the mathematical study of the second section is generalized in two directions: general vector bundles and Lagrange spaces. The latter is particularly applicable to dynamics of mechanical systems.

In "Diffusion on the Total Space of a Vector Bundle" the Finslerian *hv*-Diffusion Theory is generalized, in regard to its formalism, to vector bundles. Few probabilistic calculations are given here, for they are just the same as those of the original Finslerian Theory. The related *h*-Diffusion Theory is not so easily generalized. In "Diffusions and Laplacians on Lagrange Manifolds", *h*-stochastic parallel transport for some special Lagrange spaces is investigated. Generators for diffusions on the slit tangent bundle are obtained. The existence of a new nonlinear connection for this special class of manifolds is crucial for the construction.

In "φ-Lagrange Laplacians", the special class of φ-Lagrange manifolds (i.e. those whose geodesics are actually Finslerian) are candidates for the construction of explicit *hv*-diffusions. This is accomplished after which the Stochastic Imbedding Theorem for φ-Lagrange manifolds is proved and explicit examples with importance in biology are described.

Section IV

In this section, as in Section V, the emphasis changes from the Finslerian Laplacian constructions on the total space of various bundles (tangent, indicatrix, general vector bundle) in the previous sections to their construction on the base manifold. In "Diffusion, Laplacian and Hodge Decomposition on Finsler Spaces", the isotropic transport process on a Riemannian manifold is generalized to a Finsler manifold. A piecewise geodesic flow with speed $1/\varepsilon$, and jumps of direction at Poisson distributed times with rate parameter $1/\varepsilon^2$, is proved to converge to a Markov diffusion on the base manifold, as $\varepsilon \to 0$. Using the symbol of the generator and the Hodge star construction a Finslerian Laplacian acting on forms, Δ_{AZ}, is obtained and a complete version of the Hodge-de Rham Theorem is proved after the method of D. Bao and B. Lackey.

In "A Mean-Value Laplacian for Finsler Spaces", the idea employed is that a Finslerian Laplacian, at least infinitesimally, should be a measure of the average value around a point, of any function on which it acts. Beautiful arguments ensue, producing a Laplacian which acts on functions. Using the symbol and the Hodge star, a Finslerian Laplacian on the de Rham complex is obtained. Although there are superficial similarities with, Δ_{AZ}, this theory of local averaging leads to a different Laplacian and to a different Hodge-de Rham decomposition theorem.

Section V

The fifth and final section of this book is dedicated to other approaches to the construction of a Finslerian Laplacian. As one can see in the Prologue of this text, the Laplace operator is central to diffusion theory, but it is also significant in a number of other constructions.

In "The Non-Linear Laplacian for Finsler Manifolds", one considers the canonical Dirichlet integral constructed from the Finsler function. The integrand is quadratic homogeneous, but not quadratic in the properly Finslerian case. The resulting equation for the minimizer has the appearance of the divergence of a gradient but is not linear. Smoothness of the eigenfunctions of this nonlinear Laplacian are considered, as well as its relationship to various forms of curvature.

"A Bochner Vanishing Theorem for Elliptic Complices" is not about Finsler geometry proper, but rather an introduction to the theory of elliptic complices, and their connection and curvature theory. The main result is that if the Laplacian of a complex satisfies the "complete positivity" condition, then each grade of the complex has a unique connection yielding a Weitzenböck formula. Bochner's vanishing theorem is then generalized to every grade of any elliptic complex.

"A Lichnerowicz Vanishing Theorem for Finsler Spaces" is the natural extension of "A Bochner Vanishing Theorem for Elliptic Complices", to the famous signature complex of Hirzebruch. A Finsler structure induces in a natural way metrics on differential forms, and thus a Hodge star on the exterior algebra. Unfortunately, not all the useful properties of the Hodge star in Riemannian geometry are retained. Yet, it does contain enough of the essential structure to define the spaces of self-dual and anti-self-dual forms. If the associated Laplacian has the "complete positivity" condition, then the results of the previous paper apply, and there are unique self-dual and anti-self-dual connections and curvatures for the Finsler space. Lichnerowicz's vanishing theorem is then an immediate consequence.

In "A Geometric Inequality and a Weitzenböck Formula for Finsler Surfaces", the "complete positivity" condition is explored for Finsler surfaces. A long introduction summarizes the work in, A Hodge Decomposition

Theorem for Finsler Space, *C.R. Acad. Sc. Paris*, **323**, 51-56, by Bao and Lackey, and translates the work of "A Bochner Vanishing Theorem for Elliptic Complices" into the framework of Finsler geometry. The surface case is then examined in detail, and the "complete positivity" condition is reduced to the ratio of two determinants. The example of Rander spaces is considered. An appendix follows, exhibiting that this technique generalizes to arbitrary fibre bundles (with the proper geometric structures). Also a comparison of this Laplacian and the one proposed in Sections I - III is offered.

In paper "Spinors on Finsler Spaces", a method is developed for constructing spinors and the Dirac operator associated to a Finsler space. Lichnerowicz's famous Weitzenböck formula for the square of the Dirac operator is generalized to this context. Emphasis is placed on the fact that metric connections (in the properly Finslerian case) must have torsion, and this torsion contributes to the Weitzenböck formula.

This collection of papers is unique in that it contains all known work on Laplacians in Finsler spaces. Of the sixteen articles, nine constitute the proceedings of the first *Finsler Laplacians Conference*, held August 20 - 22, 1997, at the University of Alberta. Of the remainder, only three required permission from publishers and/or journals where they appeared first. All are dated 1993, or later, and are here included in modified form, but with appropriate citations. All deal with the stochastic Laplacian and its applications in concrete biological problems.

In addition to the authors of papers in this collection, all of whom attended the above mentioned conference, active participation was solicited from Prof. John Bland of the University of Toronto and from Prof. Dante Giarusso of St. Lawrence University.

Acknowledgements. Funding for our conference was provided by the Dean of Science of the University of Alberta and NSERC. The editors would also like to thank Vivian Spak for her excellent typesetting of this book.

INTRODUCTION TO DIFFUSION ON FINSLER MANIFOLDS

P.L. ANTONELLI * and T.J. ZASTAWNIAK

Abstract

Diffusion theory on Finsler manifolds is briefly presented, including generalizations of the notions of stochastic parallel transport, stochastic development (rolling), and Brownian motion from the well-known Riemannian case. The results discussed cover the case of an arbitrary $h-$ and $v-$metrical deflection-free $v-$symmetric Finsler connection, which proves important in applications, as can be seen from our second paper in this issue involving the so-called Wagner connection.

1 Introduction

Let M be an $n-$dimensional Finsler manifold with *metric function* $L : \mathring{T}M \to (0, \infty)$ defined and smooth on the slit tangent bundle $\mathring{T}M = TM \setminus \{0_x : x \in M\}$, where 0_x denotes the null vector in TM_x. It is assumed that $L(x, ky) = kL(x, y)$ for every $x \in M$, $0 \neq y \in TM_x$, and $k > 0$, that is, $L(x, y)$ is positively homogeneous of degree one in y. The Finsler metric tensor

$$g_{ij}(x, y) = 2^{-1}\partial_{(i)}\partial_{(j)}L^2(x, y),$$

where $\partial_{(i)} = \partial/\partial y^i$, will be assumed to be positive definite for all $x \in M$ and $0 \neq y \in TM_x$. Let M be equipped with a Finsler connection $(N^i_j, F^i_{jk}, C^i_{jk})$, N^i_j being the *nonlinear*, F^i_{jk} the *horizontal*, and C^i_{jk} the *vertical connection coefficients*. We set $\partial_i = \partial/\partial x^i$ and $\delta_i = \partial_i - N^j_i\partial_{(j)}$. It is assumed that the Finsler connection is $h-metrical$ and $v-metrical$, i.e.,

$$\delta_i g_{jk} = F^l_{ij}g_{lk} + F^l_{ik}g_{jl}, \qquad \partial_{(i)}g_{jk} = C^l_{ij}g_{lk} + C^l_{ik}g_{jl}, \tag{1.1}$$

*This work was partially supported by NSERC A-7667. This article appeared in *Mathl. Comput. Modelling* Vol. 20, pp. 109-116, 1994. It is included here, with permission from Pergamon Press.

deflection-free, i.e.,

$$N^i_j = F^i_{jk} y^k, \tag{1.2}$$

and $v-$*symmetric*, i.e.,

$$C^i_{jk} = C^i_{kj}. \tag{1.3}$$

It follows, in particular, that $C^i_{jk} = 2^{-1} g^{il} \partial_{(j)} g_{lk}$ are the standard vertical connection coefficients. Above and throughout the article we use the standard summation convention on upper and lower indices.

Perhaps the most important example of an $h-$ and $v-$metrical deflection-free Finsler connection is the *Cartan connection* with coefficients defined by

$$\begin{aligned}
N^i_j &= \partial_{(j)} \left(\gamma^i_{kl} y^k y^l \right), \\
F^i_{jk} &= 2^{-1} g^{il} \left(\delta_j g_{lk} + \delta_k g_{jl} - \delta_l g_{jk} \right), \\
C^i_{jk} &= 2^{-1} g^{il} \left(\partial_{(j)} g_{lk} + \partial_{(k)} g_{jl} - \partial_{(l)} g_{jk} \right) = 2^{-1} g^{il} \partial_{(j)} g_{lk},
\end{aligned}$$

where g^{ij} is the inverse matrix to g_{ij}, and $\gamma^i_{jk} = 2^{-1} g^{il} \left(\partial_j g_{lk} + \partial_k g_{jl} - \partial_l g_{jk} \right)$. Please note that the above formulae for F^i_{jk} and C^i_{jk} resemble the well-known expression for the Levi-Cività connection coefficients in Riemannian geometry. Moreover, the horizontal connection coefficients F^i_{jk} are symmetric in j, k. The Cartan connection can, in fact, be regarded as a generalization of the Levi-Cività connection. There is a large family of other $h-$ and $v-$metrical deflection-free $v-$symmetric Finsler connections, for which the coefficients F^i_{jk} are no longer symmetric in j, k. The so-called Wagner connection used in our article [1] in the present issue serves as an important example of such a Finsler connection.

We say that $A : \mathring{T}M \to TM$ is a Finsler vector field if $A(x,y) \in TM_x$ for every $x \in M$ and $0 \neq y \in TM_x$. The *horizontal* and *vertical covariant derivatives* of such a field will be denoted by

$$\nabla^h_j A^i = \delta_j A^i + F^i_{jk} A^k, \qquad \nabla^v_j A^i = \partial_{(j)} A^i + C^i_{jk} A^k.$$

These formulae can be extended to arbitrary Finsler tensor fields in the usual way. For example, conditions (1.1) can be written as $\nabla^h_i g_{jk} = 0$ and $\nabla^v_i g_{jk} = 0$.

A systematic presentation of Finsler geometry can be found in the classical monograph [2] by Rund. For a modern approach to Finsler manifolds see Matsumoto's monograph [3]. We also refer to Cartan's original work [4].

In the present article we shall describe our generalization of the theory of diffusion on Riemannian manifolds to the case of Finslerian manifolds [5], [6], [7]. The theory of diffusion processes on Riemannian manifolds goes back to the pioneering articles [8], [9], [10], [11] by Itô and has now become a classical branch of stochastic calculus with various applications in mathematics and physics (see, for example, [12], [13], [14]). Our work on Finslerian diffusions is motivated mainly by biological models in which the deterministic evolution of a system can be represented by a geodesic on a Finsler manifold [15], [16], [17]. Finslerian diffusions provide the much needed tool for adding noise to such systems (see [1] in this issue and [5]). It might well be that Finslerian diffusions will also prove useful in physics, for example, to tackle random walk problems in anisotropic spaces or to construct a version of Nelson's famous stochastic mechanics [18] in anisotropic spaces.

Throughout the article we use the standard probabilistic setup (see, for example, [13], [19]). We fix a complete *probability space* $(\Omega, \mathcal{F}, P)$ equipped with a right-continuous *filtration* $(\mathcal{F}_t)_{t\geq 0}$ such that each $\mathcal{F}_t$ contains all the events of probability zero from the σ-algebra $\mathcal{F}$. We denote by $Ef = \int_\Omega f\, dP$ the *expectation* of a *random variable* $f : \Omega \to \mathbb{R}$. The *conditional expectation* of f given a sub-σ-field $\mathcal{G} \subset \mathcal{F}$ will be denoted by $E(f|\mathcal{G})$. A *stochastic process* $X : [0,\infty)\times\Omega \to \mathbb{R}$ is said to be *continuous* if its sample paths $t \mapsto X(t,\omega)$ are continuous for P-almost all $\omega \in \Omega$. A stochastic process X is *adapted* if the random variable $X(t)$ is $\mathcal{F}_t$-measurable for every $t \geq 0$. A *martingale* is an adapted process X such that $E|X(t)| < \infty$ for every $t \geq 0$ and $E(X(t)|\mathcal{F}_s) = X(s)$ for every $t > s \geq 0$. A *local martingale* is an adapted process X such that the process $(t,\omega) \mapsto X(\tau_n \wedge t, \omega)\chi_{\{0<\tau_n\}}(\omega)$ is a martingale for every n, where $\tau_n = \inf\{t \geq 0 : |X(t)| \geq n\}$ (as usual, we set $\inf \emptyset = \infty$) and where χ_A is the indicator function of $A \subset \Omega$. A *semimartingale* is the sum of a continuous local martingale and a continuous adapted process whose sample paths have P–almost surely finite variation on each compact interval in $[0,\infty)$.

By a *standard Brownian motion* on $\mathbb{R}^n$ (also known as the *Wiener process*) we understand a semimartingale $W : [0,\infty) \times \Omega \to \mathbb{R}^n$ such that $(t,\omega) \mapsto W^i(t,\omega)$ and $(t,\omega) \mapsto W^i(t,\omega)W^j(t,\omega) - t\delta^{ij}$ are martingales for all $i,j \in \{1,\dots,n\}$.

For a real-valued semimartingale X and a continuous adapted process Y, we denote by $\int Y\, dX$ the *Itô stochastic integral* of Y with respect to X. Given two real-valued semimartingales X and Y, the process $[X,Y] = XY - X(0)Y(0) - \int X\, dY - \int Y\, dX$ is called the *joint quadratic variation* of X and Y, and the process $\int Y \circ dX = \int Y\, dX + 2^{-1}[X,Y]$ is called the *Stratonovich integral* of Y with respect to X. We shall use the differential notation: if $Z = \int Y\, dX$, we shall write $dZ = Y\, dX$; if $Z = [X,Y]$, we shall write $dZ = dXdY$; and if $Z = \int Y \circ dX$, then we shall write $dZ = Y \circ dX$. Thus, in particular, $Y \circ dX = Y\, dX + 2^{-1} dXdY$. It is well known that a standard Brownian motion W on $\mathbb{R}^n$ satisfies the *Itô rules*

$$dW^i dW^j = \delta^{ij} dt, \qquad dW^i dt = 0, \qquad dtdt = 0.$$

Here dt stands for dX, where X is the deterministic process $X(t,\omega) = t$.

If $X^1, \dots, X^n$ are real-valued semimartingales and $f : \mathbb{R}^n \to \mathbb{R}$ is a function of class C^2, then we have the Itô change-of-variables formula

$$df(X) = \partial_i f(X)\, dX^i + 2^{-1}\partial_i\partial_j f(X)\, dX^i dX^j,$$

where $X = (X^1, \dots, X^n)$. If f is of class C^3, then the Itô formula can be written as

$$df(X) = \partial_i f(X) \circ dX^i.$$

A continuous stochastic process $X : [0,\infty) \times \Omega \to M$ with values on a manifold M is called an *M–valued semimartingale* if, for every smooth function $f : M \to \mathbb{R}$, the process $(t,\omega) \mapsto f(X(t,\omega))$ is a real-valued semimartingale. We shall also admit M–valued semimartingales defined on a *random time interval* $[0,\tau)$, where $\tau : \Omega \to [0,\infty]$ is a *stopping time*, i.e., a random variable such that $\{t < \tau\} \in \mathcal{F}_t$ for every $t \geq 0$.

Given two manifolds M and N, we shall consider *Stratonovich stochastic differential equations* (Stratonovich SDEs) of the form $dY = e(X,Y) \circ dX$, where $e(x,y) : TM_x \to TN_y$ are linear maps depending smoothly on $x \in M$ and $y \in N$, and where X is an M–valued semimartingale defined on a random time interval $[0,\tau)$. In local coordinates the above SDE reads $dY^i = e^i_j(X,Y) \circ dX^j$. The solution Y is an N–valued semimartingale defined on a random time interval $[0,\sigma)$. It is called a *maximal solution* if $P\{\sigma = \tau \text{ or } Y(t) \to \infty \text{ as } t \nearrow \sigma\} = 1$, where ∞ is the infinity of the one-point compactification of M. (If M is a compact manifold, then the condition reduces to $P\{\sigma = \tau\} = 1$.) If $\sigma < \tau$ with non-zero probability, then we say that the maximal solution explodes. From now on, by a solution of an SDE we shall always understand a maximal solution.

Let D be an elliptic second-order operator on a manifold M. An M–valued semimartingale X is called a *diffusion* on M with *generator* D if the process $(t,\omega) \mapsto f(X(t,\omega)) - f(X(0,\omega)) - \int_0^t Df(X(s,\omega))ds$ is a local martingale for every compactly supported smooth function $f : M \to \mathbb{R}$.

The simplest example of a diffusion is a standard Brownian motion on $\mathbb{R}^n$. In this case $D = 2^{-1}\Delta$, where Δ is the Laplace operator on $\mathbb{R}^n$. More generally, a Brownian motion on a Riemannian manifold M is a an M–valued diffusion with generator $D = 2^{-1}\Delta$, where Δ is the Laplace-Beltrami operator on M. The standard construction of a Brownian motion on a Riemannian manifold is based on the concept of stochastic parallel transport of vectors governed by the Levi-Cività connection and the so-called stochastic development or 'rolling' the manifold along a standard

Brownian motion on $\mathbb{R}^n$ (see [10], [11], [12], [13], [14]). In the following sections we shall demonstrate how to generalize these concepts to the case of a Finsler manifold. The Cartan connection will play a role analogous to the Levi-Cività connection. Using any other $h-$ and $v-$metrical deflection-free v-symmetric Finsler connection will result in some additional drift, similarly as in the Riemannian case, where using a non-symmetric metrical connection in place of the Levi-Cività connection is known to produce Brownian motion with drift (see, for example, [14]). As in the Riemannian case, the additional drift is related to the torsion tensor of the connection.

Possible generalizations of the theory of Finslerian diffusions, which is outlined below, may involve Lagrange and/or Hamilton spaces presented in other articles in this issue. The existence of $h-$ and $v-$metrical deflection-free connections for spaces of this kind may facilitate such a generalization. For the time being, however, this remains an open problem.

The Finslerian diffusion described below is used in our article [1] in this issue to model random environmental perturbations in a host/parasite system of Rothschild type.

2 Finslerian Stochastic Parallel Transport

We begin with the notion of deterministic Finslerian parallelism. In the case of a Finsler manifold M the metric tensor $g_{ij}(x, y)$ and the connection coefficients $N^i_j(x, y), F^i_{jk}(x, y), C^i_{jk}(x, y)$ depend both on $x \in M$ and a tangent vector $0 \neq y \in TM_x$. Thus, it is not enough to specify a curve $x(t) \in M$ along which is defined parallel transport of a tangent vector $u(t) \in TM_{x(t)}$. One must, in addition, specify a field of directions $0 \neq y(t) \in TM_{x(t)}$ along $x(t)$ and consider parallelism along the curve $x(t) \in M$ given $y(t)$. Namely, we say that a vector field $u(t) \in TM_{x(t)}$ is *hv−parallel* along $x(t)$ given $y(t)$ if

$$\frac{du^i}{dt} = -F^i_{jk}(x, y)u^k \frac{dx^j}{dt} - C^i_{jk}(x, y)u^k \frac{\delta y^j}{dt}, \tag{2.1}$$

where

$$\frac{\delta y^j}{dt} = \frac{dy^j}{dt} + N^i_j(x, y)\frac{dx^j}{dt} \,.$$

If the Finsler connection $(N^i_j, F^i_{jk}, C^i_{jk})$ is both h- and $v-$metrical, then it follows that

$$g_{ij}(x(t), y(t))u^i(t)v^j(t) = \text{const}$$

for arbitrary vector fields $u(t), v(t) \in TM_{x(t)}$ that satisfy (2.1). In particular, if $u(t)$ and $v(t)$ are orthogonal to one another at $t = 0$, then they remain orthogonal for all t.

We use an approximation procedure to extend the notion of hv- parallelism to the case when the deterministic curve $(x(t), y(t))$ on $\mathring{T}M$ is replaced by a semimartingale $(X(t), Y(t))$ on $\mathring{T}M$. A piecewise smooth approximation of $(X(t), Y(t))$ can be obtained by taking a division π : $0 = t_0 < t_1 < \ldots < t_n = T$ of the time interval and joining the points $(X(t_i), Y(t_i))$ by pieces of suitable smooth curves on $\mathring{T}M$. This can be achieved by means of a smooth *interpolation rule* (see [13]). By this we understand a smooth mapping $I : V \times [0,1] \to \mathring{T}M$, where $V \subset \mathring{T}M \times \mathring{T}M$ is a neighbourhood of the diagonal in $\mathring{T}M \times \mathring{T}M$ such that $I((x,y),(x,y),t) = (x,y)$ for all $t \in [0,1]$, $I((x,y),(x',y'),0) = (x,y)$, $I((x,y),(x',y'),1) = (x',y')$, and $\partial^m I((x,y),(x',y'),t)/\partial t^m = O(d^m((x,y),(x',y')))$ for $m = 1,2,3$ uniformly on compact sets. Here $d((x,y),(x',y'))$ denotes the distance between (x,y) and (x',y') in the *diagonal lift metric*

$$G(x,y) = g_{ij}(x,y)dx^i \otimes dx^j + g_{ij}(x,y)\delta y^i \otimes \delta y^j \tag{2.2}$$

on $\mathring{T}M$, where $\delta y^i = dy^i + N^i_j dx^j$.

Given an interpolation rule I and a division π of the time interval, we define a piecewise smooth approximation $(X_\pi(t), Y_\pi(t))$ of the semimartingale $(X(t), Y(t))$ by

$$(X_\pi(t), Y_\pi(t)) = I\left(\left(X(t_{i-1}), Y(t_{i-1})\right), \left(X(t_i), Y(t_i)\right), \frac{t - t_{i-1}}{t_i - t_{i-1}}\right)$$

for $t \in [t_{i-1}, t_i]$, where $i = 1, 2, \ldots, n$. If $(X(t_{i-1}), Y(t_{i-1}))$ and $(X(t_i), Y(t_i))$ are too far away from one another, so that $((X(t_{i-1}), Y(t_{i-1})), (X(t_i), Y(t_i))) \notin V$, then we can leave $X_\pi(t)$ undefined for $t \geq t_{i-1}$. It is only the behaviour of I in the vicinity of the diagonal in $\mathring{T}M \times \mathring{T}M$ that matters in the limit as $|\pi| = \max_i |t_i - t_{i-1}|$ tends to zero.

There is a multitude of interpolation rules to choose form. A natural one is the *geodesic interpolation rule* such that $t \mapsto I((x,y),(x',y'),t)$ is the shortest geodesic in the diagonal lift metric (2.2) on $\mathring{T}M$ that stars from (x,y) at $t = 0$ and arrives at (x',y') at time $t = 1$. The proof of all the required properties of I is exactly the same as in the case treated in Proposition 7.13 of [13].

Let $U_\pi(t) \in TM_{X_\pi(t)}$ be a process whose sample paths satisfy the differential equation (2.1) for the hv-parallel transport along $X_\pi(t)$ given $Y_\pi(t)$, i.e.,

$$\frac{dU^i_\pi}{dt} = -F^i_{jk}(X_\pi, Y_\pi)U^k_\pi \frac{dX^j_\pi}{dt} - C^i_{jk}(X_\pi, Y_\pi)U^k_\pi \frac{\delta Y^j_\pi}{dt},$$

where

$$\frac{\delta Y^i_\pi}{dt} = \frac{dY^i_\pi}{dt} + N^i_j(X_\pi, Y_\pi)\frac{dX^j_\pi}{dt}.$$

We set $x_0 = X(0)$ and $y_0 = Y(0)$, which implies that $x_0 = X_\pi(0)$ and $y_0 = Y_\pi(0)$ for every partition π, and assume that $U_\pi(t)$ satisfies the initial condition $U_\pi(0) = u_0 \in TM_{x_0}$ for every partition π, where u_0 is an $\mathcal{F}_0$-measurable random vector. Then, by Theorem 7.24 of [13], $U_\pi(t)$ converges in probability, uniformly with respect to t, in any compact set as $|\pi| \to 0$, to a semimartingale $U(t) \in TM_{X(t)}$ that satisfies the Stratonovich SDE

$$dU^i = -F^i_{jk}(X,Y)U^k \circ dX^j - C^i_{jk}(X,Y)U^k \circ \delta Y^j, \tag{2.3}$$

where

$$\delta Y^i = dY^i + N^i_j(X,Y) \circ dY^j,$$

with the initial condition $U(0) = u_0$. We then say that $U(t)$ is the *hv-stochastic parallel transport* of u_0 along $X(t)$ given $Y(t)$, or simply that $U(t)$ is *hv-parallel* along $X(t)$ given $Y(t)$.

Similarly as in the deterministic case,

$$g_{ij}(X(t),Y(t))U^i(t)V^j(t) = \text{const} \tag{2.4}$$

almost surely for any semimartingales $U(t)$ and $V(t)$ that satisfy (2.3), provided that the Finsler connection $(N^i_j, F^i_{jk}, C^i_{jk})$ is both h- and v-metrical. It follows that if $U(t)$ and $V(t)$ are orthogonal at $t = 0$, then they remain orthogonal *almost surely* for all t. It also follows that a solution $U(t)$ of (2.3) *cannot explode*, i.e., it can almost surely be extended to all t for which the original semimartingale $(X(t), Y(t))$ is defined. Moreover, if $U(0) \neq 0$ almost surely, then $U(t) \neq 0$ almost surely for all such t.

Since the SDE (2.3) is linear in U, the notion of stochastic hv-parallel transport can be extended to tensor-valued processes $U(t) \in T^r_s M_{X(t)}$ of any type (r,s). For example, we say that a process $U(t) \in T^*M_{X(t)} = T^1_0 M_{X(t)}$ is hv-parallel along $X(t)$ given $Y(t)$ if

$$dU_i = F^k_{ji}(X,Y)U_k \circ dX^j + C^k_{ji}(X,Y)U_k \circ \delta Y^j.$$

3 Lifting a Semimartingale from $\overset{\circ}{T}M$ to the Orthonormal Finsler Bundle $F\overset{\circ}{T}M$

The *orthonormal Finsler bundle* $F\overset{\circ}{T}M$ is by definition a principal bundle over $\overset{\circ}{T}M$ whose total space consists of elements (x,y,z), where $x \in M$, $0 \neq y \in TM_x$, and $z = (z_1, \dots, z_n)$ is an orthonormal frame in TM_x relative to the quadratic form $g(x,y) \in T^0_2 M_x$, i.e.,

$$g_{ij}(x,y)z^i_k z^j_l = \delta_{kl},$$

with projection $\pi : F\mathring{T}M \ni (x, y, z) \mapsto (x, y) \in \mathring{T}M$. The differential structure of $F\mathring{T}M$ can be obtained from that of M. Namely, if (x^i) are local coordinates on M, then we can write $y = y^i \partial_i$ for any $y \in TM_x$ and $z_j = z^i_j \partial_i$ for each vector z_j of any linear frame $z = (z_1, \ldots, z_n)$ in TM_x, where $x \in M$, and use (x^i, y^i, z^i_j) as local coordinates on $F\mathring{T}M$. The structure group of $F\mathring{T}M$ is the group $O(n)$ of orthogonal transformations of $\mathbb{R}^n$ with the right action $R_h(x, y, z) = (x, y, zh)$, where $(zh)_i = h^j_i z_j$, for any $h \in O(n)$ and $(x, y, z) \in F\mathring{T}M$.

Let (X, Y) be a semimartingale on $\mathring{T}M$. A semimartingale of the form (X, Y, Z) on $F\mathring{T}M$, where $Z(t) = (Z_1(t), \ldots, Z_n(t))$ is a frame in $TM_{X(t)}$ orthonormal with respect to $g(X(t), Y(t))$ for every t, is then called an *hv-lift* of (X, Y) if each of the frame vectors $Z_j(t)$ is hv-parallel along $X(t)$ given $Y(t)$, that is, by (2.3),

$$dZ^i_j = -F^i_{kl}(X, Y) Z^l_j \circ dX^k - C^i_{kl}(X, Y) Z^l_j \circ \delta Y^k. \tag{3.1}$$

The existence of an hv-lift follows from (2.4). It suffices to specify an initial frame $Z(0)$ in $TM_{X(0)}$ orthonormal with respect to $g(X(0), Y(0))$ and solve (3.1) for Z with $Z(0)$ as the initial condition. Then (2.4) ensures that the *solution does not explode* and $(X(t), Y(t), Z(t))$ stays on $F\mathring{T}M$ for all t for which $(X(t), Y(t))$ is defined.

4 Finslerian Stochastic Development (Rolling); Finslerian Brownian Motion

Let (X, Y) be a semimartingale on $\mathring{T}M$ starting at a point $(x_0, y_0) = (X(0), Y(0)) \in M$. We choose local coordinates so that $g_{ij}(x_0, y_0) = \delta_{ij}$. Let (X, Y, Z) be the hv-lift of (X, Y) to the orthogonal Finsler bundle $F\mathring{T}M$, as described in the previous section, with the initial condition $Z^i_j(0) = \delta^i_j$. Consider the tangent space TM_{x_0} as an n-dimensional Euclidean vector space, which can be identified with $\mathbb{R}^n$. We say that a semimartingale $\tilde{X}(t)$ on TM_{x_0} is obtained by *hv-rolling* along $X(t)$ given the directions $Y(t)$ if

$$dX^i = Z^i_j \circ d\tilde{X}^j.$$

This definition is a direct generalization of the notion of rolling from the Riemannian case (see [12], [13], [14]), in which there is of course no need to specify the directions $Y(t)$. We also say that $\tilde{X}(t)$ is the *hv-lift* of $X(t)$ to TM_{x_0}. One can also define the lift $\tilde{Y}(t)$ of $Y(t)$. Namely, $\tilde{Y}(t)$ is the tangent vector at $\tilde{X}(t)$ defined by

$$Y^i(t) = Z^i_j(t) \tilde{Y}^j(t), \tag{4.1}$$

i.e., $\tilde{Y}^j(t)$ are the coordinates of $Y(t)$ in the frame $Z(t)$. Using (1.2) and (1.3), one can demonstrate that (4.1) is equivalent to the SDE

$$\delta Y^i = Z^i_j \circ d\tilde{Y}^j$$

with the initial condition $Y^i(0) = \tilde{Y}^i(0)$, where $\delta Y^i = dY^i + N^i_j \circ dX^j$. Since TM_{x_0} can be identified with $\mathbb{R}^n$ and the tangent bundle over TM_{x_0} can be identified with $\mathbb{R}^n \times \mathbb{R}^n$, $(\tilde{X}, \tilde{Y})$ can be considered as a process in $\mathbb{R}^n \times \mathbb{R}^n$.

One can reverse the situation by prescribing a semimartingale $(\tilde{X}, \tilde{Y})$ in $\mathbb{R}^n \times \mathbb{R}^n$ and solving the appropriate SDEs for X, Y, and Z. In particular, if $\tilde{X} = W$ and $\tilde{Y} = V$ are independent Brownian motions in $\mathbb{R}^n$, then the corresponding semimartingale (X, Y, Z) on $F\mathring{T}M$ is defined by the system of SDEs

$$\begin{cases} dX^i &= Z^i_j \circ dW^j \\ \delta Y^i &= Z^i_j \circ dV^j \\ dZ^i_j &= -F^i_{kl}(X,Y) Z^l_j \circ dX^k - C^i_{kl}(X,Y) Z^l_j \circ \delta Y^k, \end{cases} \tag{4.2}$$

where

$$\delta Y^i = dY^i + N^i_j \circ dX^j.$$

We then call (X, Y, Z) an *hv-Finslerian stochastic development*.

Theorem 1. *The semimartingale (X, Y) defined by (4.2) is a diffusion on $\mathring{T}M$ with generator*

$$\begin{aligned} D &= 2^{-1} g^{ij} \nabla^h_i \nabla^h_j + 2^{-1} g^{ij} \nabla^v_i \nabla^v_j \\ &= 2^{-1} g^{ij} (\delta_i \delta_j - F^k_{ij} \delta_k) + 2^{-1} g^{ij} (\partial_{(i)} \partial_{(j)} - C^k_{ij} \partial_{(k)}). \end{aligned} \tag{4.3}$$

The proof of this theorem can be found in [5], [6], and [7].

Definition 4.2. If $(N^i_j, F^i_{jk}, C^i_{jk})$ is the Cartan connection, then a diffusion (X, Y) on $\mathring{T}M$ with generator (4.3) is called an *hv-Finslerian Brownian motion.*

The theorems below contain two interesting results for the semimartingale (X, Y) defined by (4.2) when $(N^i_j, F^i_{jk}, C^i_{jk})$ is an arbitrary h- and v-metrical deflection-free v-symmetric Finsler connection. The first theorem shows that (X, Y) can be represented as a Riemannian Brownian motion with drift on $\mathring{T}M$ equipped with a suitable Riemannian metric. The drift is related to the torsion tensors $T^i_{jk} = F^i_{jk} - F^i_{kj}$, $P^i_{jk} = \partial_{(j)} N^i_k - F^i_{kj}$, and C^i_{jk} of the Finsler connection $(N^i_j, F^i_{jk}, C^i_{jk})$. (As is well known, C^i_{jk} play the double role of the components of a torsion tensor and the vertical coefficients of the connection.) The second result below is concerned with the behaviour of the metric function $L(X,Y)$ for the diffusion (X,Y) defined by (4.2). We refer to it as the theorem on the radial behaviour of Finslerian diffusion, because it generalizes the well-known theorem that the radial part $\sqrt{(W^1)^2+\ldots+(W^n)^2}$ of a Brownian motion $W=(W^1,\ldots,W^n)$ in $\mathbf{R}^n$ is a Bessel process with index n. (On Bessel processes see, for example, [14].)

Theorem 4.3. *A semimartingale* (X,Y) *on* $\mathring{T}M$ *is a diffusion with generator* (4.3) *if and only if it is a Riemannian Brownian motion with drift* B *on* $\mathring{T}M$ *equipped with the Riemannian metric* G *defined by* (2.2), *where* B *is a vector field on* $\mathring{T}M$ *defined by*

$$B=2^{-1}g^{ik}T^j_{jk}\delta_i+2^{-1}g^{ik}P^j_{jk}\delta_i-2^{-1}g^{ik}C^j_{kj}\partial_{(i)} \tag{4.4}$$

in terms of the torsion tensors $T^i_{jk}=F^i_{jk}-F^i_{kj}$, $P^i_{jk}=\partial_{(j)}N^i_k-F^i_{kj}$, *and* C^i_{jk} *of the Cartan connection.*

Theorem 4.4. *If* (X,Y) *is the Finslerian diffusion defined by* (4.2), *then* $L(X,Y)$ *is a Bessel process with index* $n=\dim M$.

The proofs of Theorems 4.3 and 4.4 in the case when $(N^i_j,F^i_{jk},C^i_{jk})$ is the Cartan connection can be found in [5], [6], and [7]. It is not hard to extend these proofs to cover the case of an arbitrary h- and v-metrical deflection-free v-symmetric Finsler connection.

Compared with the case of the Cartan connection studied in [7], the drift in (4.4) contains the additional term $2^{-1}g^{ik}T^j_{jk}\delta_i$ in the general case of an arbitrary Finsler connection that satisfies (1.1), (1.2), and (1.3). The additional drift is related to the torsion tensor $T^i_{jk}=F^i_{jk}-F^i_{kj}$ in close analogy with the Riemannian case, in which replacing the Levi-Cività connection by an arbitrary metrical connection results in a similar effect (see [14]). In our article [1] in this issue the torsion tensor T^i_{jk} appears as a consequence of using the so-called Wagner connection intrinsically related to our model of a Rothschild host/parasite system.

References

[1] Antonelli, P.L. and Zastawniak, T.J. (1994) Density-Dependent Host-Parasite Systems of Rothschild Type and Finslerian Diffusion, *Mathl. Comput. Modelling,* **20**, 117-129.

[2] Rund, H. (1959) *The Differential Geometry of Finsler Spaces*, Springer-Verlag, Berlin.

[3] Matsumoto, M. (1986) *Foundations of Finsler Geometry and Special Finsler Spaces*, Kasheisha Press, Saikawa 3–23–2, Otsushi, Shigaken.

[4] Cartan, E. (1971) Les Espaces de Finsler, *Actualités*, **79**, Paris, 1934, 2nd ed.

[5] Antonelli, P.L. and Zastawniak, T.J. (1993) Stochastic Calculus on Finsler Manifolds and an Application in Biology, *Nonlinear World*, **1**, 149-171.

[6] Antonelli, P.L. and Zastawniak, T.J. (1992) Diffusions on Finsler Manifolds, *Proc. of XXVth Symposium on Math. Phys.*, Toruń, Poland.

[7] Antonelli, P.L. and Zastawniak, T.J. (1995) Diffusion on the Tangent and Indicatrix Bundles of a Finsler Manifold, *Tensor, N. S.*, **56**, 233-247.

[8] Itô, K. (1950) Stochastic Differential Equations in a Differentiable Manifold, *Nagoya Math. J.*, **1**, 35–47.

[9] Itô, K. (1953) Stochastic Differential Equations in a Differentiable Manifold (2), *Mem. Coll. Sci. Univ. Kyoto Math.*, **28**, 81–85.

[10] Itô, K. (1962, 1963) The Brownian Motion and Tensor Fields on a Riemannian Manifold, *Proc. Intern. Congr. Math., Stockholm, Inst. Mittag-Leffler, Djursholm*, 536–539.

[11] Itô, K. (!975) Stochastic Parallel Displacement, in: M. A. Pinsky (ed.), *Probabilistic Methods in Differential Equations*, Lect. Notes in Math., Vol. 451, Springer- Verlag, Berlin-Heidelberg-New York, 1–7.

[12] Elworthy, K.D. (1982) *Stochastic Differential Equations on Manifolds*, Cambridge University Press, Cambridge.

[13] Emery, M. (1989) *Stochastic Calculus in Manifolds*, Springer-Verlag, Berlin, Heidelberg.

[14] Ikeda, N. and Watanabe, S. (1981, 1989) *Stochastic Differential Equations and Diffusion Processes*, North Holland, Amsterdam, Kodansha, Tokyo.

[15] Antonelli, P.L. (ed.), (1985) *Mathematical Essays on Growth and the Emergence of Form*, Univ. of Alberta Press, Edmonton.

[16] Antonelli, P.L. (1993) On y-Berwald Connections and Hutchinson's Ecology of Social Interactions, *Tensor, N. S.*, **52**, 27-36.

[17] Antonelli, P.L., Bradbury, R., Krivan, V. and Shimada, H. (1993) A Dynamical Theory of Heterochrony: Time-Sequencing Changes in Ecology, Development and Evolution, *J. Biol. Syst.*, **1**, 451-487.

[18] Nelson, E. (1985) *Quantum Fluctuations*, Princeton Univ. Press, Princeton, N.J.

[19] Kunita, H. (1990) *Stochastic Flows and Stochastic Differential Equations*, Cambridge University Press, Cambridge.

DENSITY DEPENDENT HOST/PARASITE SYSTEMS OF ROTHSCHILD TYPE AND FINSLERIAN DIFFUSION*

P.L. ANTONELLI and T.J. ZASTAWNIAK

1 Biological Background

The 2000 odd species of fleas are prime examples of parasites which are modest in their physiological demands on their hosts and unobtrusive in their behavior. Generally, host/parasite relationships are stable and this is true of the host/fleas relationships in particular, making their appearance in the fossil record some 50 million years ago [1]. Occasionally, however, circumstances arise which can turn these benign coevolved relationships into powerfully destructive forces, as, for example, fleas of genus *Xsenopsylla* carrying the bacillus *Yersinia pestis*, which were responsible for the deaths of 25 million people in 14th century Europe, or, in a less humanly traumatic example, the rabbit flea, *Spilopsyllus cuniculi* carrying the myxoma virus, which killed millions of its European wild rabbit (*Oryctolagus cuniculus*) hosts in the 1950's in Great Britain. This epizootic was man-induced, the reason being that *O. cuniculus* was a great pest costing upwards of $40 million annually [2].

From an academic perspective, the relationship between this rabbit species and its flea parasite is very interesting, for *S. cuniculi* has completely given over control of its reproductive cycle to its host, *O. cuniculus*. This control is at the level of hydrocortisone and other related sexual hormones produced by fertilized female rabbits and transported to the fleas as a portion of their normal blood meal [3]. M. Rothschild has demonstrated

*This work was partially supported by NSERC A-7667. This article appeared in *Mathl. Comput. Modelling*, Vol. **20**, 1994, pp. 117-129. It is republished here with permission from Pergamon Press.

that *S. cuniculi* will become sexually mature only on fertilized female rabbits and will copulate preferentially on newly born rabbit kittens up to 12 days old. She showed that increased levels of corticosteroids initiate egg maturation in female fleas about 10 days prior to the birth of the rabbit kittens. A few hours after birth the fleas jump to the kittens, feed off their blood voraciously, mate, lay their flea eggs, and return to their adult rabbit hosts, all in a period of 12 days. Mathematically, the above mechanism will be encoded in equation (2.5) below.

Recently, Rothschild and Ford [4] have discovered that other flea species and their hosts have evolved this type of chemical control of reproduction. In fact, Rothschild supposes that in the future many instances will be discovered in which hormones of vertebrate animals play some unobtrusive but definite role in the development of their parasites [1].

There is a general consensus that parasites have undergone evolution towards smaller body size and simplified adult morphology. This type of evolutionary change is called progenesis and is closely linked to life cycles in a noisy (i.e., uncertain) environment (termed r-selective). *Progenetic evolution* implies nonlinear change or shift in the time-sequence of developmental events of an organism relative to the sequence of its ancestors [5]. Thus, it is a special case of a *heterochronic process* ([5], [6]) in which formally juvenile characters are retained in adult descendants and are produced by *precocious sexual maturation* of the organism still in the juvenile state, morphologically speaking.

It is one of our goals to show how the Rothschild hormonally controlled reproduction of the parasite by its host displays, mathematically, progenetic heterochrony and that the myxomatosis epizootic is a consequence of this and the Rothschild hormonal control mechanism (2.5) for the special case of *O. cuniculus*. In particular, it will be shown that there is a close connection between density-dependent reproduction and progenetic evolution in this Rothschild class of systems.

It is essential that a model of progenesis display its dynamics in a noisy or uncertain environment. The system of deterministic differential equations defining our model must therefore be perturbed by noise in a suitable way. The problem is especially difficult to treat because of the appearance of the ratio of host to parasite densities (describing the so-called density-dependent effect), which specify a direction in the tangent plane at a point of the physiological space describing caloric intake. Such spaces equipped with a suitable metric are examples of what are called *Finsler spaces*.

In the present article we solve the question of how to introduce noise into our model of Rothschild dynamics in the density-dependent case. We do this by implementing our theory of Finslerian diffusions (see our introductory article in this issue). As a result, we obtain a system of stochastic differential equations, namely, the deterministic differential equations governing the evolution of the host/parasite system perturbed by noise. We find weak solutions to the system of stochastic differential equations by means of the Feynman-Kac formula, which lends itself well to biologically relevant qualitative analysis. In this way, we discover a strictly Finslerian (that is density-dependent) phenomenon dominating, for example, the evolution of the rabbit/flea system during the final stages of a myxomatosis epizootic. We also use the Feynman-Kac formula coupled with the Cameron-Martin-Girsanov theorem to study the dependence of the solution on two important parameters, the so-called *average metabolic rate* and *efficiency* of the host/parasite system.

In the density-independent case such stochastic differential equations were introduced in [7], [8] and used to solve relevant problems like *nonlinear filtering* for myxomatosis. In such practical problems the *curvature* of the underlying space enters the final expression for the *signal estimates* [9], [10]. Solutions to density-dependent myxomatosis filtering problems will be given elsewhere.

2 Deterministic Host/Parasite Dynamics of Rothschild Type

Let $y^P(t)$ and $y^H(t)$ denote the flea and rabbit, or, more generally, Rothschild parasite and host population densities at time t. Let l_P and l_H denote the (positive) per capita energy intake rates, so that

$$\frac{dx^P}{dt} = l_P y^P, \qquad \frac{dx^H}{dt} = l_H y^H \tag{2.1}$$

hold, where x^P and x^H denote the (log) total caloric intake up to time t for the y^P and y^H populations, respectively. As will be seen, neither x^P nor x^H enter the final equations for dy^P/dt and dy^H/dt (in fact, the stochastic version of the differential equations derived below will also have a similar property), so that we may use l_P and l_H to rescale x^P and x^H. Consequently, we take $l_P = l_H = 1$ in all that follows. We develop the model in a series of steps arriving at (2.4) and then finally (2.8).

The simplest dynamics is ($\epsilon_P \geq \epsilon_H > 0$)

$$\frac{dy^P}{dt} = -\epsilon_P y^P, \qquad \frac{dy^H}{dt} = -\epsilon_H y^H,$$

which together with (2.1) yields

$$\frac{d^2x^P}{dt^2} + \epsilon_P \frac{dx^P}{dt} = 0, \qquad \frac{d^2x^H}{dt^2} + \epsilon_H \frac{dx^H}{dt} = 0, \tag{2.2}$$

where ϵ_P and ϵ_H are the *metabolic rates* (greater for the parasite). These equations mean that x^P and x^H increase over a fixed period (longer than the host reproductive cycle including gestation period). They express the so-called log Gompertz growth law if positive initial conditions are taken in (2.2).

We now include the host/parasite terms with *density-dependence* as follows:

$$\begin{cases} \frac{dy^P}{dt} &= -\epsilon_P y^P + \gamma_P \left(\frac{y^H}{y^P}\right)^{m-2} (y^H)^2 \\ \frac{dy^H}{dt} &= -\epsilon_H y^H - \gamma_H \left(\frac{y^P}{y^H}\right)^{m-2} (y^P)^2. \end{cases} \tag{2.3}$$

Here γ_P and γ_H are positive constants and m is a positive integer ≥ 2. Specifically, by *density-dependence* we mean that the coefficients multiplying $(y^P)^2$ and $(y^H)^2$ above depend on y^P and y^H. If $m = 2$, then, in fact, the coefficients are constant, i.e., *density-independent*. Note that if y^H is large relative to y^P in the first equation of (2.3), then dy^P/dt is relatively larger, while if y^P is large relative to y^H in the second equation, dy^H/dt is relatively more negative. We term γ_H the *virulence parameter*. It expresses the negative effect of the parasite on its host. It is large when the parasite is a disease carrier, as in myxomatosis. We call γ_P the *hormonal parameter*. It expresses the positive effect of host sexual hormone production on the parasites, as we shall see below in (2.4) and (2.5).

We now introduce the full dynamics in general form:

$$\begin{cases} \frac{dy^P}{dt} &= -\epsilon_P y^P - p_P(y^P)^2 + q_P y^P y^H + \gamma_P \left(\frac{y^H}{y^P}\right)^{m-2} (y^H)^2 \\ \frac{dy^H}{dt} &= -\epsilon_H y^H + p_H(y^H)^2 - q_H y^P y^H - \gamma_H \left(\frac{y^P}{y^H}\right)^{m-2} (y^P)^2. \end{cases} \tag{2.4}$$

Note that p_H, p_P, q_H, q_P are positive functions of y^P, y^H, t (and even, more generally, x^H and x^P), and that if q_P and γ_P are formally set to zero, then the parasite population crashes. On the other hand, if q_H and γ_H are formally set to zero, the host population can increase without bound. The $y^P y^H$ terms are the classical host/parasite interaction terms and, accordingly, occur with opposite signs. The function p_H is the *host reproductivity*, while p_P has a similar interpretation as the *parasite reproductivity* and is expected to be small in Rothschild type dynamics.

We now define the *Rothschild parameter* r $(0 < r < 1)$ by

$$\gamma_P = r p_H. \tag{2.5}$$

The latter equality expresses the Rothschild mechanism of parasite reproduction hormonally controlled by the host and will be called the *Rothschild constraint*. The parameter r indicates that a certain percentage of the host reproductivity p_H enters the expression for the parasite's rate of increase dy^P/dt and this is due to the ingestion of hormones in the parasite's regular blood meal.

It may be reasonable to consider the Rothschild type dynamics, if not as truly efficient metabolically, at least as quasi-efficient, as regards the joint caloric intake. For, in a much earlier evolutionary epoch, a Rothschild host had not yet evolved its special hormonal relationship with its parasite. At that time, the host's metabolism may well have been better adapted or more efficient. But, even if the parasite's metabolic rate ϵ_P was not very different from its host's ϵ_H, progenetic evolution is often towards smaller body size and simpler morphology and hence towards higher metabolic rates, so that in the present $\epsilon_P > \epsilon_H$. But let us extrapolate back in evolutionary time to the situation when approximately $\epsilon_P = \epsilon_H = \epsilon$. What does this imply about the dynamics (2.4)? Suppose all the coefficients are constants and $m = 2$, to begin with, in (2.1) and (2.4).

We invoke the so-called fundamental theorem of Volterra-Hamilton systems [11], [12]. This theorem states that there is only one set of constant coefficients possible, given γ_P and γ_H, which arise as those of constant coefficient Euler-Lagrange equations for an optimality principle based on a positive convex Lagrangian, C^∞ in x^P, x^H, y^P, y^H, t, and quadratic in y^P and y^H. (The quadratic condition really amounts to assuming a Riemannian metric). In fact, the Lagrangian for this problem is unique up to multiplication by a positive constant and is expressed by

$$L_2 = e^{2(\gamma_P x^P - \gamma_H x^H)+\epsilon t}\left[(y^P)^2 + (y^H)^2\right].$$

The coefficients are uniquely determined as

$$q_P = 2\gamma_H, \quad q_H = 2\gamma_P, \quad p_P = \gamma_P, \quad p_H = \gamma_H. \tag{2.6}$$

Let us now relax the condition $m = 2$ by replacing 2 by $m \geq 2$ to obtain the regular variational problem

$$\delta \int_{t_0}^{t_1} L_m dt = \delta \int_{t_0}^{t_1} e^{2(\gamma_P x^P - \gamma_H x^H)+\epsilon t}\left[(y^P)^m + (y^H)^m\right]^{2/m} dt = 0, \tag{2.7}$$

from which we obtain

$$q_P = \frac{m}{m-1}\gamma_H, \quad q_H = \frac{m}{m-1}\gamma_P, \quad p_P = \gamma_P, \quad p_H = \gamma_H.$$

The equations describing host/parasite dynamics are now readily seen to be

$$\begin{aligned}
\frac{dx^P}{dt} &= y^P \\
\frac{dx^H}{dt} &= y^H \\
\frac{dy^P}{dt} &= -\epsilon_P y^P - \gamma_P (y^P)^2 + \gamma_H \left(\frac{m}{m-1}\right) y^P y^H \\
&\quad + \frac{\gamma_P}{m-1} \left(\frac{y^H}{y^P}\right)^{m-2} (y^H)^2 \\
\frac{dy^H}{dt} &= -\epsilon_H y^H + \gamma_H (y^H)^2 - \gamma_P \left(\frac{m}{m-1}\right) y^P y^H \\
&\quad - \frac{\gamma_H}{m-1} \left(\frac{y^P}{y^H}\right)^{m-2} (y^P)^2.
\end{aligned} \tag{2.8}$$

Here $m \geq 2$, the condition $\epsilon_P = \epsilon_H = \epsilon$ is no longer enforced, and the Rothschild constraint expressing the host's hormonal control of the parasite reproductivity is now expressed as

$$\gamma_P = r\gamma_H,$$

instead of (2.5). The system (2.8) expresses *intrinsic social interactions* [13], [14].

If $y^H \sim 1$ in (2.8) and $y^P = \eta < 1$, then note that $dy^P/dt \sim \gamma_P/\eta^{m-2} \gg 1$ when $m \geq 3$. Similarly, $dy^H/dt \sim -\gamma_H/\eta_{m-2} \ll -1$ if $y^P \sim 1$, $y^H = \eta < 1$ and $m \geq 3$. The first estimate shows that parasites increase rapidly if the first density-dependence term $(y^H/y^P)^{m-2}$ in not unity, even when the host's population density y^H is small. The second estimate shows that the parasite can exterminate the host, but it need not be so, as a general rule. The case of myxomatosis is special. Generally, Darwinian natural selection will favor *attenuation of virulence* γ_H. The smaller the value, the better for both the host and the parasite.

We have seen that during the course of progenesis smaller size of the parasite is selected for, and that in our model increasing ϵ_P over ϵ_H in (2.8) expresses this in terms of metabolic rates. But what role do the density-dependent terms like $(y^H/y^P)^{m-2}$ play in progenesis? In an effort to separate effects, let us again return to $\epsilon_P = \epsilon_H = \epsilon$ by metaphorically going back in evolutionary time. It is easy to see that for fixed x^P, x^H, and t, the integrand L_m in (2.7) decreases with increasing m. This expression has the interpretation of *total intake rate* and its integral over a time interval is the total caloric intake for that interval. *To increase m is to decrease the total intake rate.* But this is consistent with smaller body size, too.

A smaller body size needs less calories, in absolute terms, under optimal conditions.

But what does progenesis mean for the host? Certainly not smaller body size or simpler morphology as in the parasite. Another aspect of progenesis is *precocious sexual maturation* [5]. One result of this would be increased reproductivity p_H through a time-sequencing change in host development. The optimality criterion ($\epsilon_P \sim \epsilon_H$) implies that the optimal intake history of the host population reflects the total growth (i.e., development) history very closely. Thus, $p_H = \gamma_H$ increments in (2.7) express heterochronic change as in [14] and the result is a decrease in L_m, the total intake rate, just as with increasing m. Recall that the optimality criterion forces $p_H = \gamma_H$ in our model and that if progenesis tends to increase p_H, attenuation of virulence γ_H tends to decrease it. Therefore, the effects tend to balance each other over evolutionary time scales, at least when the parasite becomes the disease vector of a powerful pathogen, as in the case of the myxoma virus.

This, then, is the meaning of m and γ_H for *progenetic evolution*. The effects of increasing m, γ_H, and $\epsilon_P - \epsilon_H$ are compounded in our model of host/parasite progenesis. Our model dynamics is only *quasi-efficient* and depends in a major way on $\epsilon_P > \epsilon_H$. In fact, we set

$$\xi = \frac{1}{\epsilon_P - \epsilon_H}$$

and call this the *efficiency* of the system. We also set $\epsilon = (\epsilon_P + \epsilon_H)/2$ and refer to it as the *average metabolic rate* (see [7]).

In the density-independent case, when $m = 2$, there is a unique positive equilibrium

$$y_0^P = \frac{r}{2\xi\gamma_H(1+r^2)}, \quad y_0^H = \frac{y_0^P}{r},$$

which is stable if $r > r_0$ (r_0 being the so-called *Hopf value*). It will bifurcate to a periodic solution of small amplitude as r is decreased through r_0. The Floquet *stability exponent* β_2 is given by

$$\xi\beta_2 = \frac{-8a_0(1-a_0^2)}{(3-4a_0^2)}, \qquad a_0 = \frac{1-r_0^2}{1+r_0^2},$$

and indicates a *stable limit cycle*. Furthermore, the *period adjustment* τ_2 to the basic period $2\pi/\omega$ with $\omega = 2\sqrt{(3-4a_0^2)}/\xi$ is

$$\tau_2 = 4\xi^2(7 + 8a_0^2 - 16a_0^4)\frac{1-q_0^2}{(3-4a_0^2)^2}\,.$$

The reader is invited to consult [7] for more subtle properties of the Rothschild ($m = 2$) dynamics. Again, note the important role played by ξ in these formulas. *It remains an open problem to determine the* (y^P, y^H)*–dynamics for* $m \geq 3$.

Let us conclude this section with a few remarks on the relations (2.6). First, note that $\gamma_P = r\gamma_H$, so that $\gamma_P < \gamma_H$, and $p_P = \gamma_P$. Thus, the negative effect of p_P is relatively small. For the special case of myxomatosis, the myxoma virus has less effect on the fleas than on the rabbits, and our *model predicts this*. Secondly, note that the mixed product terms $2\gamma_H y^P y^H$ and $-2\gamma_P y^P y^H$ indicate that host/parasite encounters are more beneficial to the parasite then they are a negative influence on the host. Finally, the host's reproductivity p_H is seen to precisely balance the effect of the virulence parameter γ_H in the case $m = 2$. Elsewhere (see [7]) it has been shown that for suitable choices of $\epsilon_P, \epsilon_H, \gamma_H, r_0$, and r, there is a stable limit cycle of amplitude proportional to

$$|r - r_0|^{1/2} + O(|r - r_0|).$$

Increasing γ_H moves (y_0^P, y_0^H) towards the origin $(0,0)$. Since $\gamma_P = r\gamma_H$, increasing γ_H allows us to select γ_P so that r_0 is unchanged. Consequently, the amplitude is unchanged as γ_H is increased and (y_0^P, y_0^H) moved towards the origin. Since the host population y_0^H is larger than y_0^P, the result of increasing γ_H must be total extinction of the host. Therefore, Darwinian natural selection will act to keep γ_H at moderate levels over the long time-scale, because dead hosts do not leave offspring. For myxomatosis, it has been observed that virulence does indeed attenuate. Therefore, our model is consistent with the evidence in this case.

3 Introduction to Noise for Rothschild Dynamics

As we have already mentioned, coevolution of host and parasite entails a noisy or uncertain environment [5]. Clearly, part of the environment of the parasite is the uncertain host population undergoing random fluctuations. We add noise effects by perturbing the deterministic dynamical equations (2.8). Our noise terms arise intrinsically from the dynamics itself, or, more precisely, from the metric structure of the physiological space prescribed by the ecological interactions. The metric function intrinsic to the host/parasite system (2.8) is given by

$$\bar{L}(x, y) = e^{\phi(x)} L(y), \tag{3.1}$$

where

$$L(y) = \left[(y^P)^m + (y^H)^m\right]^{1/m}, \tag{3.2}$$

$$\phi(x) = \gamma_P x^P - \gamma_H x^H,$$

with $x = (x^P, x^H)$ and $y = (y^P, y^H)$. The metric function $\bar{L}(x, y)$ is positively homogeneous of degree one in y and gives the physiological space M spanned by x the structure of a Finsler manifold. If $\epsilon_P = \epsilon_H$, then the solutions to (2.8) are geodesics in the Finsler metric (3.1). But even if $\epsilon_P > \epsilon_H$, the metric remains biologically relevant, for it captures the ability of the host/parasite system to respond to external perturbations from the environment. Namely, *a state (represented by a point in the physiological space M) will be displaced by a distance (measured with the Finsler metric (3.1)) proportional to the magnitude of the perturbation.* In the case of random perturbations this can be achieved if the noise terms to be added to (2.8) depend on the metric in such a way that the covariance matrix of the solution is equal to $\bar{g}^{ij}(x, y)$, the reciprocal tensor to the Finsler metric tensor $\bar{g}_{ij}(x, y) = 2^{-1}\partial_{(i)}\partial_{(j)}\bar{L}(x, y)$, where $\partial_{(i)} = \partial/\partial y^i$. This is our *general rule for noise addition.*

In the density-independent case when $m = 2$ the system (2.8) perturbed by noise has been studied in [8] and [10]. In that case (3.1) defines a Riemannian metric on M, the metric tensor

$$\bar{g}_{ij}(x) = e^{2\phi(x)}\delta_{ij} \tag{3.3}$$

being independent of y.

For $m = 2$ the perturbed system (2.8) has the general form

$$\begin{aligned} dx^i &= y^i dt + du^i \\ dy^i &= E^i dt - \Gamma^i_{jk} y^j y^k dt + dv^i, \end{aligned} \tag{3.4}$$

where $i, j, k \in \{P, H\}$, $E^P = -\epsilon_P y^P$, $E^H = -\epsilon_H y^H$, Γ^i_{jk} are the Levi-Civitá connection coefficients of the Riemannian metric (3.3), and u, v are independent perturbations. Here and throughout the present section we use the standard summation convention with respect to lower-case Latin indices $i, j, k, \ldots$, each of which can assume two values P and H. Since $\bar{g}_{ij}(x)$ is independent of y, the fibres of the tangent bundle TM have the structure of a *flat* Euclidean manifold. Therefore, a natural choice is that v be a standard Brownian motion in $\mathbb{R}^2$ (that is, formally, dv^i/dt are white noises). Such a perturbation dv^i added to the right-hand sides of the population equations (the last two equations in (2.8)) is not only consistent with the metric, but also follows the standard procedures in mathematical ecology. In the case of the physiological equations (the first two equations in (2.8)) the same external fluctuations may affect the dynamics in a different way if the system is at a different point $x = (x^P, x^H)$ in the *curved* physiological space M describing caloric intake. The perturbations du^i should therefore have the form

$$du^i = F^i_j(x) \circ dw^j,$$

where w is a standard Brownian motion in $\mathbb{R}^2$. The circle notation indicates that Stratonovich stochastic calculus is utilized. This is necessary in order that the dynamical equations be covariant not only before, but also after the perturbations are added. The requirement that the covariance matrix of dx^i be equal to $\bar{g}^{ij}(x)$ determines $F^i_j(x)$. Namely, the noise terms du^i must be of the form

$$du^i = e^{-\phi(x)} \circ dw^i. \tag{3.5}$$

The system (3.4) with u given by (3.5) and v, w independent Brownian motions in $\mathbb{R}^2$ was successfully used to solve several filtering problems for the case of myxomatosis. The reader is invited to consult the references [8] and [10] for more details.

Before proceeding to the density-dependent case when $m > 2$, we observe that the perturbation u is closely related to the so-called *Wagner connection* on M with coefficients $\bar{F}^i_{jk} = \phi_j \delta^i_k$, where $\phi_j = \partial\phi/\partial x^j$. Namely, it follows from (3.5) that

$$u^i(t) - u^i(0) - \frac{1}{2}\int_0^t \bar{g}^{jk}(x(s))\bar{F}^i_{jk} ds \tag{3.6}$$

is a martingale. (If the term $y^i dt$ is removed from the first equation of (3.4), then, in fact, u becomes a Markov diffusion on M with generator $2^{-1}\bar{g}^{jk}(\partial_j\partial_k - \bar{F}^i_{jk}\partial_i)$.) The Wagner connection is metrical relative to $\bar{g}_{ij}(x) = e^{2\phi(x)}\delta_{ij}$ and has torsion $\bar{T}^i_{jk} = \phi_j\delta^i_k - \phi_k\delta^i_j$. Its appearance proves crucial in the density-dependent case when $m > 2$ considered below.

We now relax the condition $m = 2$ and consider any integer $m \geq 2$. Then the physiological space M equipped with metric function $\bar{L}(x, y)$ defined by (3.1) has the structure of a Finsler space conformally equivalent to a Minkowski space with the m-th root metric given by (3.2), the conformal exponent $\phi(x) = \gamma_P x^P - \gamma_H x^H$ being linear in x. (On conformal transformations of Finsler spaces see [15] and [16].) As a general rule, all objects corresponding to the Minkowski metric $L(y)$ will be written without a bar, while those corresponding to the conformally transformed metric $\bar{L}(x, y)$ will be indicated by a bar. The metric tensors $g_{ij}(y)$ and $\bar{g}_{ij}(x, y)$ of the Minkowski metric and the conformally transformed metric are related by

$$\bar{g}_{ij}(x, y) = e^{2\phi(x)} g_{ij}(y).$$

The physiological manifold M with metric function $\bar{L}(x,y)$ has the structure of a Wagner space (see [16], [17]) with the natural Wagner connection $(\bar{N}^i_j, \bar{F}^i_{jk}, C^i_{jk})$, where

$$\bar{N}^i_j = \phi_j y^i, \quad \bar{F}^i_{jk} = \phi_j \delta^i_k, \quad C^i_{jk} = 2^{-1}\bar{g}^{il}\partial_{(l)}\bar{g}_{jk} = 2^{-1}g^{il}\partial_{(l)}g_{jk} \qquad (3.7)$$

are the nonlinear, horizontal, and vertical connection coefficients, respectively. (The vertical coefficients C^i_{jk} are written without a bar, since they are conformal invariants.) As has been seen in the density-independent case ($m = 2$), the Wagner connection is intrinsically related to the noise perturbations added to the deterministic equations (2.8). It is a deflection-free, $h-$ and $v-$metrical Finsler connection relative to the metric $\bar{L}(x,y)$ with non-vanishing torsion tensors $\bar{T}^i_{jk} = \phi_j\delta^i_k - \phi_k\delta^i_j$ and C^i_{jk}. (As is usual in Finsler geometry, C^i_{jk} play the double role of coefficients of one of the five Finsler torsion tensors and the vertical coefficients of the connection.)

The deterministic equations (2.8) for the host/parasite dynamics can be written in terms of the Wagner connection as follows (see [9] and note that here we use a different convention for the ordering of the indices j, k in $\bar{F}^i_{jk}$ and $\bar{T}^i_{jk}$) :

$$\begin{aligned} dx^i &= y^i dt \\ \bar{\delta}y^i &= \bar{g}^{ij}\bar{T}^l_{jk}\bar{g}_{lm}y^k y^m dt + E^i dt, \end{aligned} \qquad (3.8)$$

where

$$\bar{\delta}y^i = dy^i + \bar{N}^i_j dx^j, \qquad (3.9)$$

and where E^i are defined as before. The perturbed equations (3.8) have the form

$$\begin{aligned} dx^i &= y^i dt + du^i \\ \bar{\delta}y^i &= \bar{g}^{ij}\bar{T}^l_{jk}\bar{g}_{lm}y^k y^m dt + E^i dt + ds^i, \end{aligned} \qquad (3.10)$$

where

$$\bar{\delta}y^i = dy^i + \bar{N}^i_j \circ dx^j. \qquad (3.11)$$

The noise terms du^i and ds^i must transform like vectors and the Stratonovich circle notation must be used again to make sure that not only the deterministic equations (3.8), but also the stochastic differential equations (3.10) are covariant.

According to our *general rule for noise addition*, the perturbation terms du^i and ds^i are determined by the metric. Because the fibres of TM are non-Euclidean when $m > 2$ (each fibre is isometrically isomorphic to a Minkowski space with the $m^{-\text{th}}$ root metric; see [9]), the perturbations ds^i added to the population equations are no longer simple white noises. To find the expressions for the random perturbations du^i and ds^i in terms of the

Finsler metric $\bar{L}(x,y)$, we employ our stochastic version of Cartan's concept of parallel transport and moving frames, or, equivalently, the concept of hv–rolling a Finsler manifold along a curve introduced in [18], [19] and presented in our introductory article [20] on Finslerian diffusions in the present issue. The process of 'rolling' (or, equivalently, the transport of frame vectors) is controlled by the Wagner connection $(\bar{N}^i_j, \bar{F}^i_{jk}, C^i_{jk})$ defined by (3.7). As we shall see, this ensures, in particular, that

$$u^i(t) - u^i(0) - \frac{1}{2}\int_0^t \bar{g}^{jk}(x(s),y(s))\bar{F}^i_{jk}ds \tag{3.12}$$

is a martingale, in close analogy with (3.6).

We begin with hv–rolling TM on $\mathbb{R}^2 \times \mathbb{R}^2$ along a given solution $(x(t),y(t))$ to the deterministic equations (3.8). The curve $(\mu(t),\nu(t))$ traced on $\mathbb{R}^2 \times \mathbb{R}^2$ is defined by the system of differential equations

$$\begin{aligned} dx^i &= z^i_j d\mu^j \\ \bar{\delta} y^i &= z^i_j d\nu^j \\ dz^i_j &= -\bar{F}^i_{kl} z^l_j dx^k - C^i_{kl} z^l_j \bar{\delta} y^k, \end{aligned} \tag{3.13}$$

where $z(t)$ is an orthonormal frame in $TM_{x(t)}$ with respect to $\bar{g}(x(t),y(t))$. We now perturb $(\mu(t),\nu(t))$ by two independent Brownian motions w and v on $\mathbb{R}^2$, which represent the environmental perturbations, and use our theory of Finslerian hv–stochastic development (see [18], [19], and [20] in this issue) to roll the perturbed curve, or rather the stochastic process $(\mu(t)+w(t),\nu(t)+v(t))$ on $\mathbb{R}^2 \times \mathbb{R}^2$ back onto TM. The resulting diffusion $(x(t),y(t))$ on TM is defined by the system

$$\begin{aligned} dx^i &= y^i dt + z^i_j \circ dw^j \\ \bar{\delta} y^i &= \bar{g}^{ij}\bar{T}^l_{jk}\bar{g}_{lm} y^k y^m dt + E^i dt + z^i_j \circ dv^j \\ dz^i_j &= -\bar{F}^i_{kl} z^l_j \circ dx^k - C^i_{kl} z^l_j \circ \bar{\delta} y^k, \end{aligned} \tag{3.14}$$

of stochastic differential equations with the same initial conditions as in the deterministic case. Note that $\bar{\delta} y^i$ in (3.14) is given by (3.11), while $\bar{\delta} y^i$ in (3.13) is given by (3.9). Please also note that by an abuse of notation, we use the same symbols for the deterministic functions x, y, z in (3.13) and the stochastic processes x, y, z in (3.14).

The noise terms in (3.10) are therefore expressed as

$$du^i = z^i_j \circ dw^j, \qquad ds^i = z^i_j \circ dv^j,$$

where z_j^i satisfy the last equation of (3.14). In particular, since

$$\begin{aligned} du^i &= z_j^i \circ dw^j = z_j^i dw^j + 2^{-1} dz_j^i dw^j = z_j^i dw^j - 2^{-1} \bar{F}_{kl}^i z_j^l dx^k dw^j \\ &= z_j^i dw^j - 2^{-1} \bar{F}_{kl}^i z_m^k z_j^l \delta^{mj} dt = z_j^i dw^j - 2^{-1} \bar{g}^{kl}(x,y) \bar{F}_{kl}^i dt, \end{aligned}$$

it follows that (3.12) is indeed a martingale, as required to ensure analogy with the $m = 2$ case.

The process (x, y) defined by (3.14) is a Markov diffusion on TM with generator

$$\begin{aligned} D = 2^{-1} \bar{g}^{jk} (\bar{\delta}_j \bar{\delta}_k - \bar{F}_{jk}^i \bar{\delta}_i) + 2^{-1} \bar{g}^{jk} (\partial_{(j)} \partial_{(k)} - C_{jk}^i \partial_{(i)}) \\ + y^i \bar{\delta}_i + \bar{g}^{ij} \bar{T}_{jk}^l \bar{g}_{lm} y^k y^m \partial_{(i)} + E^i \partial_{(i)} \end{aligned} \tag{3.15}$$

expressed in terms of the *adapted frame* $(\bar{\delta}_i, \partial_{(i)})$, where $\bar{\delta}_i = \partial_i - \bar{N}_i^j \partial_{(j)}$, $\partial_i = \partial/\partial x^i$, and $\partial_{(i)} = \partial/\partial y^i$.

The orthonormal frame $z(t)$ in (3.13) and (3.14) plays an auxiliary role and has no particular biological meaning attached to it. However, the frame provides the mathematical machinery needed to define the notion of rolling, which is biologically relevant, as it embodies the *general principle* that the displacement of a biological state measured by the intrinsic metric is proportional to the magnitude of external perturbations. Here this principle is extended to the tangent bundle TM to account for the fact that the fibres of TM are non-Euclidean if $m > 2$. A natural metric on TM is given by the so-called *diagonal lift* $\bar{G}(x, y)$ of the Finsler metric tensor $\bar{g}(x, y)$ (see [21]). The matrix of the lifted metric tensor $\bar{G}(x, y)$ in the adapted frame $(\bar{\delta}_i, \partial_{(i)})$ is defined to be

$$\begin{pmatrix} \bar{g}_{ij}(x,y) & 0 \\ 0 & \bar{g}_{ij}(x,y) \end{pmatrix}.$$

For the diffusion (x, y) defined by (3.14) we have

$$\begin{pmatrix} dx^i dx^j & dx^i dy^j \\ dy^i dx^j & dy^i dy^j \end{pmatrix} = \begin{pmatrix} \bar{g}^{ij} & -\bar{g}^{ik} \bar{N}_k^j \\ -\bar{g}^{jk} \bar{N}_k^i & \bar{g}^{ij} + \bar{g}^{kl} \bar{N}_k^i \bar{N}_l^j \end{pmatrix} dt.$$

The covariance matrix of (x, y) standing on the right-hand side is equal to the matrix of the reciprocal tensor to $\bar{G}(x, y)$ in the *natural frame* $(\partial_i, \partial_{(i)})$, which expresses our *general rule for noise addition*.

The diagonal lift metric tensor $\bar{G}(x, y)$ defines a Riemannian metric on TM. In fact, it follows from the results of [22] that the diffusion (x, y) defined by (3.14) can be regarded as a Riemannian Brownian motion on TM with a certain drift, and the generator D of (x, y) given by (3.15) can be expressed as the Laplace-Beltrami operator on TM plus a certain vector field.

Because of the interpretation of $x^i(t)$ as the total caloric intake up to time t and $y^i(t)$ as the population densities at time t, the process $(x(t), y(t))$ must be restricted to the region TM^+ of TM where all the coordinates x^i and y^i are positive, i.e., the process must start at a point (x_0, y_0) in TM^+ and will be considered up to the first moment τ of hitting the boundary $\partial(TM^+)$.

Let $G_{\alpha\beta}(y)$ denote the Riemannian tensor on TM obtained by the diagonal lift of the Minkowski metric tensor $g_{ij}(y)$ and let

$$g = \det(g_{ij}), \qquad G = \det(G_{\alpha\beta}),$$

which implies that $g^2 = G$. We denote by $p(t, x, y)$ the probability density of the process (x, y) defined by (3.14) starting in TM^+ and stopped at time τ to be in a region $A \subset TM^+$ at time t relative to the measure $\sqrt{G(y)}dxdy$ on TM :

$$P\{(x(t), y(t)) \in A\} = \int_A p(t, x, y)\sqrt{G(y)}dxdy.$$

The probability density $p(t, x, y)$ satisfies the forward initial boundary value problem

$$\begin{aligned} D^*p &= \frac{\partial p}{\partial t} \\ \lim_{t \searrow 0} p(t, x, y) &= p_0(x, y) \\ p(t, x, y)\Big|_{(x,y)\in\partial(TM^+)} &= 0, \end{aligned} \tag{3.16}$$

where the initial density $p_0(x, y)$ is supported on TM^+ and D^* is the adjoint operator to D relative to the measure $\sqrt{G(y)}dxdy$. We compute D^* to find that

$$D^* = D + A^i\bar{\delta}_i + B^i\partial_{(i)} - 2E^i\partial_{(i)} + V$$

with

$$A^i = -3\bar{g}^{ij}\phi_j - 2y^i, \tag{3.17a}$$

$$B^i = \bar{g}^{ij}C^k_{kj} - 2\bar{L}^2\bar{g}^{ij}\phi_j + 2\phi_j y^j y^i, \tag{3.17}$$

$$E^i = -\epsilon y^i - \tilde{y}^i/2\xi, \tag{3.18}$$

$$V = 6\bar{g}^{ij}\phi_i\phi_j - I_\theta/2\bar{L}^2 + I^2/2\bar{L}^2 + 5y^i\phi_i + 2\epsilon + \tilde{y}^i C^j_{ji}/\xi, \tag{3.19}$$

where $\tilde{y}^P = y^P$, $\tilde{y}^H = -y^H$, $\epsilon = (\epsilon_P + \epsilon_H)/2$ is the average metabolic rate, and $\xi = 1/(\epsilon_P - \epsilon_H)$ is the efficiency of the system. Moreover,

$$I = \frac{(m-2)}{2\sqrt{(m-1)}} \frac{(1-z^m)}{\sqrt{z^m}}, \tag{3.20}$$

$$I_\theta = -\frac{m(m-2)}{4(m-1)} \frac{(1+z^m)^2}{z^m}, \tag{3.21}$$

where $z = y^H/y^P$, are the so-called *main scalar* and the derivative of the main scalar I with respect to the *Landsberg angle* θ (for definitions, see, for example, [23], [9]).

The forward initial boundary value problem (3.16) for the probability density $p(t, x, y)$ can be solved with the aid of the Feynman-Kac formula. To this end we introduce an auxiliary diffusion (X, Y) on TM defined by the following system of stochastic differential equations:

$$\begin{aligned} dX^i &= Z^i_j \circ dw^j + A^i dt \\ \bar{\delta} Y^i &= Z^i_j \circ dv^j + B^i dt - 2E^i dt \\ dZ^i_j &= -\bar{F}^i_{kl} Z^l_j \circ dX^k - C^i_{kl} Z^l_j \circ \bar{\delta} Y^k, \end{aligned} \tag{3.22}$$

where w and v are independent standard Brownian motions on $\mathbb{R}^2$ with respect to a fixed probability measure P.

We denote by $(X_{x,y}, Y_{x,y})$ the diffusion on TM defined by (3.22) with initial conditions $X(0) = x$ and $Y(0) = y$, where $(x, y) \in TM^+$. Moreover, let $\tau_{x,y}$ be the first hitting time of the boundary $\partial(TM^+)$ by $(X_{x,y}, Y_{x,y})$. Then the solution $p(t, x, y)$ to (3.16) can be expressed by means of the Feynman-Kac formula

$$\begin{aligned} p(t, x, y) = E\Big\{ &\chi_{\{\tau_{x,y} > t\}} p_0\big(X_{x,y}(t), Y_{x,y}(t)\big) \\ &\times \exp\left[\int_0^t V\big(X_{x,y}(s), Y_{x,y}(s)\big) ds \right] \Big\}, \end{aligned} \tag{3.23}$$

E being the expectation with respect to P and χ being the indicator function of a set.

The above Feynman-Kac formula can be used to study the behaviour of $p(t, x, y)$ depending on the parameters characterizing the host/parasite system. Below we demonstrate, as an example, how to study the dependence on the average metabolic rate ϵ and the efficiency ξ of the system. Then we discuss the role of density-dependence.

One can see form (3.18), (3.19), and (3.22) that the auxiliary diffusion $(X_{x,y}, Y_{x,y})$ and the 'potential' V in (3.23) both depend on ϵ and ξ.

The Cameron-Martin-Girsanov theorem can be used to extract the joint dependence on these parameters. We set

$$\psi^i = 2\zeta^i_j E^j, \quad d\tilde{w}^i = dw^i, \quad d\tilde{v}^i = dv^i - \psi^i dt, \tag{3.24}$$

where ζ^i_j is the dual frame to Z^i_j. Then the auxiliary diffusion $(X_{x,y}, Y_{x,y})$ defined by (3.22) satisfies the following system of stochastic differential equations:

$$\begin{aligned} dX^i &= Z^i_j \circ d\tilde{w}^j + A^i dt \\ \delta Y^i &= Z^i_j \circ d\tilde{v}^j + B^i dt \\ dZ^i_j &= -\bar{F}^i_{kl} Z^l_j \circ dX^k - C^i_{kl} Z^l_j \circ \delta Y^k, \end{aligned} \tag{3.25}$$

where, by the Cameron-Martin-Girsanov theorem, $\tilde{w}$ and $\tilde{v}$ are independent standard Brownian motions on $\mathbb{R}^2$ with respect to the probability measure $\tilde{P}$ defined by

$$d\tilde{P} = \exp\left(\int_0^t \delta_{ij}\psi^i dw^j - \frac{1}{2}\int_0^t \delta_{ij}\psi^i\psi^j ds\right) dP.$$

The Feynman-Kac formula can be written in terms of the expectation $\tilde{E}$ with respect to $\tilde{P}$ as

$$\begin{aligned} p(t,x,y) = \tilde{E}\Big\{ &\chi_{\{\tau_{x,y}>t\}} p_0\big(X_{x,y}(t), Y_{x,y}(t)\big) \\ &\times \exp\Big[-\int_0^t \delta_{ij}\psi^i dw^j + \frac{1}{2}\int_0^t \delta_{ij}\psi^i\psi^j ds \\ &+ \int_0^t V\big(X_{x,y}, Y_{x,y}\big) ds\Big]\Big\}. \end{aligned} \tag{3.26}$$

The dependence on ϵ and ξ is now only in the exponent, because, by (3.25), the probability law of $(X_{x,y}, Y_{x,y})$ with respect to the new probability measure $\tilde{P}$ is independent of ϵ and ξ. From (3.24) and (3.22) we find that

$$\delta_{ij}\psi^i dw^j - 2^{-1}\delta_{ij}\psi^i\psi^j dt = 2\bar{g}_{kl}E^k \circ d\bar{Y}^l - 2\bar{g}_{kl}E^k B^l dt + 2\bar{g}_{kl}E^k E^l dt - \nabla^v_k E^k dt,$$

where $\nabla^v_j A^i = \partial_{(j)} A^i + C^i_{jk} A^k$ is the vertical covariant derivative (often denoted by $A^i\big|_j$ and called the long-bar derivative). Thus, substituting for B^i and E^i from (3.17) and (3.18) and using (3.19), we can write the entire exponent in (3.26) in the form

$$\exp\Big\{2\epsilon\int_0^t d\bar{L} + (1/\xi)\int_0^t \bar{g}_{kl}\tilde{Y}^k \circ \bar{\delta}Y^l + \int_0^t \Big[2\bar{L}^2\phi_k/\xi - 2\phi_j Y^j Y^l \bar{g}_{kl}/\xi - C^i_{ki}/(2\xi)\Big]\tilde{Y}^k ds - \int_0^t 2\bar{g}_{kl}\big(\epsilon Y^k + \tilde{Y}^k/2\xi\big)\big(\epsilon Y^l + \tilde{Y}^l/2\xi\big)ds + \int_0^t \big[6\bar{g}^{ij}\phi_i\phi_j - I_\theta/(2\bar{L}^2) + I^2/(2\bar{L}^2) + 5Y^i\phi_i\big]ds\Big\}.$$

The above expression has readily derived consequences. For example, consider the case $m \geq 2$ with ϵ_P large compared to ϵ_H. This means that ϵ is large and ξ is small. Then the above exponent is dominated by

$$-2\int_0^t \bar{g}_{kl}(X,Y)\big(\epsilon Y^k + \tilde{Y}^k/2\xi\big)\big(\epsilon Y^l + \tilde{Y}^l/2\xi\big)ds,$$

which is negative when Y^P and Y^H are both positive. We conclude that progenesis degrades the transition density of stochastic Rothschild dynamics. This means that on average the process is speeded up, relatively speaking.

On the other hand, take $m > 2$ and consider fixed values of ϵ_H and ϵ_P (or, equivalently, of ϵ and ξ). Furthermore, suppose that y^H is small, but $y^P \approx C > 0$, where C is some constant. It follows from (3.20) and (3.21) that $I^2/(2\overline{L}^2)$ and $-I_\theta/(2\overline{L}^2)$ both tend to $+\infty$ as $y^H \to 0$ and $y^P \to C > 0$. All the remaining terms being bounded, these two strictly Finslerian terms involving the main scalar I will dominate the above exponent, resulting in increased probability density for a transition into the vicinity of the $y^H = 0$ axis, as one would expect in the case of the myxomatosis epizootic when the hosts (rabbits) die out. Increased transition density into a neighbourhood of the $y^H = 0$ boundary also means that the stochastic Rothschild dynamics will be slowed down, there. This slowdown makes good biological sense for the end of myxomatosis epizootic, for instance. The major point here is that it is due entirely to density-dependent (i.e., purely Finsler) effects, for if $m = 2$, then $I \equiv I_\theta \equiv 0$.

The above two conclusions are analogous to the Lyapunov stability results of the deterministic model, where $\epsilon_P - \epsilon_H$ is of moderate size or vanishing [9], [13]. The influence of $\epsilon_P - \epsilon_H$ is to destabilize the system, while density-dependence has the opposite effect.

References

[1] Rothschild, M. (1965) Fleas, *Sci. Am.*, **213**, 44-53. (1965).

[2] Lockley, R.M. (1964) *The Private Life of the Rabbit*, MacMillan.

[3] Rothschild, M. and Ford, B. (1964) Maturation and Egg Laying of the Rabbit Flea (Spilopsyllus Cuniculi DALE) Induced by the External Application of Hydrocortisone, *Nature*, **303**, 210-211.

[4] Rothschild, M. and Ford, B. (1972) Breeding Cycle of the Flea *Cedisopsylla Simplex* Controlled by the Breeding Cycle of Host, *Science*, **178**, 625-626.

[5] Gould, S.J. (1977) *Ontogeny and Phylogeny*, Belknap Press, Harvard Univ., Cambridge, Mass.

[6] Albrecht, P., Gould, S.J., Oster, G.F. and Wake, D.B. (1979) Size and Shape of Ontogeny and Phylogeny, *Paleobiology*, **5**, 296-317.

[7] Antonelli, P.L. and Seymour, R.M. (1988) A Model of Myxomatosis Based on Hormonal Control of Rabbit-Flea Reproduction, *IMA J. Math. Appl. in Med. & Biol.*, **5**, 65-80,

[8] Antonelli, P.L. and Seymour, R.M. (1987) Nonlinear Filtering of Myxomatosis Based on Hormonal Control of Flea Reproduction by the Rabbit, *Stoch. Anal. and Appl.*, **5**, 245-265.

[9] Antonelli, P.L. and Shimada, H. (1991) On 1-Form Finsler Connections with Constant Coefficients, *Tensor, N.S*, **50**, 263-275.

[10] Antonelli, P.L. (1990) Filtering the n-Dimensional Logistic Growth Model, *Stoch. Anal. and Appl.*, **8**, 263-292.

[11] Antonelli, P.L. (ed.) (1985) *Mathematical Essays on Growth and the Emergence of Form*, Univ. of Alberta Press, Edmonton.

[12] Antonelli, P.L. (1990) A Brief Introduction to Volterra-Hamilton Theory in Ecological Modelling, in Proc. of Int. Workshop on the Population Dynamics of Outbreaks, *Math. Comp. Mod.*, **13**, 19-23.

[13] Antonelli, P.L. (1993) On y-Berwald Connections and Hutchinson's Ecology of Social Interactions, *Tensor, N. S.*, **52**, 27-36.

[14] Antonelli, P.L., Bradbury, R., Krivan, V. and Shimada, H. (1993) A Dynamical Theory of Heterochrony: Time-Sequencing Changes in Ecology, Development and Evolution, *J. Biol. Syst.*, **1**, 451-487.

[15] Hashiguchi, M. (1976) On Conformal Transformations of Finsler Spaces , *J. Math. Kyoto Univ.*, **16** (1), 25-50.

[16] Hashiguchi, M. and Ichijyō, Y. (1977) On Conformal Transformations of Wagner Spaces, *Rep. of the Faculty of Sci., Kagoshima Univ.*, **10**, 19-25.

[17] M. Matsumoto, M. (1982) On Wagner Generalized Berwald Spaces of Dimension Two, *Tensor, N. S.*, **36**, 303-311.

[18] Antonelli, P.L. and Zastawniak, J.T. (1992) Diffusions on Finsler Manifolds, *Proc. of XXVth Symposium on Math. Phys*, Toruń, Poland.

[19] Antonelli, P.L. and Zastawniak, T.J. (1993) Stochastic Calculus on Finsler Manifolds and An Application in Biology, *Nonlinear World*, **1**, 149-171.

[20] Antonelli, P.L. and Zastawniak, T.J. (1994) Introduction to Diffusion on Finsler Manifolds, *Mathl. Comput. Modelling*, **20**, 109-116.

[21] Yano, K. and Davies, E.T. (1963) On the Tangent Bundles of Finsler and Riemannian Manifolds, *Rend. Circ. Mat. Palermo*, **12**, 211-218.

[22] Antonelli, P.L. and Zastawniak, T.J (1995) Diffusion on the Tangent and Indicatrix Bundles of a Finsler Manifold, *Tensor, N. S.*, **56**, 233-247.

[23] Matsumoto, M. (1986) *Foundations of Finsler Geometry and Special Finsler Spaces*, Kasheisha Press, Saikawa 3–23–2, Otsushi, Shigaken.

STOCHASTIC FINSLER GEOMETRY IN THE THEORY OF EVOLUTION BY SYMBIOSIS

P.L. ANTONELLI * and T.J. ZASTAWNIAK

1 Introduction

In this article we present an outline of our theory of Volterra-Hamilton systems of Finsler type with random perturbations. Such systems provide a mathematical model of colonial animals with social interactions. They involve a *cost functional* to be optimized by the system, giving rise to the mathematical structure of a Finsler manifold. Random perturbations of two sorts are realized as hv−diffusions and h−diffusions on this manifold. Apart from this biological motivation, Finslerian diffusion is of interest in its own right, since it generalizes the extensively studied notion of diffusion on Riemannian manifolds, which proved useful in various areas, particularly in mathematical physics. Of special interest to us is the relationship between diffusion and curvature. For example, curvature appears in the Feynman-Kac solution in our stochastic model of a two-species colony using hv−diffusions, see Section 5. In Section 7 we prove our recent result linking the quadratic dispersion of a diffusion on a Finsler manifold of Berwald type with the *Onsager-Machlup term* $R/12$, where R is the Cartan (or, equivalently, Berwald) scalar curvature of the Finsler manifold. An elementary introduction to our modelling approach can be found in *Volterra-Hamilton Models in the Ecology and Evolution of Colonial Organisms*, by P. Antonelli and R. Bradbury, World Scientific Press Ser. in Mathematical Biology and Medicine, New York, 1996. The basic references for Finslerian diffusion are [4], [5].

The renowned biologist L. Margulis has argued convincingly that body cells (soma) of modern plants and animals (eukaryotes) evolved by (endo)

*Supported in part by NSERC-A-7667. This paper was published in Vol. 3, 1997, p. 1-18 of *Dynamics of Cont., Discrete and Impulsive Systems*, Watam Press. It appears here in modified form with permission of the publisher.

symbiosis from separately living bacterial species some two billion years ago [1]. A necessary quantitative condition for this evolutionary change is that the average proliferation rate of the cells in separate closely living populations should be equal or nearly so, to allow their DNA, RNA replication machinery an association in which all members of the smaller cell type (eg. mitochondrian) can live inside the larger in one-to-one fashion. On the other hand, from a purely population perspective, exponentially growing bacterial populations living in a constrained niche must have equal growth rates, if both are to survive indefinitely, [1]. This equality of growth rates plays a very important role in our mathematical models of evolution by symbiosis. It is a prerequisite we call *the presymbiont condition*. The Volterra-Hamilton model of Section 5 shows that it is possible, *even in a noisy background defined by hv-diffusion*, for a colony formed by symbiosis of separately living, chemically exchanging, bacterial species to form a stable relationship, but that for large growth rates, this stability is only linearly dependent on chemical exchange, growth rate and population ratio.

For the model of evolution by (endo) symbiosis in which a colony of two cell-types near a steady-state, N_0^i, the addition of noise, by the $h-$diffusion method, results in a model organism displaying noisy chemical exchanges of small *vigor* (see Section 6), when the chemical exchange parameter, ν_3, is positive. Moreover, the vigor is linear in ν_3 and λ, when the growth rate λ is large. Of course, one expects large λ in a noisy (biochemical) environment, for this is the scenario of the classical *r-selective regime.*

2 Biological Motivation. Volterra-Hamilton Systems

By a *Volterra-Hamilton system* we mean a system of ordinary differential equations [1]

$$\begin{cases} \frac{dx^i}{dt} = N^i, \\ \frac{dN^i}{dt} + \sum_{j,k=1}^{n} \gamma_{jk}^i N^j N^k = \lambda N^i. \end{cases} \tag{1}$$

Here x^i denotes the amount of product, measured as total log biomass, due to all N^i individuals of kind $i = 1, \dots, n$ in an eco-developmental system consisting of n subpopulations, also called *producers*. These two equations are known, respectively, as the *production* and *population equations.* The coefficients γ_{jk}^i governing the dynamics can depend on all the x^i and also on the ratios for the N^i, that is, N^i/N^j. The constant growth rate $\lambda > 0$ is assumed to be the same for all subpopulations (this is the above mentioned pre-symbiont condition). The quantities ax^i, with $a > 0$, denote *chemicals produced in allometric proportion to biomass.*

The substitution $s = e^{\lambda t}$ defines an *intrinsic time scale,* turning the Volterra-Hamilton system (1) into

$$\frac{d^2x^i}{ds^2} + \sum_{j,k=1}^{n} \gamma^i_{jk} \frac{dx^j}{ds}\frac{dx^k}{ds} = 0. \tag{2}$$

The coefficients γ^i_{jk} are (smooth) C^∞ functions of the x^i and $y^i = dx^i/ds$ ($y^i \neq 0\ \forall i$) homogeneous of degree zero in y^i. We further require (2) to be the Euler-Lagrange equations for a convex (smooth) C^∞ Lagrangian or *cost functional* $F(x, y)$ that is homogeneous of degree one in y^i. This leads to *Finsler differential geometry* (which includes Riemannian geometry as a special case), see Section 3.

Example. For an *endosymbiotic bacterial colony allometrically exchanging chemicals* the Volterra-Hamilton System (2) becomes Euler-Lagrange equations (geodesics)

$$\begin{aligned} &\frac{d^2x^1}{ds^2} + \lambda(\alpha_1 - \nu_3 x^2)\cdot\left(\frac{dx^1}{ds}\right)^2 = 0,\\ &\frac{d^2x^2}{ds^2} + \lambda\Big(\alpha_2 + \frac{\nu_3}{\lambda+1}\,x^1\Big)\cdot\left(\frac{dx^2}{ds}\right)^2 = 0, \end{aligned} \tag{3}$$

where the cost functional $F(x, y)$ is given by

$$F(x,y) = \frac{(y^2)^{1+\frac{1}{\lambda}}}{(y^1)^{\frac{1}{\lambda}}}\cdot \exp[-\alpha_1 x^1 + (\lambda+1)\alpha_2 x^2 + \nu_3 x^1 x^2], \tag{4}$$

with constants $\lambda > 0$, $\alpha_1 > 0$, $\alpha_2 > 0$, $\nu_3 \neq 0$. The latter is called the *exchange parameter.* This $F(x, y)$ is clearly homogeneous of degree 1 in y. Furthermore, we can prove that replacing $-\nu_3 x^2$ by a general smooth function $\Phi_1(x^1, x^2)$ and $\frac{\nu_3}{\lambda+1}\,x^1$ by $\Phi_2(x^1, x^2)$, there are *no Riemannian* metric functions with geodesic equations which are of the form (above) required for our biological assumptions. Further, g_{ij} (see (8)) can be shown to be *positive-definite* on $y^1y^2 \neq 0$. In particular, if $\alpha_1 = \alpha_2 = \nu_3 = 0$,

$$g_{11} = \Big(1+\frac{2}{\lambda}\Big)\cdot\Big(\frac{1}{\lambda}\Big)\cdot\Big(\frac{y^2}{y^1}\Big)^{2+\frac{2}{\lambda}}; \quad g_{22} = \Big(1+\frac{2}{\lambda}\Big)\cdot\Big(1+\frac{1}{\lambda}\Big)\cdot\Big(\frac{y^2}{y^1}\Big)^{\frac{2}{\lambda}}$$

$$g_{12} = -\Big(\frac{2(\lambda+1)}{\lambda^2}\Big)\cdot\Big(\frac{y^2}{y^1}\Big)^{1+\frac{2}{\lambda}} = g_{21}.$$

From these expressions and the fact that for the *principle scalar*, I, of this Finsler geometry, $I^2 = \frac{(\lambda+2)^2}{\lambda+1} > 4$, the proof that g_{ij} is positive-definite follows, as does the specific classification of this 2-dimensional Finsler space, according to the classical scheme of L. Berwald, [1]. *Different values of, I, yield different geometries. What is novel, here, is that the growth rate, λ, determines this geometry uniquely* (see [1]). This is not true if $\nu_3 \neq 0$, for then curvature plays a role.

3 Finsler Manifolds

By M we denote a *smooth manifold* of finite dimension n and by TM the *tangent bundle* over M. Let $TM\backslash\{0\}$ be the so-called *slit tangent bundle* over M, that is, the tangent bundle with the zero section removed. We say that $F : TM\backslash\{0\} \to R$ is a *Finsler metric function* if it is smooth and

$$F(x, ky) = kF(x, y), \quad k > 0, \quad x \in M, \quad y \in TM_x, \tag{5}$$

that is, $F(x, y)$ is homogeneous of degree 1 in y.

Example. The cost functional (4) is a Finsler metric function.

Example. If M is a Riemannian manifold with metric tensor $g_{ij}(x)$, $x \in M$, then the Riemannian length $F(x, y)$ of a tangent vector $y \in TM_x$ given by

$$F^2(x, y) = g_{ij}(x)y^i y^j \tag{6}$$

is a Finsler metric function. From now on, we employ *the Einstein summation convention in repeated upper and lower indices.*

The assertion below is a simple consequence of Euler's theorem on homogeneous functions.

Proposition 1. *Every Finsler metric function $F(x, y)$ can be represented as*

$$F^2(x, y) = g_{ij}(x, y)y^i y^j, \tag{7}$$

where

$$g_{ij}(x, y) = 2^{-1}\dot{\partial}_i\dot{\partial}_j F^2(x, y) \tag{8}$$

is homogeneous of degree 0 in y and transforms like a tensor.

Here $\dot{\partial}_i = \partial/\partial y^i$. Later we shall also write $\partial_i = \partial/\partial x^i$. We call $g_{ij}(x, y)$ the *Finsler metric tensor.*

Definition. The pair (M, F) is called a *Finsler manifold* if

(a) $F(x, y)$ is a Finsler metric function on M,

(b) the Finsler metric tensor $g_{ij}(x,y)$ is non-degenerate. For diffusion theories, this must of course be strengthened to positive definiteness.

Finslerian differential geometry can be developed in much the same way as Riemannian geometry, even if the dependence on y involves a few necessary extensions such as the introduction of a *non-linear connection.* Namely, the coefficients of the *Cartan non-linear connection* are defined by

$$N^i_j = \dot{\partial}_j(\gamma^i_{k\ell}y^k y^\ell), \tag{9}$$

where $\gamma^i_{jk} = 2^{-1}g^{i\ell}(\partial_j g_{\ell k} + \partial_k g_{j\ell} - \partial_\ell g_{jk})$ are the Christoffel symbols. Since neither ∂_i nor dy^i transform like a vector or one-form, respectively, the non-linear connection is used to define the following objects, which satisfy the appropriate transformation laws:

$$\delta_i = \partial_i - N^j_i\dot{\partial}_j, \quad \delta y^i = dy^i + N^i_j dx^j. \tag{10}$$

Furthermore, the coefficients of the *horizontal and vertical Cartan connection* are defined in close analogy with the Levi-Civitá connection coefficients:

$$\begin{aligned} F^i_{jk} &= 2^{-1}g^{i\ell}(\delta_j g_{\ell k} + \delta_k g_{j\ell} - \delta_i g_{jk}), \\ C^i_{jk} &= 2^{-1}g^{i\ell}(\dot{\partial}_j g_{\ell k} + \dot{\partial}_k g_{j\ell} - \dot{\partial}_\ell g_{jk}). \end{aligned} \tag{11}$$

Definition. The triple $(N^i_j(x,y), F^i_{jk}(x,y), C^i_{jk}(x,y))$, N^i_j being the non-linear, F^i_{jk} the horizontal, and C^i_{jk} the vertical connection coefficients, is called the *Cartan connection.*

It is now possible to define the *horizontal and vertical covariant derivatives* of a Finsler vector field $A^i(x,y)$ by

$$\nabla^h_j A^i = \delta_j A^i + F^i_{jk}A^k, \quad \nabla^v_j A^i = \dot{\partial}_j A^i + C^i_{jk}A^k. \tag{12}$$

These formulae can be extended to arbitrary Finsler tensor fields in the usual way. The *Cartan connection is metrical*, that is,

$$\nabla^h_i g_{jk} = 0, \quad \nabla^v_i g_{jk} = 0. \tag{13}$$

The analogy with Riemannian differential geometry can be carried further, leading to the notions of *Finslerian torsion and curvature tensors.* Here the definitions are omitted for brevity. The interested reader is referred, for example, to the monographs [2,3]. Finslerian curvature plays an important role in studying the stability of deterministic Volterra-Hamilton systems [1].

4 Random Perturbations of Volterra-Hamilton Systems. Stochastic Finsler Geometry

The Volterra-Hamilton system (1) or (2) can be rewritten in the equivalent form

$$\begin{cases} \frac{dx^i}{ds} &= y^i, \\ \frac{\delta y^i}{ds} &= 0, \end{cases} \tag{14}$$

with δy^i given by (10). Any solution of this system is a geodesic on M under the Finsler metric $F(x, y)$. Random perturbations of this system amount to perturbing Finsler geodesics. This can be achieved by the following geometric construction (see [4,5] for details in the Finsler case and [6] for the general notions of stochastic calculus):

1. First M is *rolled isometrically* (without slipping) on a flat Euclidean space along the geodesic (14). This can be described by a system of ordinary differential equations. As a result, we obtain a straight line in the Euclidean space.
2. This straight line is perturbed by adding the standard Euclidean Brownian motion.
3. The resulting diffusion is transferred back to the manifold M by isometrical rolling. This rolling along the trajectories of a diffusion is described by a system of *stochastic differential equations* (SDEs).

As a result, we obtain a hv-diffusion on M defined by the system of SDEs

$$\begin{cases} dx^i &= y^i ds + z^i_j \circ dv^j, \\ \delta y^i &= z^i_j \circ dw^j, \\ dz^i_j &= -F^i_{k\ell}(x,y) z^\ell_j \circ dx^k - C^i_{k\ell}(x,y) z^\ell_j \circ \delta y^k. \end{cases} \tag{15}$$

Here v_s and w_s are *two independent standard Brownian motions* in R^n, the circle $\circ$ indicates that these are *Stratonovich* SDEs and

$$\delta y^i = dy^i + N^i_j(x,y) \circ dx^j. \tag{16}$$

Furthermore, $g_{ij}(x, y)$ is assumed positive definite on $y^1 y^2 \neq 0$, and z^i_j is an auxiliary process taking values in the space of orthonormal frames over M, which is necessary to control the position of M relative to the Euclidean space in the course of rolling. It turns out that the solution (x_s, y_s) is a Markov diffusion on TM with infinitesimal generator Δ_F, which is independent of the choice of the initial orthonormal frame z^i_j.

Definition. If the drift term $y^i ds$ is removed from the first equation in (15), then the corresponding infinitesimal generator is called the hv – *Laplace–Beltrami operator* on the Finsler manifold M. Accordingly, the diffusion defined by (15) is called an *hv-Brownian motion* with drift y^i.

5 Application to the Growth of a Colonial Animal

The cost functional $F(x, y)$ for a two-species colony is given by (4). This will play the role of our Finsler metric function. Consider the corresponding Markov diffusion (x_s, y_s) defined by (15). This can be regarded as the Volterra-Hamilton system (3) perturbed by noise. Let Δ_F and the corresponding Finslerian hv–Laplace – Beltrami operator.

The *probability density* $\rho(s, x, y)$ of (x_s, y_s) satisfies the forward initial-boundary value problem

$$\begin{cases} \Delta_F^* \rho = \frac{\partial \rho}{\partial s}, \\ \lim_{s \searrow 0} \rho(s, x, y) = \rho_0(x, y), \\ \rho(s, x, y)\big|_{y^1 y^2 = 0} = 0. \end{cases} \tag{17}$$

Here $\rho_0(x, y)$ is a given initial probability distribution and Δ_F^* is the adjoint operator to Δ_F. The boundary condition in the third equation is imposed because of our interpretation of y^1 and y^2 as production rates (intrinsic time). These must have non-negative values and therefore cannot cross the boundary $y^1 y^2 = 0$.

The solution of this initial-boundary value problem can be represented by the Feynman-Kac formula:

$$\rho(s, x, y) = E\Big\{\rho_0(X_s^{x,y}, Y_s^{x,y}) e^{\int_0^s V(X_u^{x,y}, Y_u^{x,y}) du}\Big\},$$

where $(X_s^{x,y}, Y_s^{x,y})$ is the *backward diffusion* corresponding to (15) that starts from (x, y) at $s = 0$ and E is the mathematical expectation. The 'potential' V in the Feynman-Kac formula can be evaluated in terms of the so-called *Berwald scalar curvature* K of the Finsler metric (4). This makes it possible to study the behaviour of the probability density $\rho(s, x, y)$ as it depends on the curvature. Namely, *large positive curvature will cause ρ to increase locally for small times s, while large negative curvature will have the opposite effect* (see [4] for more details). This is consistent with the well-known results on the stability of solutions of a *deterministic system* depending on the sign of the curvature (see [3, Chap. 4.4]). For metric (4) above, computations lead to

$$K = \frac{\lambda^2}{\lambda+1} \cdot \nu_3 \cdot \left(\frac{N^1}{N^2}\right)^{1+\frac{2}{\lambda}} \cdot \mathrm{Exp}\,\{-2[-\alpha_1 x^1 + (\lambda+1)\alpha_2 x^2 + \nu_3 x^1 x^2]\}. \quad (18)$$

This quantity is positive if and only if $\nu_3 > 0$, and ρ will increase locally for small times. *For large growth rates* λ, K *is linear in* λ, *in the exchange parameter* ν_3, *and linear in the ratio*, N^1/N^2.

6 Relationship Between Finslerian Diffusion and Curvature

As in the well-known Riemannian case, one would expect Finslerian diffusion to be closely related to the curvature of the manifold. The Feynman-Kac solution above involving the Berwald scalar curvature K provides a confirmation of this relationship for hv–diffusions.

Here, we state our recent result on a direct link between *Finslerian h-diffusion and curvature.* The h–Laplace – Betrami operator is generator of (15) with $\delta y^i = 0$, in (16), (see (21) and (23)). The result applies only to a class of Finsler manifolds that admit normal coordinates, i.e. so-called Berwald spaces (see [1,2,3]). The proof is quite complicated. The main step involves a generalization from the Riemannian to the Finslerian case of the Cartan formula expressing the metric tensor g_{ij} in terms of the curvature tensor and its derivatives in normal coordinates and is given in the next section.

Note that the drift term, $y^i ds$, of (15) has no counterpart in (21). The reason is that *these* x*–variables are, in the case* $\nu_3 = 0$, *the* so-called *real growth* variables, which require the population densities N^i to exhibit a *unique positive equilibrium*, N_0^i, so that, in real time,

$$x^i(t) := \int_0^t (N^i(\tau) - N_0^i)d\tau + D^i$$

holds (see the Antonelli/Bradbury book mentioned in the introduction). Thus, dx^i/dt vanishes at this equilibrium and so does $y^i := dx^i/ds$, in terms of the intrinsic time s. The relations (3) and (4) are assumed to hold when $\nu_3 \neq 0$. *The noise, then, reflects stochastic effects at the level of maintenance and repair close to the steady-state.*

Let (x_s, y_s) be the Finslerian diffusion defined by (21) below. For $h > 0$ we define the *Quadratic Dispersion* to be (in normal coordinates)

$$\mathcal{U}_h(x_s, y_s) = \tfrac{1}{2}\, g_{ij}(x_s, y_s) E_s\{x^i_{s+h} x^j_{s+h}\}, \quad (19)$$

where $E_s\{\cdot\}$ is the conditional expectation given x_s, y_s. Then we define the *relative vigor*, $\mathcal{V}$, of the diffusion to be the *negative of the coefficient multiplying* h^2 in the expansion $\mathcal{U}_h = A + Bh + Ch^2 + o(h^2)$.

Theorem 2. *For every Finsler space* (M, F) *that is Berwald*

$$\mathcal{U}_h = \frac{\dim M}{2}\, h + \frac{R}{12}\, h^2 + o(h^2), \tag{20}$$

that is, the relative vigor, $\mathcal{V}$, *of the Finsler* h–diffusion (x_s, y_s) *is equal to* $-R/12$, *where* R *is the Cartan scalar curvature (which is equal to the Berwald scalar curvature,* $R = K$, *in the case of Berwald spaces.)*

The term $R/12$ is clearly the Finslerian analogue of the Onsager-Machlup term well known in stochastic calculus on Riemannian manifolds [7,8,9]. If $K << 0$, the relative vigor is large, which implies *weak chaos* for $n = 2$, [2]. If $K >> 0$, $\mathcal{V}$ is small, in fact negative. From (18), the vigor is small when $\nu_3 > 0$. Yet $\nu_3 < 0$ implies chaos. This is a completely unexpected consequence.

7 Computation of the Quadratic Dispersion Formula

We split this difficult section into three subsections.

7.1 Notation, assumptions, and auxiliary results

M a Berwald space with metric function $L(x, y)$

$g_{ij} = \frac{1}{2}\, \dot{\partial}_j \dot{\partial}_j L^2$ metric tensor, assumed to be positive definite

$\gamma^i_{jk} = \frac{1}{2}\, g^{im}(\partial_j g_{mk} + \partial_k g_{jm} - \partial_m g_{jk})$ Levi-Civitá symbols

$G^i = \gamma^i_{jk} y^j y^k$

$N^i_j = G^i_j = \dot{\partial}_j G^i$ nonlinear coefficients of the Cartan and Berwald connections

$F^i_{jk} = G^i_{jk} = \dot{\partial}_j G^i_k = \dot{\partial}_j \dot{\partial}_k G^i$ horizontal coefficients of the Cartan and Berwald connections (equal because M is a Berwald space)

$C^i_{jk} = \frac{1}{2}\, g^{im}\dot{\partial}_m g_{jk}$ vertical coefficients of the Cartan connection

$R = K$ scalar h-curvature of the Cartan connection,equal to scalar K-curvature.

$\exp_x^{-1}$ inverse exponential map from a neighbourhood of $x \in M$ to TM_x

NC_x normal coordinates at $x \in M$, they exist because M is a Berwald space.

We will now employ, t, as arc-length parameter i.e. *in this section, t replaces s, of the previous sections.*

Lemma 3. *Fix any $x \in M$. Then*

(a) $N^i_j(x,y) = G^i_j(x,y) = 0 \quad \text{in} \quad NC_x \quad \forall\, y \in TM_x,$

(b) $F^i_{jk}(x,y) = G^i_{jk}(x,y) = 0 \quad \text{in} \quad NC_x \quad \forall\, y \in TM_x,$

(c) $\delta_i g_{jk}(x,y) = 0 \quad \text{in} \quad NC_x \quad \forall\, y \in TM_x,$

(d) $\delta_i \dot{\partial}_j f(x,y) = \dot{\partial}_j \delta_i f(x,y) \quad \text{in} \quad NC_x \quad \forall\, y \in TM_x \quad \forall\, f,$

(e) $\delta_h F^i_{jk}(x,y) + \delta_k F^i_{hj}(x,y) + \delta_j F^i_{kh}(x,y) = 0 \quad \text{in} \quad NC_x \quad \forall\, y \in TM_x.$

Proof: Since the geodesic equation in normal coordinates reads $\frac{d^2x^i}{dt^2} = 0$, it follows that

$$G^i(x,y) = 0 \quad \text{in} \quad NC_x \quad \forall\, y \in TM_x.$$

This immediately yields (a) and (b), since $N^i_j = G^i_j = \dot{\partial}_j G^i$ and $F^i_{jk} = G^i_{jk} = \dot{\partial}_j G^i_k = \dot{\partial}_j \dot{\partial}_k G^i$. Now (c) follows from (b) because of the $h-$metricity condition $\delta_i g_{jk} = F^m_{ij} g_{mk} + F^m_{ik} g_{jm}$. Obviously, (d) follows from (a). To prove (e) write down the geodesic equation as

$$\frac{d^2x^i}{dt^2} + F^i_{jk}\left(x, \frac{dx}{dt}\right)\frac{dx^j}{dt}\frac{dx^k}{dt} = 0$$

and differentiate both sides with respect to t to get

$$y^h y^i y^j \delta_h F^i_{jk}(x,y) = 0 \quad \text{in} \quad NC_x \quad \forall\, y \in TM_x.$$

Differentiate this three times with respect to y to get (e). This is very much like the Riemannian case, except now one has to deal with the dependence of $\delta_h F^i_{jk}$ on y when differentiating. But, by (d) and (b),

$$\dot{\partial}_m \delta_h F^i_{jk}(x,y) = \delta_h \dot{\partial}_m F^i_{jk}(x,y) = 0 \quad \text{in} \quad NC_x \quad \forall\, y \in TM_x,$$

so it works out, as required.

7.2 Finslerian version of Cartan's lemma for Berwald spaces

In this section the arguments (x, y) will be omitted for brevity.

Lemma 4. (a) $g^{hk}g^{ij}\delta_h\delta_k g_{ij} = \frac{2}{3}\,R$ in NC_x $\forall\, y \in TM_x$,

(b) $\delta_i F^i = -\,\frac{2}{3}\,R$ in NC_x $\forall\, y \in TM_x$.

Proof: By contracting indices in formula (e) of the Lemma in Sect. 7.1, we get

$$0 = 2g^{ij}\delta_j F^k_{ki} + \delta_i F^i \quad \text{in} \quad NC_x \quad \forall\, y \in TM_x.$$

Furthermore,

$$\begin{aligned} R = K = g^{hk}g^{ij}K_{hijk} &= g^{hk}g^{ij}(\delta_j F_{khi} - \delta_k F_{jhi}) \\ &= g^{ij}\delta_j F^k_{ki} - \delta_i F^i \quad \text{in} \quad NC_x \quad \forall\, y \in TM_x. \end{aligned}$$

By adding these two equalities, we obtain (a). Subtracting the second equality twice from the first one gives (b).

Remark. In the Riemannian version of Cartan's lemma there is no contraction of indices, so the full curvature tensor features there. This may well be possible in the Finslerian case too, but would be harder to prove and the above is just what is needed for our purposes.

7.3 Quadratic dispersion and relative vigor

Finslerian h–diffusion x_t, y_t is defined by the system of SDEs

$$\begin{cases} dx^i &= z^i_j \circ dw^j, \\ \delta y^i &= 0, \\ dz^i_j &= -F^i_{k\ell}(x,y)z^\ell_j \circ dx^k, \end{cases} \tag{21}$$

where $\delta y^i = dy^i + N^i_j(x,y) \circ dx^j$ and z^i_j is an auxiliary orthonormal frame process (ref. [4]). By the Itô formula, for any smooth function f

$$df(x,y) = (\delta_i f)z^i_j dw^j + (Df)dt, \tag{22}$$

where

$$D = \tfrac{1}{2}\, g^{ij}(\delta_i\delta_j - F^k_{ij}\delta_k) \tag{23}$$

is the generator of the diffusion x_t, y_t. Formula (23) defines the h–*Laplace–Beltrami* operator on the Finsler manifold and (21) defines the corresponding h-*Brownian motion*.

Let us consider the *quadratic dispersion*

$$\mathcal{U}_h(x_t, y_t) := \tfrac{1}{2}\, g_{ij}(x_t, y_t) E_t\{[\exp_{x_t}^{-1} x_{t+h}]^i [\exp_{x_t}^{-1} x_{t+h}]^j\},$$

where $E_t\{\cdot\}$ is the conditional expectation given x_t, y_t. Thus

$$\mathcal{U}_h(x_t, y_t) = \tfrac{1}{2}\, g_{ij}(x_t, y_t) E_t\{x_{t+h}^i x_{t+h}^j\} \quad \text{in} \quad NC_x. \tag{24}$$

We define the *relative vigor*, $\mathcal{V}$ to be the negative of the coefficient of h^2 in the Taylor formula for $\mathcal{U}_h$.

Remark. The reader may compare our definition of relative vigor $\mathcal{V}$ of a diffusion on a Finsler manifold of Berwald type with the notion of kinetic energy $U = \frac{1}{2} g_{ij}(x)\dot{x}^i \dot{x}^j$ of a smooth trajectory x_t on a Riemannian manifold as used in classical mechanics. The latter can be defined by

$$\mathcal{U}_h(x_t) = \tfrac{1}{2}\, g_{ij}(x_t)[\exp_{x_t}^{-1} x_{t+h}]^i [\exp_{x_t}^{-1} x_{t+h}]^j = h^2 U + o(h^2).$$

Theorem 5. *In the case of Finslerian h-Brownian motion x_t, y_t defined by* (21)

$$\mathcal{U}_h = \frac{\dim\, M}{2}\, h + \frac{R}{12}\, h^2 + o(h^2), \tag{25}$$

that is, for the relative vigor,

$$\mathcal{V} = -\, \frac{R}{12}\, .$$

Proof: For an arbitrary smooth function f the Itô formula (22) implies that

$$\begin{aligned} E_t f(x_{t+h}, y_{t+h}) &= f(x_t, y_t) + \int_t^{t+h} E_t Df(x_s, y_s) ds \\ &= f(x_t, y_t) + hDf(x_t, y_t) + o(h). \end{aligned} \tag{26}$$

Applying the Itô formula once more, this time to the integrand $Df(x_s, y_s)$ in (26), we obtain

$$\begin{aligned} E_t f(x_{t+h}, y_{t+h}) &= f(x_t, y_t) + \int_t^{t+h} \left[Df(x_t, y_t) + \int_t^s E_t DDf(x_r, y_r) dr\right] ds \\ &= f(x_t, y_t) + hDf(x_t, y_t) + \frac{h^2}{2}\, DDf(x_t, y_t) + o(h^2). \end{aligned} \tag{27}$$

This is a stochastic version of Taylor formula. However, D is a second order operator here.

Now, for any fixed $x_0 \in M$ and $y_0 \in TM_{x_0}$ we take the diffusion x_t, y_t defined by (21) that starts from x_0, y_0 and set

$$f(x,y) := x^i x^j \quad \text{in} \quad NC_{x_0}.$$

Then, by (24) and (27),

$$\begin{aligned}\mathcal{U}(x_0,y_0) = &\ \frac{h}{2}\, g_{ij}(x_0,y_0)D(x^i x^j)\\ &+ \frac{h^2}{4}\, g_{ij}(x_0,y_0)DD(x^i x^j) + o(h^2) \quad \text{in} \quad NC_{x_0}.\end{aligned} \tag{28}$$

Below the arguments (x_0, y_0) will be omitted for brevity. From (23) we have

$$D(x^i x^j) = g^{ij} - \tfrac{1}{2}\, x^i F^j - \tfrac{1}{2}\, x^j F^i.$$

The last two terms vanish in NC_{x_0}, so that

$$g_{ij}D(x^i x^j) = g_{ij}g^{ij} = \dim M \quad \text{in} \quad NC_{x_0},$$

which by (28) gives the first term in (25). To find the second term, i.e. the term U, we compute

$$\begin{aligned} g_{ij}DD(x^i x^j) &= g_{ij}D(g^{ij} - x^i F^j) = \tfrac{1}{2}\, g_{ij}g^{hk}\delta_h\delta_k g^{ij} - \delta_i F^i \\ &= -\, \tfrac{1}{2} g^{ij}g^{hk}\delta_h\delta_k g_{ij} - \delta_i F^i = \tfrac{1}{3}\, R \quad \text{in} \quad NC_{x_0}.\end{aligned}$$

Here we used (23) and the fact that $x_0^i = 0$ and $F^i_{jk}(x_0, y_0) = 0$ in NC_{x_0}. The identity $g_{ij}g^{ij} = \dim M$ was used in the second equality and (a) and (b) of the Lemma in Sect. 7.2 were used in the last equality. Hence (25) follows from (28). In particular, we find that $-\mathcal{V} = R/12$, as required.

8 Acknowledgements

The authors would like to thank Vivian Spak for her excellent preparation of this manuscript.

References

[1] Antonelli, P.L., Ingarden, R.S. and Matsumoto, M. (1993) *The Theory of Sprays and Finsler Spaces with Applications in Physics and Biology*, Kluwer Academic Publishers, Dordrecht-Boston-London.

[2] Rund, H. (1959) *The Differential Geometry of Finsler Spaces*, Springer-Verlag, Berlin.

[3] Matsumoto, M. (1986) *Foundations of Finsler Geometry and Special Finsler Spaces*, Kaiseisha Press, Saikawa, Shigaken, Japan.

[4] Antonelli, P.L. and Zastawniak, T.J. (1994) Stochastic Calculus on Finsler Manifolds and An Application in Biology, *Nonlinear World,* **1**, 149-171.

[5] Antonelli, P.L. and Zastawniak, T.J. (1994) Density-Dependent Host/Parasite Systems of Rothschild Type and Finslerian Diffusion, *Mathl. Comput. Modelling*, **20**, 117-129.

[6] Ikeda, N. and Watanabe, S. (1981, 1989) *Stochastic Differential Equations and Diffusion Processes*, North Holland, Amsterdam, Kodansha, Tokyo.

[7] Onsager, L. and Machlup, S. (1953) Fluctuations of Irreversible Processes. I, II, *Phys. Rev.*, **91**, 1505-1512, 1512-1515.

[8] Takahashi, Y. and Watanabe, S. (1981) The Probability Functions (Onsager-Machlup Functions) of Diffusion Processes, *Lecture Notes Math.*, **851**, 434-463.

[9] Fujita, T. and Kotani, S. (1982) The Onsager-Machlup Function For Diffusion Processes, *J. Math. Kyoto Univ.*, **22**, 115-130.

[10] Molchanov, S.A. (1975) Diffusion Processes and Riemannian Geometry, *Russian Math. Surveys*, **30**, 1-63.

DIFFUSIONS ON FINSLER MANIFOLDS

P.L. ANTONELLI * and T.J. ZASTAWNIAK

1 Introduction

Let M be a finite-dimensional Finsler manifold with arc element $ds = L(x^1, \dots, x^n, dx^1, \dots, dx^n)$, L being a non-negative smooth homogeneous function of degree one in dx^i. The Finsler metric tensor

$$g_{ij}(x,y) = 2^{-1}\dot{\partial}_i\dot{\partial}_j L^2(x,y),$$

where $\dot{\partial}_i = \partial/\partial y^i$, will be assumed to be positive definite for all $x \in M$ and $0 \neq y \in TM_x$. (For some examples, $y^1 \dots y^n \neq 0$, is required.) Let M be equipped with the Cartan connection $\left(N^i_j(x,y), F^i_{jk}(x,y), C^i_{jk}(x,y)\right)$, N^i_j being the nonlinear, F^i_{jk} the horizontal, and C^i_{jk} the vertical connection coefficients, which can be expressed as

$$\begin{aligned} N^i_j &= \dot{\partial}_j\left(\gamma^i_{kl}y^k y^l\right), \\ F^i_{jk} &= 2^{-1}g^{il}\left(\delta_j g_{lk} + \delta_k g_{jl} - \delta_l g_{jk}\right), \\ C^i_{jk} &= 2^{-1}g^{il}\left(\dot{\partial}_j g_{lk} + \dot{\partial}_k g_{jl} - \dot{\partial}_l g_{jk}\right) = 2^{-1}g^{il}\dot{\partial}_j g_{lk}, \end{aligned}$$

where $\dot{\partial}_i = \partial/\partial y^i$ as above, $\delta_i = \partial_i - N^j_i\dot{\partial}_j$ with $\partial_i = \partial/\partial x^i$, and $\gamma^i_{jk} = 2^{-1}g^{il}\left(\partial_j g_{lk} + \partial_k g_{jl} - \partial_l g_{jk}\right)$. Here and throughout the paper we use the standard summation convention. The horizontal and vertical covariant derivatives of a Finsler vector field $A^i(x,y)$ will be denoted by

$$\nabla^h_j A^i = \delta_j A^i + F^i_{jk}A^k, \qquad \nabla^v_j A^i = \dot{\partial}_j A^i + C^i_{jk}A^k.$$

These formulae can be extended to any Finsler tensor fields in the usual way. The Cartan connection is metrical, i.e.,

$$\delta_i g_{jk} = F^l_{ij}g_{lk} + F^l_{ik}g_{jl}, \qquad \dot{\partial}_i g_{jk} = C^l_{ij}g_{lk} + C^l_{ik}g_{jl}, \tag{1.1}$$

*Partially supported by NSERC A-7667. This article appeared in *Rep. Math. Phys.*, **33**, (1993), 303-315.

and deflection-free, i.e., $N^i_j = F^i_{jk}y^k$. A systematic presentation of Finsler geometry can be found in the classical monograph [13] by Rund. For a modern approach to Finsler manifolds see Matsumoto's monograph [11]. We also refer to Cartan's original work [3].

The dependence of the metric of a Finsler manifold on the direction of the tangent vector y offers more flexibility than a Riemannian metric, providing a suitable geometric framework for applications in areas such as variational calculus, mechanics, theoretical physics (e.g., it facilitates the description of anisotropic properties of the physical space; see, for example, [7]), and developmental and population biology [1]. Our results on diffusions on a Finsler manifold reported in this paper, which generalize the well-known results on diffusions on a Riemannian manifold [4], [8], [9], [10], [5], [12] have been developed with a view to applications in the above areas.

Throughout the paper we shall use the standard differential notation of stochastic calculus. Instead of writing the Itô stochastic integral with respect to a diffusion (or, more generally, a semimartingale) $x(t)$ in the form $\int y\,dx$, we shall write simply $y\,dx$. By $dx\,dy = d(xy) - x\,dy - y\,dx$ we denote the joint quadratic variation of x and y. The Itô change-of-variables formula can then be written as

$$df(x) = \partial_i f(x)\,dx^i + 2^{-1}\partial_i\partial_j f(x)\,dx^i dx^j \tag{1.2}$$

for a sufficiently smooth function $f : \mathbf{R}^n \to \mathbf{R}$. We shall often use the Stratonovich stochastic integral $\int y \circ dx$, or simply $y \circ dx$ in the differential notation, defined by $y \circ dx = y\,dx + 2^{-1}dy\,dx$. The Itô formula (1.2) can be expressed in terms of the Stratonovich integral as

$$df(x) = \partial_i f(x) \circ dx^i,$$

which has the same form as in ordinary calculus. By a diffusion $x(t)$ on a manifold M we shall mean a solution to a stochastic differential equation of the form

$$dx^i = A^i_j(x) \circ dw^j + B^i(x)\,dt, \tag{1.3}$$

where $w(t)$ is a Brownian motion in $\mathbf{R}^n$ and $A_1, \ldots, A_n, B$ are vector fields on M.

2 Finslerian Stochastic Parallelism

In the present section the concept of stochastic parallel transport along a trajectory of a diffusion process on a Riemannian manifold as introduced by Itô [8], [9] and Dynkin [4] will be extended to a Finsler manifold with the Cartan connection. In fact, our construction can readily be carried out along similar lines using any deflection-free metrical Finsler connection. In the Finslerian case there are two types of stochastic parallel transports emerging from the geometric setting, the h–transport along a diffusion on M and the hv–transport along a diffusion on the tangent bundle TM. In the next section these two types of parallelism will be combined to obtain the fundamental concept of Finslerian stochastic development. Because of the non-differentiability of the sample paths of a diffusion we cannot use ordinary differential equations of the usual form to define the stochastic parallel transport. Instead, we use an approximation of the diffusion by piecewise smooth paths similar to that in [8], [9] and [4] leading to a stochastic differential equation for the limiting process.

A. The h-stochastic parallel transport

Let $x(t)$, $t \in [0, T]$ be a diffusion on M starting at $x(0) = x_0 \in M$. Given a partition $\pi : 0 = t_0 < t_1 < \ldots < t_n = T$, we take a piecewise smooth curve $x_\pi(t)$ on M with bounded first, second, and third-order derivatives such that $x_\pi(t_\alpha) = x(t_\alpha)$ for $\alpha = 0, \ldots, n$. For instance, one can take $x_\pi(t)$ consisting of pieces of geodesics joining $x(t_{\alpha-1})$ with $x(t_\alpha)$, or a piecewise linear approximation in a fixed coordinate system.

The parallel transport $y_\pi(t) \in TM_{x_\pi(t)}$ of a tangent vector $y_0 \in TM_{x_0}$ along $x_\pi(t)$ can be defined as a piecewise smooth function that satisfies the ordinary differential equation

$$\frac{dy_\pi^i}{dt} = -N_j^i(x_\pi, y_\pi)\,\frac{dx_\pi^j}{dt} \tag{2.1}$$

with initial condition $y_\pi(0) = y_0$. We have the following result

Theorem 2.1. mesh $\pi = \max_\alpha |t_\alpha - t_{\alpha-1}|$ *tends to zero,* $y_\pi(t)$ *tends in probability for any* $t \in [0, T]$ *to the solution* $y(t)$ *of the Stratonovich stochastic differential equation*

$$dy^i = -N_j^i(x, y) \circ dx^j \tag{2.2}$$

with initial condition $y(0) = y_0$.

We shall call $y(t)$ the *h-stochastic parallel transport.*

2.1 Outline of the proof

For any function $f(t)$ on $[0,T]$ and $\alpha = 1,\dots,n$, we set $\Delta_\alpha f = f(t_\alpha) - f(t_{\alpha-1})$. From (2.1) we have

$$\begin{aligned}\Delta_\alpha y^i_\pi &= \frac{dy^i_\pi}{dt}\Delta_\alpha t + \frac{1}{2}\frac{d^2y^i_\pi}{dt^2}(\Delta_\alpha t)^2 + o(\Delta_\alpha t)^2 = \\ &= -N^i_j\frac{dx^j_\pi}{dt}\Delta_\alpha t - \frac{1}{2}(\delta_j N^i_k)\frac{dx^j_\pi}{dt}\frac{dx^k_\pi}{dt}(\Delta_\alpha t)^2 + o(\Delta_\alpha t)^2.\end{aligned}$$

Since

$$\begin{aligned}\Delta_\alpha x^i = \Delta_\alpha x^i_\pi &= \frac{dx^i_\pi}{dt}\Delta_\alpha t + \frac{1}{2}\frac{d^2x^i_\pi}{dt^2}(\Delta_\alpha t)^2 + o(\Delta_\alpha t)^2,\\ \Delta_\alpha x^i\Delta_\alpha x^j &= \frac{dx^i_\pi}{dt}\frac{dx^j_\pi}{dt}(\Delta_\alpha t)^2 + o(\Delta_\alpha t)^2,\end{aligned}$$

it follows that

$$\Delta_\alpha y^i_\pi = -N^i_j\Delta_\alpha x^j - \frac{1}{2}(\delta_j N^i_k)\Delta_\alpha x^j\Delta_\alpha x^k + o(\Delta_\alpha t)^2.$$

Please note that there is no summation on α in these formulae. In the limit as mesh $\pi \longrightarrow 0$ we obtain

$$dy^i = -N^i_j dx^j - \frac{1}{2}(\delta_j N^i_k)dx^j dx^k = -N^i_j \circ dx^j,$$

as required. A precise treatment of the limiting passage can be extracted from Chap. 7 of [6] or Chap. 7 of [5]. □

An important property of the limiting process $y(t)$, which satisfies (2.2), is that

$$L\big(x(t), y(t)\big) = \text{const} \tag{2.3}$$

almost surely. This can be verified by evaluating the stochastic differential $dL(x,y)$ with the aid of the Itô formula. The result is $\delta_i L(x,y)\circ dx^i$, which is equal to zero because of the identity $\delta_i L = 0$. From (2.3) it follows that if $y_0 \in TM_{x_0}$ is a non-zero vector, then $y(t) \neq 0$ almost surely for all t. This is important because many functions in Finsler geometry, for instance the metric tensor $g_{ij}(x,y)$, are not defined for $y = 0$. It also follows from (2.3) that if y_0 is chosen so that (x_0, y_0) belongs to the indicatrix $IM = \{(x,y) \in TM : L(x,y) = 1\}$, then the process $(x(t), y(t))$ will stay on IM almost surely.

B. The *hv*-stochastic parallel transport

Let $(x(t), y(t))$, where $t \in [0,T]$, $x(t) \in M$, and $0 \neq y \in TM_{x(t)}$, be a diffusion on TM starting at (x_0, y_0). Given a partition $\pi : 0 = t_0 < t_1 < \ldots < t_n = T$, we take a piecewise smooth curve $(x_\pi(t), y_\pi(t))$ on TM with bounded first, second, and third-order derivatives such that $x_\pi(t) \in M$, $y_\pi(t) \in TM_{x_\pi(t)}$, $x_\pi(t_\alpha) = x(t_\alpha)$, and $y_\pi(t_\alpha) = y(t_\alpha)$ for $\alpha = 0, \ldots, n$. Since $(x(t), y(t))$ 'lives' on TM rather than M, we cannot approximate by geodesics unless a suitable definition of arc length is introduced on TM, but we can still use piecewise linear approximation in a fixed coordinate system.

The parallel transport $u_\pi(t)$ of a vector $u_0 \in TM_{x_0}$ along $(x(t), y(t))$ can be defined as a piecewise smooth function that satisfies the ordinary differential equation

$$\frac{du_\pi^i}{dt} = -F^i_{jk}(x_\pi, y_\pi) u_\pi^k \frac{dx_\pi^j}{dt} - C^i_{jk}(x_\pi, y_\pi) u_\pi^k \frac{\delta y_\pi^j}{dt}, \tag{2.4}$$

where

$$\frac{\delta y_\pi^i}{dt} = \frac{dy_\pi^i}{dt} + N^i_j(x_\pi, y_\pi) \frac{dx_\pi^j}{dt},$$

with initial condition $u_\pi(0) = u_0$.

Theorem 2.2. *For each $t \in [0,T]$, $u_\pi(t)$ tends in probability as* mesh $\pi \to 0$ *to the solution $u(t)$ of the Stratonovich stochastic differential equation*

$$du^i = -F^i_{jk}(x, y) u^k \circ dx^j - C^i_{jk}(x, y) u^k \circ \delta y^j, \tag{2.5}$$

where

$$\delta y^i = dy^i + N^i_j(x, y) \circ dx^j, \tag{2.6}$$

with initial condition $u(0) = u_0$.

We shall call $u(t)$ the *hv – stochastic parallel transport.*

2.2 Outline of the proof

We set

$$\Gamma^i_{jk} = F^i_{jk} + N^l_j C^i_{lk}. \tag{2.7}$$

Then (2.4) can be written as

$$\frac{du_\pi^i}{dt} = -\Gamma^i_{jk} u_\pi^k \frac{dx_\pi^j}{dt} - C^i_{jk} u_\pi^k \frac{dy_\pi^j}{dt}. \tag{2.8}$$

Using the notation introduced in the proof of Theorem 1, we can write

$$\begin{aligned}
\Delta_\alpha x^i = \Delta_\alpha x^i_\pi &= \frac{dx^i_\pi}{dt}\,\Delta_\alpha t + \frac{1}{2}\,\frac{d^2x^i_\pi}{dt^2}\,(\Delta_\alpha t)^2 + o(\Delta_\alpha t)^2,\\
\Delta_\alpha y^i = \Delta_\alpha y^i_\pi &= \frac{dy^i_\pi}{dt}\,\Delta_\alpha t + \frac{1}{2}\,\frac{d^2y^i_\pi}{t^2}\,(\Delta_\alpha t)^2 + o(\Delta_\alpha t)^2,\\
\Delta_\alpha x^i\Delta_\alpha x^j &= \frac{dx^i_\pi}{dt}\,\frac{dx^j_\pi}{dt}\,(\Delta_\alpha t)^2 + o(\Delta_\alpha t)^2,\\
\Delta_\alpha y^i\Delta_\alpha y^j &= \frac{dy^i_\pi}{dt}\,\frac{dy^j_\pi}{dt}\,(\Delta_\alpha t)^2 + o(\Delta_\alpha t)^2,\\
\Delta_\alpha x^i\Delta_\alpha y^j &= \frac{dx^i_\pi}{dt}\,\frac{dy^j_\pi}{dt}\,(\Delta_\alpha t)^2 + o(\Delta_\alpha t)^2.
\end{aligned}$$

From this and (2.8) we find that

$$\begin{aligned}
\Delta_\alpha u^i_\pi &= \frac{du^i_\pi}{dt}\,\Delta_\alpha t + \frac{1}{2}\,\frac{d^2u^i_\pi}{dt^2}\,(\Delta_\alpha t)^2 + o(\Delta_\alpha t)^2\\
&= -\Gamma^i_{jk}u^k_\pi\Delta_\alpha x^j - C^i_{jk}u^k_\pi\Delta_\alpha y^j\\
&\quad - \frac{1}{2}\Big[(\partial_l\Gamma^i_{jk})u^k_\pi\Delta_\alpha x^l\Delta_\alpha x^j + (\dot{\partial}_l\Gamma^i_{jk})u^k_\pi\Delta_\alpha y^l\Delta_\alpha x^j\\
&\quad - \Gamma^i_{jk}(\Gamma^k_{lm}u^m_\pi\Delta_\alpha x^l + C^k_{lm}u^m_\pi\Delta_\alpha y^l)\Delta_\alpha x^j\Big]\\
&\quad - \frac{1}{2}\Big[(\partial_l C^i_{jk})u^k_\pi\Delta_\alpha x^l\Delta_\alpha y^j + (\dot{\partial}_l C^i_{jk})u^k_\pi\Delta_\alpha y^l\Delta_\alpha y^j\\
&\quad - C^i_{jk}(\Gamma^k_{lm}u^m_\pi\Delta_\alpha x^l + C^k_{lm}u^m_\pi\Delta_\alpha y^l)\Delta_\alpha y^j\Big] + o(\Delta_\alpha t)^2.
\end{aligned}$$

As in the proof of Theorem 1, there is no summation on α in the above formulae. In the limit as mesh $\pi \to 0$ we obtain

$$\begin{aligned}
du^i &= -\Gamma^i_{jk}u^k dx^j - C^i_{jk}u^k dy^j\\
&\quad - \frac{1}{2}\Big[(\partial_l\Gamma^i_{jk})u^k dx^l dx^j + (\dot{\partial}_l\Gamma^i_{jk})u^k dy^l dx^j\\
&\quad - \Gamma^i_{jk}(\Gamma^k_{lm}u^m dx^l + C^k_{lm}u^m dy^l)dx^j\Big]\\
&\quad - \frac{1}{2}\Big[(\partial_l C^i_{jk})u^k dx^l dy^j + (\dot{\partial}_l C^i_{jk})u^k dy^l dy^j\\
&\quad - C^i_{jk}(\Gamma^k_{lm}u^m dx^l + C^k_{lm}u^m dy^l)dy^j\Big]\\
&= -\Gamma^i_{jk}u^k \circ dx^j - C^i_{jk}u^k \circ dy^j. \qquad (2.9)
\end{aligned}$$

Using (2.6) and (2.7), we can write this equation in the form (2.5). As in the proof of Theorem 1, a precise justification of the limiting passage can be obtained from the results of Chap. 7 of [6] or Chap. 7 of [5]. □

In Theorem 2 we use the stochastic differential equation (2.5), rather than the equivalent equation (2.9), because the latter involves the stochastic differential $dy^i = y^i(t+dt) - y^i(t)$, which does not transform like a vector, since $y(t+dt)$ and $y(t)$ belong to the different tangent spaces $TM_{x(t+dt)}$ and $TM_{x(t)}$. On the other hand, (2.5) involves δy^i, which can be obtained by pulling $y(t+dt)$ back from $TM_{x(t+dt)}$ to $TM_{x(t)}$ along $x(t)$ with the aid of the h–stochastic parallel transport and then subtracting $y(t)$, so that the result $\delta y^i = dy^i + N^i_j \circ dx^j$ is a well-behaved vector. Because of this the stochastic differential equation (2.5) lends itself well to geometric constructions such as the Finslerian stochastic development presented in the next section. Please also note that equation (2.2) can be written as $\delta y^i = 0$.

The solutions to the stochastic differential equation (2.5) have the following important property:

$$g_{ij}(x(t), y(t))u^i(t)v^j(t) = \text{const} \tag{2.10}$$

almost surely for any $u(t)$ and $v(t)$ that satisfy (2.5). This can be proved by using the Itô formula to compute the stochastic differential $d\big[g_{ij}(x,y)u^i v^j\big]$, which is seen to be equal to zero due to (1.1). It follows from (2.10) that if $u(t)$ is a solution to (2.5) with non-zero initial condition $0 \neq u_0 \in TM_{x_0}$, then $u(t) \neq 0$ almost surely for all t. Another consequence of (2.10) is that if $u(t)$ and $v(t)$ are two solutions to (2.5) with the initial conditions $u_0, v_0 \in TM_{x_0}$ being orthogonal unit vectors relative to the quadratic form $g(x_0, y_0)$ in TM_{x_0}, then $u(t)$ and $v(t)$ are almost surely orthogonal unit vectors in $TM_{x(t)}$ relative to $g(x(t), y(t))$. The latter property facilitates the construction of the Finslerian stochastic development based on the concept of a moving orthonormal frame as presented in the next section.

We also note that $y(t) \in TM_{x(t)}$ is parallel along a diffusion $x(t) \in M$ in the sense of the h–stochastic parallel transport if an only if $y(t)$ is parallel along $(x(t), y(t))$ in the sense of the hv–stochastic parallel transport. This property depends on the formula $C^i_{jk}(x,y)y^k = 0$ and the fact that the connection is deflection-free, i.e., $N^i_j(x,y) = F^i_{jk}(x,y)y^k$.

Remark. The diffusions considered in the present and subsequent sections, which are solutions to stochastic differential equations of the form (1.3), may have explosions, i.e., they may escape to infinity in finite time with non-zero probability. Therefore it is, in general, necessary to regard $[0, T]$ as a random time interval with T a stopping time. Please note, however, that due to (2.3) and (2.10) neither the solutions to the stochastic differential equation (2.2) for the h–stochastic parallel transport nor the solutions to the stochastic differential equation (2.5) for the hv–stochastic parallel transport can have explosions.

3 Finslerian Stochastic Development

We denote by OM the so-called *orthonormal Finsler bundle*, a principal bundle over TM whose total space consists of elements (x, y, z), where $x \in M$, $y \in TM_x$, and $z = (z_1, \ldots, z_n)$ is an orthonormal frame in TM_x relative to the quadratic form $g(x, y)$, i.e.,

$$g_{ij}(x, y) z_k^i z_l^j = \delta_{kl},$$

with projection $\pi : OM \ni (x, y, z) \to (x, y) \in TM$. The differential structure on OM can be obtained from that on M. Namely, if (x^i) are local coordinates defined on an open set $U \subset M$, then we can write $y = y^i \partial_i$ for any $y \in TM_x$ and $z_j = z_j^i \partial_i$ for each vector z_j of any linear frame $z = (z_1, \ldots, z_n)$ in TM_x, where $x \in U$, and use (x^i, y^i, z_j^i) as the local coordinates on $\pi^{-1}U \subset OM$. The structure group of OM is $O(n)$, the group of orthogonal transformations of $\mathbb{R}^n$, with the right action $R_g(x, y, z) = (x, y, zg)$, where $(zg)_i = g_i^j z_j$, for any $g \in O(n)$ and $(x, y, z) \in OM$.

Given a diffusion $(x(t), y(t))$ on TM starting at (x_0, y_0) and a linear frame z_0 in TM_{x_0} orthonormal relative to the quadratic form $g(x_0, y_0)$, we can use the $hv-$stochastic parallel transport to move each of the frame vectors along $(x(t), y(t))$. We shall thus obtain a moving frame $z(t) = (z_1(t), \ldots, z_n(t))$ of vectors, each of which satisfies the stochastic differential equation (2.5), i.e.,

$$dz_l^i = -F_{jk}^i(x, y) z_l^k \circ dx^j - C_{jk}^i(x, y) z_l^k \circ \delta y^j. \tag{3.1}$$

By (2.10), the frame $z(t)$ is almost surely orthonormal relative to the quadratic form $g(x(t), y(t))$ in $TM_{x(t)}$ for all t. We can therefore regard $(x(t), y(t), z(t))$ as a process on OM.

We are now ready to extend the concept of stochastic development or 'rolling' a manifold along a flat Brownian motion from the Riemannian to Finslerian case. We recall that the process of 'rolling' an n-dimensional Riemannian manifold M with metric $g_{ij}(x)$ and Levi-Cività connection $\gamma_{jk}^i(x)$ along a Brownian motion $w(t)$, $t \geq 0$ in $\mathbb{R}^n$ can be described by the system of stochastic differential equations [10], [5]

$$\begin{cases} dx^i = z_l^i \circ dw^l, \\ dz_l^i = -\gamma_{jk}^i(x) z_l^k \circ dx^j, \end{cases} \tag{3.2}$$

where $x(t) \in M$ and $z(t) = (z_1(t), \ldots, z_n(t))$ is a linear frame in $TM_{x(t)}$. As the initial conditions for (3.2) we take $x_0 \in M$ and an orthonormal frame z_0 in TM_{x_0}. It is then seen that $z(t)$ is an orthonormal frame in

$TM_{x(t)}$ almost surely for all t and $x(t)$ is a diffusion on M with generator $2^{-1}g^{ij}\nabla_i\nabla_j = 2^{-1}g^{ij}(\partial_i\partial_j - \gamma_{ij}^k\partial_k)$, the Laplace-Beltrami operator on M.

Now, let M be an n-dimensional Finsler manifold and $w(t)$, $t \geq 0$ a Brownian motion in $\mathbb{R}^n$. To extend the system of stochastic differential equations (3.2) to this case, we replace the second equation in (3.2) by (3.1). It is then clear that the system must be supplemented with one more equation describing the evolution of the tangent vector $y(t) \in TM_{x(t)}$. We assume that $y(t)$ is obtained by means of the h–stochastic parallel transport along $x(t)$, i.e., satisfies the stochastic differential equation (2.2), which can be written as $\delta y^i = 0$. The resulting system of stochastic differential equations reads

$$\begin{cases} dx^i = z_l^i \circ dw^l, \\ \delta y^i = 0, \\ \\ dz_l^i = -F_{jk}^i(x,y)z_l^k \circ dx^j - C_{jk}^i(x,y)z_l^k \circ \delta y^j. \end{cases} \tag{3.3}$$

The second term on the right-hand side of the last equation in (3.3) can be omitted, since $\delta y^i = 0$. As the initial conditions we take $x_0 \in M$, $0 \neq y_0 \in TM_{x_0}$, and a linear frame z_0 in TM_{x_0} orthonormal relative to $g(x_0, y_0)$. By (2.10), the solution $(x(t), y(t), z(t))$ is a process on the orthogonal Finsler bundle OM. We shall call $(x(t), y(t), z(t))$ the *h-stochastic development*. Please note that $y(t) \neq 0$ almost surely for any t due to (2.3).

Theorem 3.1. *The projection* $(x(t), y(t))$ *of the above solution of* (3.3) *from* OM *to* TM *is a diffusion starting at* (x_0, y_0) *whose probability law is independent of the choice of the initial orthonormal frame* z_0 *in* TM_{x_0} *and whose generator is*

$$2^{-1}g^{ij}\nabla_i^h\nabla_j^h = 2^{-1}g^{ij}(\delta_i\delta_j - F_{ij}^k\delta_k).$$

3.1 Outline of the proof

The proof is similar to the Riemannian case. Let $\tilde{z}_0$ be another orthonormal frame in TM_{x_0} and let $(\tilde{x}(t), \tilde{y}(t), \tilde{z}(t))$ be the corresponding solution to (3.3) with initial conditions x_0, y_0, $\tilde{z}_0$. There exists $g \in O(n)$ such that $\tilde{z}_l(t) = g_l^m z_m(t)$. It is readily seen that $(\tilde{x}(t), \tilde{y}(t), \tilde{z}(t))$ is the solution to (3.3) with $w^l(t)$ replaced by $\tilde{w}^l(t) = g_m^l w^m(t)$ and initial conditions x_0, y_0, z_0. But $\tilde{w}(t)$ is also a Brownian motion in $\mathbb{R}^n$. It follows that $(\tilde{x}(t), \tilde{y}(t))$ has the same probability law as $(x(t), y(t))$. We also remark that as a consequence of the independence of the probability law of $(x(t), y(t))$ on the choice of the initial orthonormal frame $z_0 \in TM_{x_0}$, it is easy to deduce the Markov property of $(x(t), y(t))$ from that of $(x(t), y(t), z(t))$, the latter process being Markovian as a solution of the system of stochastic differential equations (3.3).

To compute the generator of $(x(t), y(t))$, we express the stochastic differential dx^i from (3.3) and the product $dx^i dx^j$ as

$$dx^i = z^i_l dw^l - 2^{-1} g^{jk}(x,y) F^i_{jk}(x,y) dt, \quad dx^i dx^j = g^{ij}(x,y) dt. \tag{3.4}$$

Next, we evaluate the stochastic differential $df(x,y)$ of any smooth function f on TM with the aid of the Itô formula:

$$\begin{aligned} df(x,y) &= (\delta_i f) dx^i + 2^{-1}(\delta_i \delta_j f) dx^i dx^j \\ &= (\delta_i f) z^i_l dw^l + 2^{-1} g^{ij} \left[\delta_i \delta_j f - F^k_{ij} \delta_k f\right] dt. \end{aligned}$$

The generator can be obtained from the expression multiplying dt above. □

One can readily verify that the second equation $\delta y^i = 0$ in (3.3) is equivalent to $y^i(t) = v^l z^i_l(t)$ with $v^l = \text{const}$ almost surely. Let us modify (3.3) so that the first and third equations remain unchanged, but the middle equation is replaced by $y^i(t) = v^l(t) z^i_l(t)$, where $v(t)$ is a Brownian motion in $\mathbb{R}^n$ independent of $w(t)$ (we assume that $v(t)$ starts from a non-zero point in $\mathbb{R}^n$). A simple computation, which depends on the fact that the Cartan connection is deflection-free, shows that the new equation for $y(t)$ can be written in the differential form $\delta y^i = z^i_l \circ dv^l$. We thus obtain the system of stochastic differential equations

$$\begin{cases} dx^i = z^i_l \circ dw^l, \\ \delta y^i = z^i_l \circ dv^l, \\ dz^i_l = -F^i_{jk}(x,y) z^k_l \circ dx^j - C^i_{jk}(x,y) z^k_l \circ \delta y^j. \end{cases} \tag{3.5}$$

As the initial conditions for this system we also take $(x_0, y_0, z_0) \in OM$ with $y_0 \neq 0$, in which case it follows from (2.10) that the solution $(x(t), y(t), z(t))$ is a process on OM. We shall call $(x(t), y(t), z(t))$ the *hv–stochastic development*. Since $y^i(t) = v^l(t) z^i_l(t)$ and the Brownian motion $v(t)$ is almost surely non-zero for any t if $n \geq 2$, it follows that $y(t) \neq 0$ almost surely for all t.

Theorem 3.2. *The projection $(x(t), y(t))$ of the above solution of (3.5) from OM to TM is a diffusion starting at (x_0, y_0) whose probability law is independent of the choice of the initial orthonormal frame z_0 in TM_{x_0} and whose generator is*

$$2^{-1} g^{ij} \nabla^h_i \nabla^h_j + 2^{-1} g^{ij} \nabla^v_i \nabla^v_j = 2^{-1} g^{ij} \left(\delta_i \delta_j - F^k_{ij} \delta_k\right) + 2^{-1} g^{ij} \left(\dot{\partial}_i \dot{\partial}_j - C^k_{ij} \dot{\partial}_k\right).$$

3.2 Outline of the proof

The independence of the probability law of $(x(t), y(t))$ on the choice of the initial orthonormal frame z_0 in TM_{x_0} can be proved in exactly the same way as in the case of Theorem 3. It also follows in the same way that $(x(t), y(t))$ has the Markov property. To compute the generator, we first express dx^i and δy^i from (3.5) as well as their products as

$$dx^i = z^i_l dw^l - 2^{-1} g^{jk}(x,y) F^i_{jk}(x,y) dt,$$
$$\delta y^i = z^i_l dv^l - 2^{-1} g^{jk}(x,y) C^i_{jk}(x,y) dt,$$
$$dx^i dx^j = g^{ij}(x,y)dt, \qquad \delta y^i \delta y^j = g^{ij}(x,y)dt, \qquad dx^i \delta y^j = 0.$$

Next, we use the Itô formula to evaluate the stochastic differential $df(x,y)$ of any smooth function f on TM :

$$\begin{aligned} df(x,y) &= (\delta_i f)dx^i + (\dot{\partial}_i f)\delta y^i + 2^{-1}(\delta_i \delta_j f) dx^i dx^j + 2^{-1}(\dot{\partial}_i \dot{\partial}_j f)\delta y^i \delta y^j \\ &= (\delta_i f) z^i_l dw^l + (\dot{\partial}_i f) z^i_l dv^l + 2^{-1} g^{ij}\left[\delta_i \delta_j f - F^k_{ij}\delta_k f\right]dt \\ &\quad + 2^{-1} g^{ij}\left[\dot{\partial}_i \dot{\partial}_j f - C^k_{ij}\dot{\partial}_k f\right]dt. \end{aligned}$$

The generator can be obtained from the expressions multiplying dt above. □

Remark. We note that due to (2.3) and (2.10) only the x part of the $h-$ or $hv-$stochastic development $(x(t), y(t), z(t))$ is subject to explosions, i.e., can escape to infinity in finite time with non-zero probability. An interesting consequence of this observation is that if M is compact, then $(x(t), y(t)) \in TM$ cannot have explosions, even though the tangent bundle TM is not compact.

4 Diffusion of Finsler Tensors

The stochastic differential equation (2.5) for the hv-stochastic parallel transport of a tangent vector $u(t) \in TM_{x(t)}$ along a diffusion $(x(t), y(t))$ on TM can be extended to describe the transport of a tensor $A(t) \in T^r_s M_{x(t)}$ of any type (r, s) along $(x(t), y(t))$. We observe that if $z(t)$ is a linear frame transported along $(x(t), y(t))$, so that each of the frame vectors $z_l(t)$ satisfies (3.1), then a necessary and sufficient condition for $u(t) \in TM_{x(t)}$ to satisfy (2.5) is that $u^i(t) = \alpha^l z^i_l(t)$ with $\alpha^l =$ const almost surely. By analogy, we shall say that a tensor $A(t) \in T^r_s M_{x(t)}$ is transported along $(x(t), y(t))$ if the components of $A(t)$ relative to the frame $z(t)$ are constant almost surely. The stochastic differential equation derived from this condition reads

$$dA(x,y) = -(F_k A)(x,y) \circ dx^k - (C_k A)(x,y) \circ \delta y^k, \tag{4.1}$$

where

$$(F_k A)^{i_1\dots i_r}_{j_1\dots j_s} = -\sum_{\rho=1}^{r} F^{i_\rho}_{kl}(x,y) A^{i_1\dots i_{\rho-1} l i_{\rho+1}\dots i_r}_{j_1\dots j_s} - \sum_{\sigma=1}^{s} F^{l}_{kj_\sigma}(x,y) A^{i_1\dots i_r}_{j_1\dots j_{\sigma-1} l j_{\sigma+1}\dots j_s},$$

$$(C_k A)^{i_1\dots i_r}_{j_1\dots j_s} = -\sum_{\rho=1}^{r} C^{i_\rho}_{kl}(x,y) A^{i_1\dots i_{\rho-1} l i_{\rho+1}\dots i_r}_{j_1\dots j_s} - \sum_{\sigma=1}^{s} C^{l}_{kj_\sigma}(x,y) A^{i_1\dots i_r}_{j_1\dots j_{\sigma-1} l j_{\sigma+1}\dots j_s}.$$

Suppose that $(x(t), y(t))$ starts at (x_0, y_0) and let $A(t)$ be the solution to (4.1) with initial condition $A_0 \in T^r_s M_{x_0}$. We denote by T_t the linear mapping

$$T_t : T^r_s M_{x_0} \ni A_0 \mapsto A(t) \in T^r_s M_{x(t)}.$$

Let us now consider a Finsler tensor field A of type (r, s). By this we mean a smooth function $A : TM \to T^r_s M$ such that $A(x, y) \in T^r_s M_x$ for any $x \in M$ and $0 \neq y \in TM_x$. For alternative equivalent approaches to the definition of Finsler tensor fields, see [13], [11], or [2]. We shall consider the process $A(x(t), y(t))$, where $(x(t), y(t), z(t))$ is either the h-stochastic development or the hv-stochastic development with initial conditions $(x_0, y_0, z_0) \in OM$ such that $y_0 \neq 0$. In the case of a scalar vector field $f : TM \to \mathbb{R}$ the generator

$$Df(x_0, y_0) = \lim_{t \searrow 0} t^{-1} \left[E\big(f(x(t), y(t))\big) - f(x_0, y_0)\right]$$

was computed in Theorems 3 and 4. However, for a Finsler tensor field A of type (r, s)., the values of $A(x(t), y(t))$ and $A(x_0, y_0)$ belong to the different spaces $T^r_s M_{x(t)}$ and $T^r_s M_{x_0}$ and cannot be subtracted from one another. We thus apply T_t^{-1} to pull $A(x(t), y(t))$ back to TM_{x_0} and set

$$DA(x_0, y_0) = \lim_{t \searrow 0} t^{-1} \left[E\big(T_t^{-1} A(x(t), y(t))\big) - A(x_0, y_0)\right].$$

Please note that DA is independent of z_0. This is clearly so because the probability law of $(x(t), y(t))$ is independent of the choice of the initial orthonormal frame z_0.

Theorem 4.1. *If $(x(t), y(t), z(t))$ is the h*–stochastic *development with initial conditions $(x_0, y_0, z_0) \in OM$, where $y_0 \neq 0$, and $A : TM \to T^r_s M$ is a Finsler tensor field, then*

$$DA^{i_1\dots i_r}_{j_1\dots j_s} = 2^{-1} g^{kl} \nabla^h_k \nabla^h_l A^{i_1\dots i_r}_{j_1\dots j_s}.$$

4.1 Outline of the proof

The proof is adapted from [4], where a similar theorem is proved in the Riemannian case. For any $x \in M$, $a \in T^r_s M_x$, and $b \in T^s_r M_x$, we set

$$\langle a, b\rangle = a^{i_1\dots i_r}_{j_1\dots j_s} b^{j_1\dots j_s}_{i_1\dots i_r}.$$

It is easily verified that

$$\langle F_k a, b\rangle = -\langle a, F_k b\rangle, \qquad \langle C_k a, b\rangle = -\langle a, C_k b\rangle.$$

We take any solution $B(t) \in T^s_r M_{x(t)}$ to the stochastic differential equation (4.1) and use (3.4) and the Itô formula to compute the stochastic differential

$$\begin{aligned} d\big\langle A(x,y), B\big\rangle &= \big\langle dA(x,y), B\big\rangle + \big\langle A(x,y), dB\big\rangle + 2^{-1}\big\langle dA(x,y), dB\big\rangle \\ &= \big\langle \delta_k A(x,y) + F_k A(x,y), B\big\rangle dx^k \\ &\quad + 2^{-1}\big\langle \delta_j \delta_k A(x,y) \\ &\quad + F_j(\delta_k A)(x,y) + (\delta_j F_k)A(x,y) + F_j F_k A(x,y), B\big\rangle dx^j dx^k \\ &= \big\langle \nabla^h_k A(x,y), B\big\rangle z^k_l dw^l + 2^{-1} g^{jk}(x,y)\big\langle \nabla^h_j \nabla^h_k A(x,y), B\big\rangle dt. \end{aligned} \tag{4.2}$$

We set $a(t) = T^{-1}A(x(t), y(t))$. It is easily verified that $\big\langle a(t), B(0)\big\rangle = \big\langle A(x(t), y(t)), B(t)\big\rangle$. Thus it follows from (4.2) that

$$da = \nabla^h_k A(x,y) z^k_l dw^l + 2^{-1} g^{jk} \nabla^h_j \nabla^h_k A(x,y) dt$$

at $t = 0$. The generator can be obtained from the expression multiplying dt above. □

Theorem 4.2. *If $(x(t), y(t), z(t))$ is the hv–*stochastic *development with initial conditions $(x_0, y_0, z_0) \in OM$, where $y_0 \neq 0$, and $A : TM \to T^r_s M$ is a Finsler tensor field, then*

$$DA^{i_1\dots i_r}_{j_1\dots j_s} = 2^{-1} g^{kl} \nabla^h_k \nabla^h_l A^{i_1\dots i_r}_{j_1\dots j_s} + 2^{-1} g^{kl} \nabla^v_k \nabla^v_l A^{i_1\dots i_r}_{j_1\dots j_s}.$$

The proof of Theorem 6 is similar to that of Theorem 5.

5 Examples

5.1 Minkowski spaces

A Minkowski space M is by definition a finite-dimensional normed vector space whose norm $L(y)$ is smooth for all $0 \neq y \in M$. Such M is clearly a special case of a Finsler manifold with arc element $ds = L(dx^1, \ldots, dx^n)$. It is easily seen that the connection coefficients N^i_j and F^i_{jk} are identically equal to zero and $\delta_i = \partial_i$ for any Minkowski space. The system of stochastic differential equations (3.3) for the h–stochastic development takes the form

$$\begin{cases} dx^i = z^i_l \circ dw^l, \\ dy^i = 0, \\ dz^i_l = 0, \end{cases}$$

which means that $x^i(t) = w^i(t)$, $y^i(t) = 0$, and $z^i_j(t) = 0$ almost surely, that is, we obtain just a standard Brownian motion in $\mathbb{R}^n$. On the other hand, the system of stochastic differential equations (3.5) for the hv–stochastic development reads

$$\begin{cases} dx^i = z^i_l \circ dw^l, \\ dy^i = z^i_l \circ dv^l, \\ dz^i_l = -C^i_{jk}(x,y) z^k_l \circ \delta y^j. \end{cases}$$

The latter two equations of this system are independent of $x(t)$ and have the form of the stochastic differential equations (3.2) for the Riemannian stochastic development. The process $y(t)$ is therefore a Brownian motion on M regarded as a Riemannian manifold with metric $g_{ij}(y) = 2^{-1}\dot{\partial}_i\dot{\partial}_j L^2(y)$ and Levi-Cività connection $C^i_{jk}(y) = 2^{-1} g^{il}(y)\dot{\partial}_l g_{jk}(y)$.

In many cases the metric tensor $g_{ij}(y)$ of a Minkowski space fails to be positive definite for all $0 \neq y \in M$. For example, this is so for the two-dimensional Minkowski space with fourth-root metric $L(y^1, y^2) = \left[(y^1)^4 + (y^2)^4\right]^{1/4}$, for which $g_{ij}(y)$ is positive definite for all $y \in M$ such that $y^1 y^2 \neq 0$ and $\det(g_{ij}) = 0$ on the coordinate axes, where $y^1 y^2 = 0$. In such a case one can still consider a stochastic development starting in a region where $g_{ij}(y)$ is positive definite and defined up to the first hitting time of the boundary of the region.

5.2 Berwald spaces

A Berwald space M can be defined as a Finsler space such that $\nabla_l^h C^i_{jk} = 0$, where C^i_{jk} are the vertical coefficients of the Cartan connection. A consequence of this is that there exists a coordinate system in which the horizontal coefficients of the Cartan connection are functions of position only, i.e., are independent of the tangent vector y. In particular, it follows that the nonlinear connection coefficients N^i_j are, in fact, linear functions of y, i.e., $N^i_j(x,y) = F^i_{jk}(x)y^k$, in this coordinate system. Then the stochastic differential equations for the $h-$stochastic development read

$$\begin{cases} dx^i = z^i_l \circ dw^l, \\ dy^i = -F^i_{jk}(x,y)y^k \circ dx^j, \\ dz^i_l = -F^i_{jk}(x,y)z^k_l \circ dx^j. \end{cases} \tag{5.1}$$

We note that the first and third equations in this system are independent of $y(t)$. By a theorem due to Szabó [14], for every Berwald space M there exists a Riemannian metric $h_{ij}(x)$ on M such that the horizontal coefficients F^i_{jk} of the Cartan connection coincide with the Levi-Cività connection coefficients for $h_{ij}(x)$ in the coordinate system in which F^i_{jk} depend on x alone.* Then the first and third equations in (5.1) are just the stochastic differential equations (3.2) for the stochastic development on M as a Riemannian manifold with metric $h_{ij}(x)$. The second equation in (5.1) describes the Riemannian stochastic parallel transport of $y(t)$ along $x(t)$. We note, however, that the initial conditions for (5.1) are different for the Riemannian and Finslerian stochastic development, even though the governing stochastic differential equations (5.1) are the same. Namely, in the Riemannian case we start from an initial frame $z_0 \in TM_{x_0}$ orthonormal relative to the Riemannian metric tensor $h_{ij}(x_0)$ (and then it is seen that the frame $z(t)$ remains orthonormal relative to $h_{ij}(x(t))$ for all t), while in the Finslerian case we start from a frame $z_0 \in TM_{x_0}$ orthonormal relative to the Finslerian metric tensor $g_{ij}(x_0, y_0)$ (in which case the frame $z(t)$ remains orthonormal relative to $g_{ij}(x(t), y(t))$ for all t).

References

[1] Antonelli, P (ed.) (1985) *Mathematical Essays on Growth and the Emergence of Form*, University of Alberta Press.

*Szabó requires the indicatrix $L = 1$ to be a smooth closed surface. However, this is impossible for a large class of Berwald spaces which are nevertheless positive definite when $y^1 \dots y^n \neq 0$.

[2] Bejancu, A. (1990) ***Finsler Geometry and Applications***, Ellis Harwood, New York.

[3] Cartan, E. (1934) 2nd ed (1971) Les Espaces de Finsler, ***Actualités***, **79**, Paris.

[4] Dynkin, E.B. (1968) Diffusion of Tensors, ***Dokl. Akad. Nauk SSSR***, **179**; English translation: ***Soviet Math. Dokl.***, **9** , 532-535.

[5] Elworthy, K.D. (1982) ***Stochastic Differential Equations on Manifolds***, Cambridge University Press, Cambridge.

[6] Emery, M. (1989) ***Stochastic Calculus in Manifolds***, Springer-Verlag, Berlin-Heidelberg-New York.

[7] Ingarden, R.S. (1976) Differential Geometry and Physics, *Tensor, N.S.*, **30**, 201–209.

[8] Itô, K. (1963) The Brownian Motion and Tensor Fields on Riemannian Manifold, *Proc. Intern. Congr. Math.*, Stockholm, 536-539.

[9] Itô, K. (1975) Stochastic Parallel Displacement, Probabilistic Methods in Differential Equations, *Lect. Notes in Math.*, **451**, Springer-Verlag, Berlin, 1-7.

[10] Ikeda, N. and Watanabe, S. (1981) ***Diffusion Processes and Stochastic Differential Equations***, North Holland, Amsterdam.

[11] Matsumoto, M. (1986) ***Foundations of Finsler Geometry and Special Finsler Spaces***, Kaiseisha Press, Otsu, Japan.

[12] Meyer, P.A. (1982) Geometrie Differentielle Stochastique, *Lect. Notes Math.*, **921**, Springer-Verlag, 165-207.

[13] Rund, H. (1959) ***The Differential Geometry of Finsler Spaces***, Springer-Verlag, Berlin.

[14] Szabó, Z.I. (1981) Positive Definite Berwald Spaces (Structure Theorems on Berwald Spaces), *Tensor, N.S.*, **35**, 25-39.

STOCHASTIC CALCULUS ON FINSLER MANIFOLDS AND AN APPLICATION IN BIOLOGY

P.L. ANTONELLI * and T.J. ZASTAWNIAK

0 Introduction

Since the pioneering works by Itô [15,16,17,18] the theory of Brownian motion and stochastic development on Riemannian manifolds has become a classical branch of stochastic calculus (see, for example [10,11,13]) with numerous applications in other areas. In the present article we extend the theory of Brownian motion and stochastic development to the case of Finsler manifolds.

The dependence of the metric tensor $g_{ij}(x, y)$ of a Finsler manifold not only on $x \in M$, but also on a tangent vector $y \in TM_x$ offers more flexibility than a Riemannian metric, providing a convenient geometric framework for applications in areas such as variational calculus, mechanics, theoretical physics (e.g., it facilitates the description of anisotropic properties of the physical space; see, for example, [6,14]), complex analysis (the Caratheodory and Kobayashi metrics), and developmental and population biology [1]. The theory presented in this article has been developed with a view to applications in the above areas.

In Section 1 we introduce the key notion of stochastic parallel transport of a vector (and/or tensor) along a semimartingale on a Finsler manifold M. This is done by approximating the sample paths of the semimartingale by piecewise smooth curves and using the deterministic parallel transport. In the Finsler case the notion of parallelism along a semimartingale splits into two concepts: the (nonlinear) $h-$parallel transport along a semimartingale on M and the (linear) $hv-$transport along a semimartingale on the tangent bundle TM. In Section 2 we develop the concept of rolling a

*Partially supported by NSERC A-7667. This paper appeared in Vol. 1, 1993, p. 149-171, of *Nonlinear World*, published by de Gruyter. It appears here with permission of the Editor-in-Chief.

Finsler manifold along a semimartingale. Section 3 is devoted to Finslerian stochastic development (i.e., rolling a Finsler manifold along a standard flat Brownian motion), leading to the definition of a Finslerian Brownian motion. In Section 4 we study the radial behaviour of a Finslerian Brownian motion. Three elementary examples are given in Section 5. Finally, in Section 6, we present a biological application of stochastic calculus on Finsler manifolds, namely, a model describing the growth of a colonial animal perturbed by developmental and/or environmental noise. The main concepts of the present article were announced in [7].

An n–dimensional manifold M without boundary is said to be a *Finsler manifold* if the arc element on M is expressed by $ds = L(x^1, \ldots, x^n, dx^1, \ldots, dx^n)$, where L is a smooth function from the so-called *slit tangent bundle* $\mathring{T}M$ (i.e. the tangent bundle TM with the zero section removed) to $[0,\infty)$ such that $L(x, ky) = kL(x,y)$ for any $x \in M$, $0 \neq y \in TM_x$, and $k > 0$, i.e., $L(x,y)$ is positively homogeneous of degree one in y. This being the case, L is called the *fundamental metric function* of M. The *Finsler metric tensor* is defined by

$$g_{ij}(x,y) = 2^{-1}\dot{\partial}_i\dot{\partial}_j L^2(x,y),$$

where $\dot{\partial}_i = \partial/\partial y^i$. We shall assume that $g_{ij}(x,y)$ is positive definite. By analogy with the Levi-Cività symbols on a Riemannian manifold, we set

$$\gamma^i_{jk} = 2^{-1}g^{i\ell}(\partial_j g_{\ell k} + \partial_k g_{j\ell} - \partial_\ell g_{jk}),$$

where $\partial_i = \partial/\partial x^i$ and g^{ij} is the inverse matrix to g_{ij}. Here g_{ij}, g^{ij}, and γ^i_{jk} are functions of $x \in M$ and $0 \neq y \in TM_x$. In the above formula and throughout the article we use the standard summation convention on upper and lower indices. We set

$$\begin{aligned}
N^i_j &= \dot{\partial}_j(\gamma^i_{k\ell}y^k y^\ell),\\
F^i_{jk} &= 2^{-1}g^{i\ell}(\delta_j g_{\ell k} + \delta_k g_{j\ell} - \delta_\ell g_{jk}),\\
C^i_{jk} &= 2^{-1}g^{i\ell}(\dot{\partial}_j g_{\ell k} + \dot{\partial}_k g_{j\ell} - \dot{\partial}_\ell g_{jk}) = 2^{-1}g^{i\ell}\dot{\partial}_j g_{\ell k},
\end{aligned}$$

where $\delta_i = \partial_i - N^j_i\dot{\partial}_j$. These formulae define the *Cartan connection*. N^i_j, F^i_{jk}, and C^i_{jk} being, respectively, the coefficients of the nonlinear, horizontal, and vertical parts of the connection.

We say that $A : \mathring{T}M \to TM$ is a Finsler vector field if $A(x,y) \in TM_x$ for every $x \in M$ and $0 \neq y \in TM_x$. The *horizontal* and *vertical covariant derivatives* of such a field will be denoted by

$$\nabla^h_j A^i = \delta_j A^i + F^i_{jk}A^k, \quad \nabla^v_j A^i = \dot{\partial}_j A^i + C^i_{jk}A^k.$$

These formulae can be extended to arbitrary Finsler tensor fields in the usual way.

The Cartan connection is both $h-$ *metrical* and $v-$*metrical*, i.e.,

$$\delta_i g_{jk} = F^{\ell}_{ij} g_{\ell k} + F^{\ell}_{ik} g_{j\ell}, \quad \dot{\partial}_i g_{jk} = C^{\ell}_{ij} g_{\ell k} + C^{\ell}_{ik} g_{j\ell},$$

and *deflection-free*, i.e.,

$$N^i_j = F^i_{jk} y^k.$$

A systematic modern approach to Finsler geometry can be found in Matsumoto's monograph [20]. We also refer to the classical monograph [21] by Rund and to Cartan's original work [8].

Throughout the article we use the standard probabilistic setup. We fix a complete *probability space* $(\Omega, \mathcal{F}, P)$ equipped with a right-continuous *filtration* $(\mathcal{F}_t)_{t\geq 0}$ such that each $\mathcal{F}_t$ contains all events of probability zero from the $\sigma-$algebra $\mathcal{F}$. We denote by $Ef = \int_\Omega f dP$ the *expectation* of a *random variable* $f : \Omega \to \mathbb{R}$. The *conditional expectation* of f given a sub$-\sigma-$field $\mathcal{G} \subset \mathcal{F}$ will be denoted by $E(f|\mathcal{G})$. A *stochastic process* $x : [0, \infty) \times \Omega \to \mathbb{R}$ is said to be *continuous* if its sample paths $t \mapsto x(t, \omega)$ are continuous for almost all $\omega \in \Omega$. A stochastic process x is *adapted* if the random variable $x(t)$ is $\mathcal{F}_t-$measurable for every $t \geq 0$. A *martingale* is an adapted process x such that $E|x(t)| < \infty$ for every $t \geq 0$ and $E(x(t)|\mathcal{F}_s) = x(s)$ for every $t > s \geq 0$. A *local martingale* is an adapted process x such that the process $(t, \omega) \mapsto x(\tau_n \wedge t, \omega)\chi_{\{0<\tau_n\}}(\omega)$ is a martingale for every n, where $\tau_n = \inf\{t \geq 0 : |x(t)| \geq n\}$ (as usual, we set $\inf \emptyset = \infty$) and χ_A is the indicator function of $A \subset \Omega$. A *semimartingale* is the sum of a continuous local martingale and a continuous adapted process whose sample paths have almost surely (a.s.) finite variation on each compact interval in $[0, \infty)$.

For a real-valued semimartingale x and a continuous adapted process y, we denote by $\int y\, dx$ the *Itô stochastic integral* of y with respect to x. Given two real-valued semimartingales x and y, the process $[x, y] = xy - x(0)y(0) - \int x dy - \int y\, dx$ is called the *joint quadratic variation* of x and y and the process $\int y \circ dx = \int y dx + 2^{-1}[x, y]$ is called the Stratonovich integral of y with respect to x. We shall use the differential notation: if $z = \int y dx$, we shall write $dz = y dx$; if $z = [x, y]$, we shall write $dz = dx dy$; and if $z = \int y \circ dx$, then we shall write $dz = y \circ dx$. Thus, in particular, $y \circ dx = y dx + 2^{-1} dx dy$. We shall also need the formula $z \circ (y \circ dx) = (zy) \circ dx$ valid for any real-valued semimartingales x, y, z.

If $x^1, \ldots, x^n$ are real-valued semimartingales and $f : \mathbb{R}^n \to \mathbb{R}$ is a function of class C^2, then we have the Itô change-or-variables formula

$$df(x) = \partial_i f(x) dx^i + 2^{-1} \partial_i \partial_j f(x) dx^i dx^j,$$

where $x = (x^1, \dots, x^n)$. If f is of class C^3, then the Itô formula can be written

$$df(x) = \partial_i f(x) \circ dx^i.$$

A continuous stochastic process $x : [0,\infty) \times \Omega \to M$ with values on a manifold M is called an *M-valued semimartingale* if, for every smooth function $f : M \to \mathbb{R}$, the process $f \circ x$ is a real-valued semimartingale. We shall also admit $M-$valued semimartingales defined on a *random time interval* $[0,\tau)$, where $\tau : \Omega \to [0,\infty]$ is a *stopping time,* i.e., a random variable such that $(t < \tau) \in \mathcal{F}_t$ for every $t \geq 0$.

Given two manifolds M and N, an $M-$valued semimartingale x defined on a random time interval $[0,\tau)$, and a family $\left(e(x,y)\right)_{x\in M, y\in N}$ of linear mappings from T_xM to T_yN, we shall consider *Stratonovich stochastic differential equations* (SDEs) of the form $dy = e(x,y) \circ dx$, which can be written as $dy^i = e^i_j(x,y) \circ dx^j$ in local coordinates. The solution y is an $N-$valued semimartingale defined on a random time interval $[0,\sigma)$. It is called a *maximal solution* if $P\{\sigma = \tau \text{ or } y(t) \to \infty \text{ as } t \nearrow \sigma\} = 1$, where ∞ is the infinity of the one-point compactification of M (if M is a compact manifold, then the condition reduces to $P\{\sigma = \tau\} = 1$). If $\sigma < \tau$ with nonzerc probability then we say that the maximal solution explodes.

For the precise definitions of stochastic integrals and solutions of SDEs as well as their standard properties see, for example, [11]. We also refer to [10,13,19].

1 Stochastic Parallel Transport

On a Finsler manifold the notion of parallelism along a given curve splits into two alternative concepts, which will be referred to as $h-$parallelism and $hv-$parallelism. We begin with introducing these two notions of parallelism along a sufficiently smooth curve in the deterministic case. The deterministic case will then be used in an approximation procedure leading to a definition of $h-$ and $hv-$parallelism along a semimartingale.

In this section M will denote an $n-$dimensional Finsler manifold with fundamental metric function L, metric tensor g_{ij}, and the Cartan connection $(N^i_j, F^i_{jk}, C^i_{jk})$. Any deflection-free $h-$ and $v-$metrical Finsler connection can, in fact, be used instead of the Cartan connection.

1.1 Deterministic h − parallel transport

Let $x(t) \in M$, where $t \in [0,T]$, be a smooth curve on M. We say that a vector field $y(t) \in TM_{x(t)}$ is *h−parallel* along $x(t)$ if

$$\frac{dy^i}{dt} = -N^i_j(x,y)\,\frac{dx^j}{dt}, \tag{1.1}$$

$N^i_j(x,y)$ being the coefficients of the nonlinear connection on M.

In particular, if M is a Riemannian manifold with the Levi-Cività connection $\gamma^i_{jk}(x)$, then $N^i_j(x,y) = \gamma^i_{jk}(x)y^k$ is a linear function of y and (1.1) is just the standard linear differential equation for a parallel vector field along $x(t)$. But in the general case (1.1) is a nonlinear differential equation in y. Because of this, the h−parallel transport governed by (1.1) will also be referred to as the *nonlinear parallel transport.*

It is easily verified that if (1.1) is satisfied, then $L\big(x(t),y(t)\big) = \text{const}$. It follows that if $y(0) \neq 0$, then $y(t) \neq 0$ for all t. This simple property is essential, because the majority of objects in Finsler geometry are not defined (or at least not differentiable) at $y = 0$ due to the homogeneity in y. Since $L\big(x(t),y(t)\big) = \text{const}$, it also follows that a solution $y(t)$ to (1.1) with an arbitrary initial condition $y(0) = y_0 \neq 0$ cannot explode in finite time (i.e. for every finite time interval, the trajectory of the solution is contained in a compact subset of $\mathring{T}M$), which means that it can be extended to all $t \in [0,T]$.

In case $0 \neq y(t) \in TM_{x(t)}$ is h−parallel along $x(t)$ we shall also say that the curve $\big(x(t),y(t)\big)$ is an *h−lift* of the curve $x(t)$ from M to the slit tangent bundle $\mathring{T}M$. The h−lift is defined uniquely, if the initial condition $0 \neq y_0(0) \in TM_{x(0)}$ is specified.

1.2 Deterministic hv−parallel transport

As opposed to the previous case, we consider an arbitrary smooth curve $\big(x(t),y(t)\big)$ on the slit tangent bundle $\mathring{T}M$, rather than on M itself. Here $x(t) \in M$ and $0 \neq y(t) \in TM_{x(t)}$ for all $t \in [0,T]$. A vector field $u(t) \in TM_{x(t)}$ is said to be *hv−parallel* along $\big(x(t),y(t)\big)$ if

$$\frac{du^i}{dt} = -F^i_{jk}(x,y)u^k\,\frac{dx^j}{dt} - C^i_{jk}(x,y)u^k\,\frac{\delta y^j}{dt}, \tag{1.2}$$

where

$$\frac{\delta y^j}{dt} = \frac{dy^j}{dt} + N^i_j(x,y)\,\frac{dx^j}{dt}.$$

A simple consequence of the fact that the connection is h− and v

-metrical and deflection-free is that if $u(t)$ and $v(t)$ are two solutions to (1.2), then

$$g_{ij}(x(t), y(t))u^i(t)v^j(t) = \text{const.} \tag{1.3}$$

It follows that a solution $u(t)$ to (1.2) cannot explode in finite time, i.e., can be extended to all $t \in [0,T]$, and if $u(0) \neq 0$, then $u(t) \neq 0$ for all $t \in [0,T]$.

The differential equation (1.2) for the hv−parallel transport is linear in u, which makes it possible to extend the notion of hv−parallelism from vector fields $u(t) \in TM_{x(t)}$ to tensor fields $u(t) \in T^r_s M_{x(t)}$ of any type (r,s). For example, we say that a field $u(t) \in T^*M_{x(t)} = T^0_1 M_{x(t)}$ is hv−parallel along $(x(t), y(t))$ if

$$\frac{du_i}{dt} = F^k_{ji}(x,y)u_k \frac{dx^j}{dt} + C^k_{ji}(x,y)u_k \frac{\delta y^j}{dt}.$$

By comparing (1.1) with (1.2) and taking into account that $F^i_{jk}(x,y)y^k = N^i_j(x,y)$ and $C^i_{jk}(x,y)y^k = 0$, one can easily see a relationship between the two kinds of parallelism: on a Finsler manifold M a vector field $0 \neq y(t) \in TM_{x(t)}$ is h−parallel along $x(t) \in M$ if and only if $y(t)$ is hv−parallel along $(x(t), y(t))$.

Suppose that for every $t \in [0,T]$, $z(t) = (z_1(t), \ldots, z_n(t))$ is a linear frame in $TM_{x(t)}$ such that each of the frame vectors $z_j(t) \in TM_{x(t)}$ is hv−parallel along $(x(t), y(t))$, i.e. satisfies (1.2). By (1.3), if the frame $z(t)$ is orthonormal relative to the quadratic form $g(x(t), y(t)) \in T^0_2 M_{x(t)}$ for any fixed $t \in [0,T]$, then this is so for all $t \in [0,T]$. We say that the curve $(x(t), y(t), z(t))$ is an *hv−lift* of $(x(t), y(t))$ from $\mathring{T}M$ to the so-called *orthonormal Finsler bundle* OM. The latter is defined as a principal bundle over $\mathring{T}M$ whose total space consists of elements (x,y,z), where $x \in M$, $0 \neq y \in TM_x$, and $z = (z_1, \ldots, z_n)$ is an orthonormal frame in TM_x relative to the quadratic form $g(x,y) \in T^0_2 M_x$, i.e.,

$$g_{ij}(x,y)z^i_k z^j_\ell = \delta_{k\ell},$$

with projection $\pi : OM \ni (x,y,z) \mapsto (x,y) \in \mathring{T}M$. The differential structure of OM can be obtained from that of M. Namely, if (x^i) are local coordinates in an open set $U \subset M$, then we can write $y = y^i\partial_i$ for any $y \in TM_x$ and $z_j = z^i_j\partial_i$ for each vector z_j of any linear frame $z = (z_1, \ldots, z_n)$ in TM_x, where $x \in U$, and use (x^i, y^i, z^i_j) as local coordinates in $\pi^{-1}U \subset OM$. The structure group of OM is the group $O(n)$ of orthogonal transformations of $\mathbb{R}^n$ with the right action $R_h(x,y,z) = (x,y,zh)$, where $(zh)_i = h^j_i z_j$, for any $h \in O(n)$ and $(x,y,z) \in OM$.

1.3 Stochastic h–parallel transport

Let $x(t)$ be a semimartingale on M defined for all $t \in [0, T]$, where $T > 0$ is a stopping time. We shall approximate $x(t)$ by processes with piecewise smooth sample paths, for which the h–parallel transport is defined by the ordinary differential equation (1.1). In the limit as the approximation tends to $x(t)$ in the sense described below we shall obtain a SDE extending (1.1) to the case of semimartingales. In doing so we shall rely on the well-known results on the approximation of SDEs on manifolds by ordinary differential equations (see, for example, [10,11]).

A piecewise smooth approximation of a semimartingale $x(t)$ on M can be obtained by taking a division $\pi : 0 = t_0 < t_1 < \cdots < t_n = T$ of the time interval and joining the points $x(t_0), x(t_1), \ldots, x(t_n)$ by pieces of suitable smooth curves. This can be achieved by means of a smooth *interpolation rule* (see [11]). By this we understand a smooth mapping $I : V \times [0,1] \to M$, where V is a neighbourhood of the diagonal $\{(x,x) : x \in M\}$ in $M \times M$ such that $I(x,x,t) = x$ for all $t \in [0,1]$, $I(x,x',0) = x$, $I(x,x',1) = x'$, and $\partial^m I(x,x',t)/\partial t^m = O(d^m(x,x'))$ for $m = 1,2,3$ uniformly on compact sets. Here $d(x,x')$ denotes the distance on M, i.e., the length of the shortest geodesic joining x and x'.

Given an interpolation rule I and a division π of the time interval, we define the approximation $x_\pi(t)$ of the process $x(t)$ by

$$x_\pi(t) = I\Big(x(t_{i-1}), x(t_i), \frac{t - t_{i-1}}{t_i - t_{i-1}}\Big) \quad \text{for} \quad t \in [t_{t-1}, t_i], \quad i = 1,2,\ldots,n. \tag{1.4}$$

If $x(t_{i-1})$ and $x(t_i)$ are too far away from one another, so that $\big(x(t_{i-1}), x(t_i)\big) \notin V$, then we can, for example, extend I to a measurable function on $M \times M \times [0,1]$, or leave $x_\pi(t)$ undefined for $t \geq t_{i-1}$. It is, anyway, only the behaviour of I in the vicinity of the diagonal that matters in the limit as $|\pi| = \max_i |t_i - t_{i-1}|$ tends to zero.

There is a multitude of interpolation rules to choose form. A natural one is the *geodesic interpolation rule* such that $t \mapsto I(x,x',t)$ is the (shortest) geodesic that starts from x at $t = 0$ and arrives at x' at time $t = 1$. The proof of all the required properties of I is exactly the same as in the Riemannian case treated in Proposition 7.13 of [11].

Each sample path of the approximation $x_\pi(t)$ is a piecewise smooth curve on M and can be h–lifted to $\mathring{T}M$, thus yielding a process $y_\pi(t) \in TM_{x_\pi(t)}$, whose sample paths satisfy the ordinary differential equation (1.1) for the h–parallel transport, which now takes the form

$$\frac{dy_\pi^i}{dt} = -N_j^i(x_\pi, y_\pi)\,\frac{dx_\pi^j}{dt}\,.$$

We set $x_0 = x(0)$ and state the same initial condition $y_\pi(0) = y_0$ for every partition π, where $0 \neq y_0 \in TM_{x_0}$ is an $\mathcal{F}_0$–measurable random vector (please note that $x_\pi(0) = x_0$ for every partition π). We are now in a position to apply Theorem 7.24 of [11]. It follows that $y_\pi(t)$ converges in probability uniformly with respect to t in any compact set as $|\pi| \to 0$ to the solution $y(t) \in TM_{x(t)}$ of the Stratonovich SDE

$$dy^i = -N_j^i(x, y) \circ dx^j \tag{1.5}$$

with the initial condition $y(0) = y_0$. The limiting process $y(t)$ will be called the *stochastic h–parallel transport* of y_0 along $x(t)$ and the semimartingale $\bigl(x(t), y(t)\bigr)$ on TM will be called an *h–lift* of the semimartingale $x(t) \in M$.

Similarly as in the deterministic case, it is easily verified that

$$L\bigl(x(t), y(t)\bigr) = \text{const} \quad \text{a.s.} \tag{1.6}$$

if $y(t)$ is a solution to (1.5). This means that if $y_0 \neq 0$ a.s., then $y(t) \neq 0$ a.s. for all $t \in [0, T]$. Moreover, $y(t)$ cannot explode, which means that it can a.s. be extended to all $t \in [0, T]$.

1.4 Stochastic *hv*–parallel transport

The construction is similar as in the case of the h–parallel transport, except that now we begin with an arbitrary semimartingale $\bigl(x(t), y(t)\bigr)$ on $\mathring{T}M$ and we use an interpolation rule I on $\mathring{T}M$, rather than on M. A natural interpolation rule can be chosen using the diagonal lift G of the Finsler metric tensor g to the slit bundle $\mathring{T}M$ defined by (see [25])

$$G(x, y) = g_{ij}(x, y) dx^i \otimes dx^j + g_{ij}(x, y) \delta y^i \otimes \delta y^j \tag{1.7}$$

for any $x \in M$ and $0 \neq y \in TM_x$. Here $\delta y^i = dy^i + N_j^i(x, y) dx^j$. The diagonal lift G is a Riemannian metric on $\mathring{T}M$, and we can take the corresponding geodesic interpolation rule I. Then, for any partition π, we take the piecewise smooth approximation $\bigl(x_\pi(t), y_\pi(t)\bigr)$ of $\bigl(x(t), y(t)\bigr)$ defined by a formula analogous to (1.4). Let $u_\pi(t) \in TM_{x_\pi(t)}$ be a process whose sample paths satisfy the differential equation (1.2) for the hv–parallel transport, i.e.,

$$\frac{du_\pi^i}{dt} = -F_{jk}^i(x_\pi, y_\pi) u_\pi^k \frac{dx_\pi^j}{dt} - C_{jk}^i(x_\pi, y_\pi) u_\pi^k \frac{\delta y_\pi^j}{dt},$$

where

$$\frac{\delta y_\pi^i}{dt} = \frac{dy_\pi^i}{dt} + N_j^i(x_\pi, y_\pi) \frac{dx_\pi^j}{dt}.$$

We also assume that $u_\pi(t)$ satisfies the initial condition $u_\pi(0) = u_0 \in TM_{x_0}$ for every partition π, where u_0 is an $\mathcal{F}_0$–measurable random vector. Then, by Theorem 7.24 of [11], $u_\pi(t)$ converges in probability uniformly with respect to t in any compact set as $|\pi| \to 0$ to the solution $u(t) \in TM_{x(t)}$ of the Stratonovich SDE

$$du^i = -F^i_{jk}(x,y)u^k \circ dx^j - C^i_{jk}(x,y)u^k \circ \delta y^j, \tag{1.8}$$

where

$$\delta y^i = dy^i + N^i_j(x,y) \circ dx^j,$$

with initial condition $u(0) = u_0$. We then say that $u(t)$ is the *stochastic hv–parallel* transport of u_0 along the semimartingale $\big(x(t), y(t)\big)$.

Similarly as in the deterministic case,

$$g_{ij}\big(x(t), y(t)\big)u^i(t)v^j(t) = \text{const} \quad \text{a.s.} \tag{1.9}$$

for any solutions $u(t)$ and $v(t)$ of (1.8). It follows that if $u(t)$ is a solution to (1.8), then it cannot explode, i.e., it can a.s be extended to all $t \in [0, T]$. Moreover, if $u_0 = u(0) \neq 0$ a.s., then $u(t) \neq 0$ a.s. for all t.

Since the SDE (1.8) is linear in u, the notion of stochastic hv–parallel transport can be extended to tensor-valued processes $u(t) \in T^r_s M_{x(t)}$ of any type (r, s). For example, we say that a process $u(t) \in T^*M_{x(t)} = T^1_0 M_{x(t)}$ is h–parallel along $\big(x(t), y(t)\big)$ if

$$du_i = F^k_{ji}(x,y)u_k \circ dx^j + C^k_{ji}(x,y)u_k \circ \delta y^j.$$

The property (1.9) makes it possible to lift a semimartingale $\big(x(t), y(t)\big)$ from $\mathring{T}M$ to OM in a similar way as in the deterministic case. We say that $\big(x(t), y(t), z(t)\big)$ is an *hv–lift* of $\big(x(t), y(t)\big)$ from $\mathring{T}M$ to OM if $z(t) = \big(z_1(t), \ldots, z_n(t)\big)$ is a frame in $TM_{x(t)}$ such that each of the frame vectors $z_j(t)$ satisfies the SDE (1.8) for the stochastic hv–parallel transport.

2 Rolling Finsler Manifolds Along Smooth Curves and Semimartingales

We consider two n–dimensional Finsler manifolds M and $\widetilde{M}$ with metric tensors g_{ij} and $\widetilde{g}_{ij}$ and Cartan connection coefficients $N^i_j, F^i_{jk}, C^i_{jk}$ and $\widetilde{N}^i_j, \widetilde{F}^i_{jk}, \widetilde{C}^i_{jk}$, respectively. Let us imagine that the manifolds roll (without slipping) along a smooth curve drawn on either of them. To make this concept precise we shall specify an appropriate system of differential equations. These differential equations will then be extended to stochastic differential equations to cover the case of rolling along a semimartingale.

The fact that the metric tensor and connection coefficients depend not only on a point on the manifold, but also on a tangent vector at that point suggests that the process of rolling is determined not only by the curve of contact $x(t)$ on M (or $\widetilde{x}(t)$ on $\widetilde{M}$), but also by a tangent field $y(t) \in TM_{x(t)}$ (or $\widetilde{y}(t) \in T\widetilde{M}_{\widetilde{x}(t)}$) along that curve. The vectors $y(t)$ can either be obtained by lifting the curve $x(t)$ form M to $\mathring{T}M$, or defined arbitrarily, which leads to two concepts of rolling.

2.1 *h*–Rolling Finsler manifolds along a smooth curve on *M*

We assume that $x(t)$, where $t \in [0,T]$, is a smooth curve on M. Similarly, $\widetilde{x}(t)$ will be a smooth curve on $\widetilde{M}$. Using the nonlinear connections N^i_j and $\widetilde{N}^i_j$, we can lift the curves from M and $\widetilde{M}$ to $\mathring{T}M$ and $\mathring{T}\widetilde{M}$, respectively. As a result, we obtain vector fields $0 \neq y(t) \in TM_{x(t)}$ and $0 \neq \widetilde{y}(t) \in T\widetilde{M}_{\widetilde{x}(t)}$ satisfying the nonlinear differential equations

$$\frac{dy^i}{dt} = -N^i_j(x,y)\,\frac{dx^j}{dt} \quad \text{and} \quad \frac{d\widetilde{y}^i}{dt} = -\widetilde{N}^i_j(\widetilde{x},\widetilde{y})\,\frac{d\widetilde{x}^j}{dt}\,. \tag{2.1}$$

Next, we lift the curves $(x(t),y(t)) \in \mathring{T}M$ and $(\widetilde{x}(t),\widetilde{y}(t)) \in \mathring{T}\widetilde{M}$ to OM and $O\widetilde{M}$, respectively. Thus we obtain orthonormal frames $z(t) = (z_1(t),\ldots,z_n(t))$ in $TM_{x(t)}$ and $\widetilde{z}(t) = (\widetilde{z}_1(t),\ldots,\widetilde{z}_n(t))$ in $T\widetilde{M}_{\widetilde{x}(t)}$ satisfying (1.2), i.e.,

$$\frac{dz^i_j}{dt} = -F^i_{k\ell}(x,y)z^\ell_j\,\frac{dx^k}{dt} \quad \text{and} \quad \frac{d\widetilde{z}^i_j}{dt} = -\widetilde{F}^i_{k\ell}(\widetilde{x},\widetilde{y})\widetilde{z}^\ell_j\,\frac{d\widetilde{x}^k}{dt}\,. \tag{2.2}$$

The terms involving $C^i_{k\ell}$ and $\widetilde{C}^i_{k\ell}$ (cf. (1.2)) disappear, since $\delta y^k/dt = 0$ and $\delta\widetilde{y}^k/dt = 0$ by (2.1). Let $\zeta(t) = (\zeta^1(t),\ldots,\zeta^n(t))$ and $\widetilde{\zeta}(t) = (\widetilde{\zeta}^1(t),\ldots,\widetilde{\zeta}^n(t))$ be the dual frames to $z(t) = (z_1(t),\ldots,z_n(t))$ and $\widetilde{z}(t) = (\widetilde{z}_1(t),\ldots,\widetilde{z}_n(t))$, respectively. Then we say that the manifolds M and $\widetilde{M}$ *h–roll along the curve* $x(t)$ *on* M (or, equivalently, $\widetilde{x}(t)$ on $\widetilde{M}$) if

$$\zeta^i_j\,\frac{dx^j}{dt} = \widetilde{\zeta}^i_j\,\frac{d\widetilde{x}^j}{dt}\,. \tag{2.3}$$

The differential equations (2.2) and (2.3) extend the concept of rolling from the case of Riemannian manifolds (as described in [13,18]) to Finsler manifolds.

2.2 hv–Rolling Finsler manifolds along a smooth curve on $\mathring{T}M$

We shall now assume that $x(t)$, where $t \in [0,T]$, is a smooth curve on M and $0 \neq y(t) \in TM_{x(t)}$ is an arbitrary smooth vector field along $x(t)$. Similarly, $\widetilde{x}(t)$ will be a smooth curve on $\widetilde{M}$ and $0 \neq \widetilde{y}(t) \in T\widetilde{M}_{\widetilde{x}(t)}$ will be an arbitrary smooth vector field along $\widetilde{x}(t)$. We lift the curves $\big(x(t), y(t)\big)$ and $\big(\widetilde{x}(t), \widetilde{y}(t)\big)$ from $\mathring{T}M$ and $\mathring{T}\widetilde{M}$ to OM and $O\widetilde{M}$, respectively. As a result, we obtain frames $z(t) = \big(z_1(t), \ldots, z_n(t)\big)$ in $TM_{x(t)}$ and $\widetilde{z}(t) = \big(\widetilde{z}_1(t), \ldots, \widetilde{z}_n(t)\big)$ in $T\widetilde{M}_{\widetilde{x}(t)}$ such that

$$\frac{dz_j^i}{dt} = -F_{k\ell}^i(x,y) z_j^\ell \,\frac{dx^k}{dt} - C_{k\ell}^i(x,y) z_j^\ell \,\frac{\delta y^k}{dt}\,, \tag{2.4}$$

$$\frac{d\widetilde{z}_j^i}{dt} = -\widetilde{F}_{k\ell}^i(\widetilde{x},\widetilde{y}) \widetilde{z}_j^\ell \,\frac{d\widetilde{x}^k}{dt} - \widetilde{C}_{k\ell}^i(\widetilde{x},\widetilde{y}) \widetilde{z}_j^\ell \,\frac{\delta \widetilde{y}^k}{dt}\,, \tag{2.5}$$

where

$$\frac{\delta y^k}{dt} = \frac{dy^k}{dt} + N_m^k(x,y)\,\frac{dx^m}{dt} \quad \text{and} \quad \frac{\delta \widetilde{y}^k}{dt} = \frac{d\widetilde{y}^k}{dt} + \widetilde{N}_m^k(\widetilde{x},\widetilde{y})\,\frac{d\widetilde{x}^m}{dt}\,.$$

We say that the manifolds M and $\widetilde{M}$ *hv–roll* along the curve $\big(x(t), y(t)\big)$ (or, equivalently, along $\big(\widetilde{x}(t), \widetilde{y}(t)\big)$ if

$$\zeta_j^i\,\frac{dx^j}{dt} = \widetilde{\zeta}_j^i\,\frac{d\widetilde{x}^j}{dt} \quad \text{and} \quad \zeta_j^i\,\frac{\delta y^j}{dt} = \widetilde{\zeta}_j^i\,\frac{\delta \widetilde{y}^j}{dt}\,, \tag{2.6}$$

where $\zeta(t) = \big(\zeta^1(t), \ldots, \zeta^n(t)\big)$ and $\widetilde{\zeta}(t) = \big(\widetilde{\zeta}^1(t), \ldots, \widetilde{\zeta}^n(t)\big)$ are the dual frames to $z(t) = \big(z_1(t), \ldots, z_n(t)\big)$ and $\widetilde{z}(t) = \big(\widetilde{z}_1(t), \ldots, \widetilde{z}_n(t)\big)$, respectively.

2.3 h–Rolling Finsler manifolds along a semimartingale on M

Suppose that $x(t)$ is a semimartingale on M and $\widetilde{x}(t)$ is a semimartingale on $\widetilde{M}$ for $t \in [0,T]$, where $T > 0$ is a stopping time. We can use the SDE (1.5) to lift $x(t)$ from M to $\mathring{T}M$ and $\widetilde{x}(t)$ from $\widetilde{M}$ to $\mathring{T}M$. The resulting SDEs, which generalize (2.1), read

$$dy^i = -N_j^i(x,y) \circ dx^j \quad \text{and} \quad d\widetilde{y}^i = -\widetilde{N}_j^i(\widetilde{x},\widetilde{y}) \circ d\widetilde{x}^j\,. \tag{2.7}$$

The semimartingales $\big(x(t), y(t)\big)$ and $\big(\widetilde{x}(t), \widetilde{y}(t)\big)$ can, in turn, be lifted from $\mathring{T}M$ and $\mathring{T}\widetilde{M}$ to OM and $O\widetilde{M}$, respectively, the lifted semimartingales

$(x(t), y(t), z(t))$ and $(\widetilde{x}(t), \widetilde{y}(t), \widetilde{z}(t))$ being defined by the SDEs

$$dz_j^i = -F_{k\ell}^i(x,y)z_j^\ell \circ dx^k \quad \text{and} \quad d\widetilde{z}_j^i = -\widetilde{F}_{k\ell}^i(\widetilde{x},\widetilde{y})\widetilde{z}_j^\ell \circ d\widetilde{x}^k, \tag{2.8}$$

which generalize (2.2). Then, in analogy with (2.3), we say that the manifolds M and $\widetilde{M}$ *h−roll along the semimartingale* $x(t)$ on M (or, equivalently, along $\widetilde{x}(t)$ on $\widetilde{M}$) if

$$\zeta_j^i \circ dx^j = \widetilde{\zeta}_j^i \circ d\widetilde{x}^j, \tag{2.9}$$

$\zeta(t) = (\zeta^1(t), \ldots, \zeta^n(t))$ and $\widetilde{\zeta}(t) = (\widetilde{\zeta}(t), \ldots, \widetilde{\zeta}^n(t))$ being the dual frames to $z(t) = (z_1(t), \ldots, z_n(t))$ and $\widetilde{z}(t) = (\widetilde{z}_1(t), \ldots, \widetilde{z}_n(t))$, respectively.

2.4 *hv*−Rolling Finsler manifolds along a semimartingale on $\mathring{T}M$

We take an arbitrary semimartingale $(x(t), y(t))$ on $\mathring{T}M$, where $x(t) \in M$ and $0 \neq y(t) \in T_{x(t)}M$ for $t \in [0,T]$, $T > 0$ being a stopping time. Similarly, $(\widetilde{x}(t), \widetilde{y}(t))$ will be a semimartingale on $\mathring{T}\widetilde{M}$. We lift the semimartingales from $\mathring{T}M$ and $\mathring{T}\widetilde{M}$ to OM and $O\widetilde{M}$, respectively. The lifted semimartingales $(x(t), y(t), z(t))$ and $(\widetilde{x}(t), \widetilde{y}(t), \widetilde{z}(t))$ are defined by the SDEs

$$dz_j^i = -F_{k\ell}^i(x,y)z_j^\ell \circ dx^k - C_{k\ell}^i(x,y)z_j^\ell \circ \delta y^k, \tag{2.10}$$

$$d\widetilde{z}_j^i = -\widetilde{F}_{k\ell}^i(\widetilde{x},\widetilde{y})\widetilde{z}_j^\ell \circ d\widetilde{x}^k - \widetilde{C}_{k\ell}^i(\widetilde{x},\widetilde{y})\widetilde{z}_j^\ell \circ \delta\widetilde{y}^k, \tag{2.11}$$

obtained from (1.8), which generalize (2.4) and (2.5). Here

$$\delta y^k = dy^k + N_m^k(x,y) \circ dx^m \quad \text{and} \quad \delta\widetilde{y}^k = d\widetilde{y}^k + \widetilde{N}_m^k(\widetilde{x},\widetilde{y}) \circ d\widetilde{x}^m.$$

Let $\zeta(t) = (\zeta^1(t), \ldots, \zeta^n(t))$ and $\widetilde{\zeta}(t) = (\widetilde{\zeta}^1(t), \ldots, \widetilde{\zeta}^n(t))$ be the dual frames to $z(t) = (z_1(t), \ldots, z_n(t))$ and $\widetilde{z}(t) = (\widetilde{z}_1(t), \ldots, \widetilde{z}_n(t))$, respectively. This being the case, we say that the manifolds *hv−roll along the semimartingale* $(x(t), y(t))$ *on* $\mathring{T}M$ (or, equivalently, $(\widetilde{x}(t), \widetilde{y}(t))$ on $\mathring{T}\widetilde{M}$) if

$$\zeta_j^i \circ dx^j = \widetilde{\zeta}_j^i \circ d\widetilde{x}^j \quad \text{and} \quad \zeta_j^i \circ \delta y^j = \widetilde{\zeta}_j^i \circ \delta\widetilde{y}^j, \tag{2.12}$$

the latter two SDEs being generalizations on (2.6).

3 Finslerian Stochastic Development and Brownian Motion

We shall apply the concepts introduced in the previous section to the case of rolling a Finsler manifold along a standard Brownian motion in a Euclidean space. Let M be an n-dimensional Finsler manifold. Then $\mathbb{R}^n$ with the ordinary Euclidean metric can play the role of another Finsler manifold. We denote by $v(t)$ and $w(t)$ two independent standard Brownian motions in $\mathbb{R}^n$. Thus $(w(t), v(t))$ is a standard Brownian motion in $\mathbb{R}^n \times \mathbb{R}^n$. According to the discussion in the previous section, M can either be h-rolled along $w(t)$, which is a semimartingale in $\mathbb{R}^n$, or hv-rolled along $(w(t), v(t))$, the latter being a semimartingale in $\mathbb{R}^n \times \mathbb{R}^n$. This leads to the following two alternative approaches to Finslerian stochastic development.

3.1 Finslerian h-stochastic development

We adapt equations (2.7), (2.8), and (2.9) to the case of h-rolling M along the semimartingale $w(t)$ in $\mathbb{R}^n$. The resulting semimartingale $(x(t), y(t), z(t))$ on OM defined by the system of SDEs

$$\begin{cases} dx^i &= z^i_j \circ dw^j, \\ dy^i &= -N^i_j(x,y) \circ dx^j, \\ dz^i_j &= -F^i_{k\ell}(x,y) z^\ell_j \circ dx^k \end{cases} \tag{3.1}$$

is called a *Finslerian h-stochastic development* over M.

3.2 Finslerian hv-stochastic development

In the case when M hv-rolls along the semimartingale $(w(t), v(t))$ in $\mathbb{R}^n \times \mathbb{R}^n$ the corresponding SDEs (2.10)-(2.12) reduce to the system

$$\begin{cases} dx^i &= z^i_j \circ dw^j, \\ \delta y^i &= z^i_j \circ dv^j, \\ dz^i_j &= -F^i_{k\ell}(x,y) z^\ell_j \circ dx^k - C^i_{k\ell}(x,y) z^\ell_j \circ \delta y^k, \end{cases} \tag{3.2}$$

which defines a semimartingale $(x(t), y(t), z(t))$ on OM called a *Finslerian hv-stochastic development.*

Remark 3.1. In general , the maximal solutions $(x(t), y(t), z(t)) \in OM$ of either of the systems (3.1) and (3.2) are subject to explosions. Therefore, in general, they are defined on a random time interval $[0, T]$, where $0 < T \leq \infty$ is a stopping time. However, due to (1.6) and (1.9), it is only the x-part of the process $(x(t), y(t), z(t))$ that can explode. As a consequence, there are no explosions if the manifold M is compact (in which case the solution $(x(t), y(t), z(t))$ exists for all $t \geq 0$).

The following theorem is concerned with the projection of a solution of (3.1) or (3.2) from OM to $\mathring{T}M$.

Theorem 1. *If $(x(t), y(t), z(t))$ is a solution (3.1) or (3.2) starting from $(x_0, y_0, z_0) \in OM$, then the probability law of $(x(t), y(t))$ is independent of the choice of the initial orthonormal frame z_0. Moreover, if $(x(t), y(t), z(t))$ is a solution of (3.1), then $(x(t), y(t))$ is a Markov diffusion on $\mathring{T}M$ with generator*

$$2^{-1} g^{ij} \nabla_i^h \nabla_j^h = 2^{-1} g^{ij} (\delta_i \delta_j - F_{ij}^k \delta_k), \tag{3.3}$$

and if $(x(t), y(t), z(t))$ is a solution of (3.2), then $(x(t), y(t))$ is a Markov diffusion on $\mathring{T}M$ with generator

$$2^{-1} g^{ij} \nabla_i^h \nabla_j^h + 2^{-1} g^{ij} \nabla_i^v \nabla_j^v = 2^{-1} g^{ij} (\delta_i \delta_j - F_{ij}^k \delta_k) + 2^{-1} g^{ij} (\dot{\partial}_i \dot{\partial}_j - C_{ij}^k \dot{\partial}_k). \tag{3.4}$$

Proof. We shall deal with the case when $(x(t), y(t), z(t))$ is a solution of (3.1). In the case of a solution of (3.2) the proof is similar. Let $\widetilde{z}_0$ be any other orthonormal frame in TM_{x_0}. It is easily seen that $(x(t), y(t), \widetilde{z}(t))$ is a solution of (3.1) with w^i replaced by $\widetilde{w}^i$ such that $w^j = h_i^j \widetilde{w}^i$ and with initial conditions $(x_0, y_0, \widetilde{z}_0)$. But $\widetilde{w}$ is also a Brownian motion on $\mathbb{R}^n$, so it has the same probability law as w. It follows that the probability law of $(x(t), y(t))$ is independent of z_0. Because of this, it follows that $(x(t), y(t))$ has the Markov property, since $(x(t), y(t), z(t))$ has the Markov property as a solution to (3.1).

To compute the generator of $(x(t), y(t))$ we express the stochastic differential dx^i from (3.1) and the product $dx^i dx^j$ as

$$\begin{aligned} dx^i &= z_j^i dw^j - 2^{-1} g^{jk}(x, y) F_{jk}^i(x, y) dt, \\ dx^i dx^j &= g^{ij}(x, y) dt. \end{aligned}$$

Next, we evaluate the stochastic differential $df(x, y)$ of any smooth function f on TM with the aid of the Itô formula:

$$df(x, y) = (\delta_i f) dx^i + 2^{-1} (\delta_i \delta_j f) dx^i dx^j = (\delta_i f) z_j^i dw^j + 2^{-1} g^{ij} [\delta_i \delta_j f - F_{ij}^k \delta_k f] dt.$$

The generator is given by the expression multiplying dt above. □

The expressions (3.3) and (3.4) can be regarded as generalizations of the Laplace-Beltrami operator to the Finslerian case. Thus, in view of Theorem 3.1, the projection $(x(t), y(t))$ of a Finslerian $h-$ or $hv-$stochastic development $(x(t), y(t), z(t))$ from OM to $\mathring{T}M$ will be called a *Finslerian $h-$ or hv-Brownian motion,* respectively.

Remark 3.2. The Finslerian $hv-$Brownian motion defined by (3.2) can also be described as a Brownian motion with drift on the slit tangent bundle

$\mathring{T}M$ with the Riemannian diagonal lift metric tensor G defined by (1.7), the drift being expressed in terms of the torsion tensors of the Finsler connection $(N^i_j, F^i_{jk}, C^i_{jk})$. This is, however, beyond the scope of the present article and will be discussed in a forthcoming publication.

4 Radial Behaviour of Finslerian Brownian Motion

The theorems below are concerned with the behaviour of the fundamental metric function L of a Finsler manifold M along a Finslerian $h-$ or $hv-$Brownian motion over M.

Theorem 2. *If* $(x(t), y(t))$ *is a Finslerian* $h-$Brownian *motion over* M, *then* $L(x(t), y(t)) =$ const *almost surely.*

Remark 4.1. An interesting consequence of Theorem 4.1 is that a Finslerian $h-$Brownian motion that starts on the *indicatrix bundle* $IM = \{(x,y) \in TM : L(x,y) = 1\}$ will stay on IM for all $t \in [0,T]$ almost surely.

Proof of Theorem 4.1. Let $(x(t), y(t), z(t))$ be a process on OM that satisfies the system (3.1) of SDEs for the Finslerian $h-$stochastic development. From the Itô formula we find that

$$dL^2(x,y) = \delta_i L^2(x,y) \circ dx^i + \dot{\partial}_i L^2(x,y) \circ \delta y^i. \tag{4.1}$$

But $\delta y^i = 0$ by the second equation in (3.1). We compute $\delta_i L^2$ using the fact that the connection is deflection-free and $h-$metrical, i.e.,$N^i_j(x,y) = F^i_{jk}(x,y)y^k$ and $\delta_i g_{jk} = F^\ell_{ij} g_{ik} + F^\ell_{ik} g_{j\ell}$:

$$\begin{aligned}\delta_i L^2 &= \delta_i[y^j y^k g_{jk}] = y^j y^k \delta_i g_{jk} - N^j_i y^k g_{jk} - y^j N^k_i g_{jk} \\ &= y^j y^k (\delta_i g_{jk} - F^\ell_{ij} g_{\ell k} - F^\ell_{ik} g_{j\ell}) = 0.\end{aligned}$$

It follows that $dL^2(x,y) = 0$, and so $L^2(x,y) =$ const a.s. which proves the theorem. □

Theorem 3. *If* $(x(t), y(t))$ *is a Finslerian* $hv-$*Brownian motion over* M, *then* $L(x(t), y(t))$ *is a Bessel process with index* $n =$ dim M *(on Bessel processes see, for example,* [13]*).*

Remark 4.2. Theorem 4.2 means that the behaviour of the fundamental function L along a Finslerian $hv-$Brownian motion is the same as the radial behaviour of a standard Brownian motion in $\mathbb{R}^n$.

Proof of Theorem 4.2. We now assume that $(x(t), y(t), z(t)) \in OM$ satisfies the system (3.2) of SDEs for the hv–stochastic development. We express the stochastic derivative $dL^2(x,y)$ using (4.1) and apply the identity $\delta_i L^2 \equiv 0$ proved above. Hence

$$dL^2 = \dot{\partial}_i L^2 \circ \delta y^i = \dot{\partial}_i L^2 \delta y^i + 2^{-1}\dot{\partial}_i\dot{\partial}_j L^2 \delta y^i \delta y^j = 2y_i \delta y^i + g_{ij}\delta y^i \delta y^j. \quad (4.2)$$

It follows from (3.2) that

$$\delta y^i = z^i_j \circ dv^j = z^i_j dv^j - 2^{-1} g^{jk} C^i_{jk} dt,$$
$$\delta y^i \delta y^j = g^{ij} dt.$$

Since $y_i C^i_{jk}(x,y) = 0$ and $g_{ij}g^{ij} = n =$ dim M, we find from (4.2) that

$$dL^2 = 2y_i z^i_j dv^j + ndt.$$

We set $n_j = \dot{\partial}_i L z^i_j$ and $du = n_j dv^j$. Since $Ln_j = L\dot{\partial}_i L z^i_j = y_i z^i_j$, we can write

$$dL^2 = 2Ldu + ndt. \quad (4.3)$$

We claim that $u(t)$ is a standard Brownian motion in $\mathbb{R}$. If this is so, then the proof is completed, because $L(x,y)$ is a Bessel process with index n as a solution of the SDE (4.3).

We can use the Lévy theorem (see, for example, [9]) to prove that $u(t)$ is a Brownian motion. Thus it suffices to verify that $u(t)$ and $u^2(t) - t$ are martingales. We first note that $L^2 n_i n_j \delta^{ij} = y_k y_\ell z^k_i z^\ell_j \delta^{ij} = y_k y_\ell g^{k\ell} = L^2$, which implies that $n_i n_j \delta^{ij} = 1$. It follows that $u = \int n_i dv^i$ is a square-integrable martingale. By the Itô formula, we find that $du^2 = 2un_i dv^i + n_i n_j \delta^{ij} dt = 2un_i dv^i + dt$. Thus $u(t)^2 - t$ is a martingale, which proves that $u(t)$ is a standard Brownian motion in $\mathbb{R}$. □

5 Examples

Example 5.1. Riemannian manifolds. Let M be a Riemannian manifold with metric tensor $g_{ij}(x)$, i.e. let $L^2(x,y) = g_{ij}(x)y^i y^j$. Then $F^i_{jk}(x,y) = \gamma^i_{jk}(x)$ are just the ordinary Levi-Cività connection coefficients, $N^i_j(x,y) = \gamma^i_{jk}(x)y^k$ are linear functions of y, and $C^i_{jk}(x,y) \equiv 0$. In this case the first and third equations of the system (3.1) of SDEs for the h–stochastic development are independent of the second equation and read

$$\begin{cases} dx^i &= z^i_j \circ dw^j, \\ dz^i_j &= -\gamma^i_{k\ell}(x) z^\ell_j \circ dx^k. \end{cases} \quad (5.1)$$

This is just the well-known system of SDEs for the Riemannian stochastic development (cf. [10,11,13,17,18]). This means that $x(t)$ is an ordinary Riemannian Brownian motion on M. The remaining equation in (3.1), which can be written as

$$dy^i = -\gamma^i_{jk}(x)y^k \circ dx^j,$$

is the SDE for the stochastic parallel transport of a tangent vector $y(t)$ along $x(t)$ on a Riemannian manifold (see the above-mentioned references).

In the Riemannian case the first and third equations of the system (3.2) for the hv–stochastic development are also independent of the second equation and take the same form as (5.1) above, thus defining a Riemannian Brownian motion $x(t)$ on M. The second equation in (3.2) can be written as $dy^i = d(v^j z^i_j)$, which means that $y^i(t) = y^i(0) + v^j z^i_j$, i.e., the coordinates of $y(t)$ in the frame $z(t)$ form a standard Brownian motion in $\mathbb{R}^n$.

Example 5.2. Minkowski spaces. A Minkowski space M is, by definition, a finite-dimensional normed vector space whose norm $L(y)$ is smooth for all $0 \neq y \in M$. Such M is clearly a special case of a Finsler manifold with arc element $ds = L(dx^1, \ldots, dx^n)$. It is easily seen that the connection coefficients N^i_j and F^i_{jk} are identically equal to zero, $\delta_i = \partial_i$, and C^i_{jk} is independent of x for any Minkowski space. The system (3.1) of SDEs for the h–stochastic development takes the form

$$\begin{cases} dx^i &= z^i_j \circ dw^j, \\ dy^i &= 0, \\ dz^i_j &= 0, \end{cases}$$

which means that $x^i(t) = x^i(0) + w^j(t)$, $y^i(t) =$ const, and $z^i_j(t) =$ const a.s., that is, we obtain just a standard Brownian motion in $\mathbb{R}^n$. On the other hand, the system (3.2) of SDEs for the hv–stochastic development reads

$$\begin{cases} dx^i &= z^i_j \circ dw^j, \\ dy^i &= z^i_j \circ dv^j, \\ dz^i_j &= -C^i_{k\ell}(y) z^\ell_j \circ dy^k. \end{cases}$$

The latter two equations of this system are independent of x and have the same form as the system of SDEs (5.1) for the Riemannian stochastic development. The process $y(t)$ is therefore a Brownian motion on M regarded as a Riemannian manifold with metric $g_{ij}(y) = 2^{-1}\dot{\partial}_i\dot{\partial}_j L^2(y)$ and Levi-Cività connection $C^i_{jk}(y) = 2^{-1}g^{i\ell}(y)\dot{\partial}_\ell g_{jk}(y)$.

In many interesting cases the metric tensor $g_{ij}(y)$ of a Minkowski space fails to be positive definite for all $0 \neq y \in M$. For example, this is so for the two-dimensional Minkowski space with fourth-root metric $L(y^1, y^2) = [(y^1)^4 + (y^2)^4]^{1/4}$, for which $g_{ij}(y)$ is positive definite for all $y \in M$ such that $y^1 y^2 \neq 0$ and $\det(g_{ij}) = 0$ on the coordinate axes, where $y^1 y^2 = 0$. In such a case one can still consider a stochastic development starting in the region where $g_{ij}(y)$ is positive definite and defined up to the first hitting time of the boundary of the region.

Example 5.3. Berwald spaces. A Berwald space M can be defined as a Finsler space such that $\nabla^h_\ell C^i_{jk} = 0$, where C^i_{jk} are the vertical coefficients of the Cartan connection. As a consequence, there exists a coordinate system in which the horizontal coefficients of the Cartan connection are functions of position only, i.e., are independent of the tangent vector y (see [20]). In particular, it follows that the nonlinear connection coefficients N^i_j are, in fact, linear functions of y, i.e., $N^i_j(x, y) = F^i_{jk}(x) y^k$, in this coordinate system. Then the SDEs for the h–stochastic development read

$$\begin{cases} dx^i &= z^i_j \circ dw^j, \\ dy^i &= -F^i_{k\ell}(x) y^\ell \circ dx^k, \\ dz^i_j &= -F^i_{k\ell}(x) z^\ell_j \circ dx^k. \end{cases} \tag{5.2}$$

We note that the first and third equations in this system are independent of y. By a theorem due to Szabó [22], for every Berwald space M there exists a Riemannian metric $\widetilde{g}_{ij}(x)$ on M such that the horizontal coefficients F^i_{jk} of the Cartan connection coincide with the Levi-Cività connection coefficients γ^i_{jk} of the Riemannian metric $\widetilde{g}_{ij}(x)$ in the coordinate system in which F^i_{jk} depend on x alone.* Then the first and third equations in (5.2) are just the SDEs (5.1) for the stochastic development on M as a Riemannian manifold with metric $\widetilde{g}_{ij}(x)$. The second equation in (5.2) describes the Riemannian stochastic parallel transport of $y(t)$ along $x(t)$. We note, however, that the initial conditions for (5.2) differ for the Riemannian and Finslerian stochastic developments, even though the governing SDEs (5.2) are the same. Namely, in the Riemannian case we start from an initial frame $z_0 \in TM_{x_0}$ orthonormal relative to the Riemannian metric tensor $\widetilde{g}_{ij}(x_0)$ (it is then seen that the frame $z(t)$ remains orthonormal relative to $\widetilde{g}_{ij}(x(t))$ for all t), while in the Finslerian case we start from a frame $z_0 \in TM_{x_0}$ orthonormal relative to the Finslerian metric tensor $g_{ij}(x_0, y_0)$ (in which case the frame $z(t)$ remains orthonormal relative to $g_{ij}(x(t), y(t))$ for all t).

---

*Szabo requires that the indicatrix be a smooth closed surface. This may be impossible, even though positive definiteness holds when $y^1 \dots y^n \neq 0$.

6 Application: Growth of a Colonial Animal

A colonial animal is both a set of populations called *castes* and a genetic individual whose growth and development exhibits the division of labour principle (specific tasks for specific castes) compatible with a globally organized physiology (cf. [24]).

6.1 Deterministic model

We postulate the fundamental metric function

$$L(y) = e^{Q\,\mathrm{Arctan}\,(\frac{y^1}{y^2})}\big((y^1)^2 + (y^2)^2\big)^{1/2}.^{\dagger} \tag{6.1}$$

where $y = (y^1, y^2) \in \mathbb{R}^2$, and where $Q \geq 0$ is a constant. This defines the structure of a Minkowski manifold M on $\mathbb{R}^2$. We also consider the Finsler fundamental function

$$\widetilde{L}(x, y) = e^{\phi(x)} L(y) \tag{6.2}$$

obtained from L by a conformal transformation (see [12]), $\phi(x)$ being a smooth function of $x = (x^1, x^2) \in \mathbb{R}^2$. The Finsler metric (6.2) was used in [4] in application to coral reef ecology. It defines a two-dimensional Finsler space of Berwald type. The geodesic equations in this Finsler metric read

$$\begin{aligned}
0 &= \frac{d^2x^1}{ds^2} + (-Q\alpha_2 + \alpha_1)\Big(\frac{dx^1}{ds}\Big)^2 \\
&\quad + (Q\alpha_2 - \alpha_1)\Big(\frac{dx^2}{ds}\Big)^2 + 2[\alpha_2 + \alpha_1 Q]\,\frac{dx^1}{ds}\,\frac{dx^2}{ds}, \\
0 &= \frac{d^2x^2}{ds^2} + (-Q\alpha_1 - \alpha_2)\Big(\frac{dx^1}{ds}\Big)^2 \\
&\quad + (Q\alpha_1 + \alpha_2)\Big(\frac{dx^2}{ds}\Big)^2 + 2[\alpha_1 - \alpha_2 Q]\,\frac{dx^1}{ds}\,\frac{dx^2}{ds},
\end{aligned} \tag{6.3}$$

where $\alpha_j = (Q^2 + 1)^{-1}\partial\phi/\partial x^j$ and s is the arc length parameter. These geodesic equations also serve as the Volterra equations for a dimorphic colonial individual. Here x^1 and x^2 are the log biomasses of the two subpopulations of polyps (i.e., castes), y^1 and y^2 are the relative population sizes, and Q measures the influence of the ratio of polyp numbers (i.e., caste ratios) on the growth of the colony. The actual time parameter t is related to the arc length s by $ds = \lambda e^{\lambda t} dt$, where λ is a constant. When $\phi(x) \equiv 0$, we obtain the Minkowski metric (6.1) and the geodesics reduce

†The indicatrix of (6.1) is not a smooth curve, although it is closed.

to straight lines, describing the ordinary allometric growth law. The conformal transformation describes the adaptation of the system to changes in the environment, the allometric curves (straight lines) being transformed into the geodesics (6.3) (cf. [5]).

6.2 Stochastic model

Our goal is to introduce noise into the equations to represent external environmental variations (e.g. weather) and/or developmental noise *a la* Waddington [23] (presumably due to thermodynamical fluctuations at the molecular level in living cells). In doing so we shall adapt the technique put forward in [3] to the Finslerian case.

We assume that the external (or developmental) noise is independent of the behaviour of populations constituting the colonial organism. As a mathematical model of such a noise we take the ordinary (Euclidean) Brownian motion in $\mathbb{R}^4$ or, equivalently, two independent Brownian motions $w(s)$ and $v(s)$ in $\mathbb{R}^2$. The log biomasses $x(s)$ will be subject to perturbation by $w(s)$, while the relative population sizes $y(s)$ will be perturbed by $v(s)$.

Even though the noise is independent of the dynamics of the colony, the effect it has on the dynamics depends on the current state of the colony, i.e., the current biomasses and population sizes, as well as the ability of the colony to react to changing circumstances, which is mathematically expressed by the form of the fundamental metric function L.

In the absence of any noise the dynamics of the colony is described by the geodesic equations (6.3), which can be written in the form

$$\begin{cases} dx^i &= y^i ds, \\ \overline{\delta} y^i &= 0, \end{cases} \tag{6.4}$$

where $\overline{\delta} y^i = dy^i + \overline{N}^i_j(x,y)dx^j$, $\overline{N}^i_j$ being the coefficients of the nonlinear connection of the Finsler metric $\overline{L}$ of the form (6.2). Please note that the latter form of the geodesic equations is independent of the choice of the Minkowski metric L in (6.2). As noted above, the way the noise enters the equations should depend on the current state $x(s)$, $y(s)$ of the colony, and the geometry of the manifold in the vicinity of $x(s)$, $y(s)$, which is determined by the metric function $\overline{L}$. This can be achieved by employing the concept of rolling as described in the earlier sections. First of all, we observe that any geodesic described by (6.4) can be obtained by hv–rolling the Finsler manifold along a curve $(\gamma(s), \eta(s))$ in $\mathbb{R}^2 \times \mathbb{R}^2$, where $\gamma(s) = as + b$ is a straight line and $\eta(s) = c$ is a constant. Here a, b, c are points in $\mathbb{R}^2$ depending on the initial conditions in the equations (2.4), (2.5), and

(2.6) for hv–rolling. We now perturb $(\gamma(s), \eta(s))$ by the noise $(w(s), v(s))$ to obtain a process $(\gamma(s) + w(s), \eta(s) + v(s))$ in $\mathbb{R}^2 \times \mathbb{R}^2$, along which we hv–roll the manifold with the same initial conditions as before. As a result, we obtain a process $(x(s), y(s), z(s))$ on OM that satisfies the system of SDEs

$$\begin{cases} dx^i &= z^i_j \circ dw^j + y^i ds, \\ \overline{\delta} y^i &= z^i_j \circ dv^j, \\ dz^i_j &= -\overline{F}^j_{k\ell}(x, y) z^\ell_j \circ dx^k - C^i_{k\ell}(y) z^\ell_j \circ \overline{\delta} y^k. \end{cases} \tag{6.5}$$

Here $\overline{\delta} y^i = dy^i + \overline{N}^i_j(x, y) \circ dx^j$ and $(\overline{N}^i_j, \overline{F}^i_{jk}, C^i_{jk})$ are the coefficients of the Cartan connection in the $\overline{L}$ metric. Notice that there is no bar over C^i_{jk}, which, being a conformal invariant, is the same for L and $\overline{L}$ (cf. [12]).

The system (6.5) is the same as the system (3.2) of SDEs for the hv–stochastic development, except for the additional drift term $y^i ds$. Similarly as in the case of the hv–stochastic development (cf. Theorem 3.1), the process $(x(s), y(s))$ is a homogeneous Markov diffusion on $\mathring{T}M$ with generator

$$\mathcal{D} = 2^{-1} \overline{g}^{ij} (\overline{\delta}_i \overline{\delta}_j - \overline{F}^k_{ij} \overline{\delta}_k) + 2^{-1} \overline{g}^{ij} (\dot{\partial}_i \dot{\partial}_j - C^k_{ij} \dot{\partial}_k) + y^j \overline{\delta}_j, \tag{6.6}$$

where $\overline{g}_{ij} = e^{2\phi} g_{ij}$ is the metric tensor of the Finsler metric $\overline{L}$, g_{ij} being the metric tensor of the Minkowski metric L.

Because of the interpretation of $y^i(s)$ as the relative population sizes, the process $(x(s), y(s))$ must be restricted to the region where both coordinates of y are positive, i.e., the process must start at a point (x_0, y_0) such that $y^j_0 > 0$ for $i = 1, 2$, and will be considered up to the first moment σ of hitting the surface $y^1 y^2 = 0$.

Let $G_{\alpha\beta}$ $(\alpha, \beta = 1, 2, 3, 4)$ be the Riemannian metric tensor on $\mathring{T}M$ obtained by the diagonal lift of g_{ij} $(i, j = 1, 2)$ (cf. (1.7)) and let

$$g = \det (g_{ij}), \qquad G = \det (G_{\alpha\beta}),$$

which implies that $g^2 = G$. We denote by $\rho(s, x, y)$ the probability density of the process with generator (6.6) stopped at time σ to be in a region $A \subset \mathring{T}M \cap \{y^1 > 0,\ y^2 > 0\}$ relative to the measure $\sqrt{G}\, dxdy$ on $\mathring{T}M$, i.e.,

$$P\{(x(s), y(s)) \in A\} = \int_A \rho(s, x, y) \sqrt{G(y)}\, dxdy.$$

The function $\rho(s,x,y)$ satisfies the forward initial boundary value problem

$$\begin{cases} \mathcal{D}^*\rho = \frac{\partial \rho}{\partial s}, \\ \lim\limits_{s\searrow 0} \rho(s,x,y) = \rho_0(x,y), \\ \rho(s,x,y)\big|_{y^1y^2=0} = 0, \end{cases} \tag{6.7}$$

where the initial density $\rho_0(x,y)$ is supported on $\mathring{T}M \cap \{y^1 > 0, y^2 > 0\}$ and $\mathcal{D}^*$ is the adjoint of $\mathcal{D}$ relative to the measure $\sqrt{G}\,dxdy$.

The solution to (6.7) can be found by means of the Feynman-Kac formula. To this end we compute $\mathcal{D}^*$ to find that

$$\mathcal{D}^* f = \mathcal{D}f - 2y^i\overline{\delta}_i f - 4\overline{g}^{ij}\phi_i\overline{\delta}_j f + \overline{g}^{ij}C^k_{ik}\dot{\partial}_j f + Vf, \tag{6.8}$$

where V is a scalar function on $\mathring{T}M$, which can be expressed in terms of Berwald's scalar curvature K of the metric (6.2), and where $\phi_i = \partial\phi/\partial x^i$. The scalar curvature is defined by $2K = \overline{g}^{jk}R^{\ell}_{jk\ell}$, where $R^i_{jk\ell}$ is the h–curvature tensor of the Cartan connection (cf. [20], formula (28.8)). We shall use the following theorem.‡

Theorem 4. *For any two-dimensional Finsler space with metric function $\overline{L}$ of the form* (6.2), *where L is an arbitrary Minkowski metric function (not necessarily of the form* (6.1)*),*

$$K = -\overline{g}^{ij}\phi_{ij} - I\overline{m}^i\overline{\ell}^j\phi_{ij} - I_\theta\overline{m}^i\overline{m}^j\phi_i\phi_j, \tag{6.9}$$

where $\phi_i = \partial\phi/\partial x^i$, $\phi_{ij} = \partial^2\phi\partial x^i\partial x^j$, I is the main scalar, $I_\theta = \partial I/\partial\theta$ is the derivative of I with respect to the Landsberg angle θ, and $\overline{\ell}^i$, $\overline{m}^i$ is the Berwald frame in the $\overline{L}$ metric (for definitions see, for example, [20]*).*

The proof of this theorem is rather involved and will be omitted. It is based on Hashiguchi's results [12] on the transformation of the curvature tensor $R^i_{jk\ell}$ under a conformal change of the metric.

In the special case of the Minkowski metric (6.1) $I = 2Q/\sqrt{Q^2+1}$ is a constant function and the last term in (6.9) vanishes. We can now state the formula for V :

$$\begin{aligned} V = {} & 2K + 2I\overline{m}^i\overline{\ell}^j\phi_{ij} - 4I\overline{m}^i\overline{l}^j\phi_i\phi_j - 2I^2\overline{m}^i\overline{m}^j\phi_i\phi_j \\ & + 8\overline{g}^{ij}\phi_i\phi_j + 4y^i\phi_i - 2^{-1}I^2/\overline{L}^2. \end{aligned} \tag{6.10}$$

‡If $\overline{L}$ is positive definite and Berwald, with indicatrix a smooth closed curve, then $K = 0$ (Szabó, 1981). Obviously, such spaces have little interest in applications.

In view of (6.8), to express the solution of the initial boundary value problem (6.7), we introduce a process $(\xi(s), \eta(s))$ on $\mathring{T}M$ being a Finslerian $hv-$Brownian motion with additional horizontal and vertical drifts

$$b^i = -2y^i - 4\overline{g}^{ij}\phi_j, \quad b^{\bar{i}} = \overline{g}^{ij}C^k_{jk}.$$

This means that $(\xi(s), \eta(s))$ satisfies the system of SDEs

$$\begin{cases} d\xi^i &= z^i_j \circ dw^j + b^i(\xi,\eta)ds, \\ \overline{\delta}\eta^i &= z^i_j \circ dv^j + b^{\bar{i}}(\xi,\eta)ds, \\ dz^i_j &= -\overline{F}^i_{k\ell}(\xi,\eta)z^\ell_j \circ d\xi^k - C^i_{k\ell}(\eta)z^\ell_j \circ \overline{\delta}\eta^k. \end{cases} \tag{6.11}$$

We claim that $\overline{L}(\xi(s), \eta(s))$ is a Bessel process with index $n = 2$ (cf. Theorem 4.2). The proof is similar to that of Theorem 4.2, except that the drifts b^i and $b^{\bar{i}}$ must be taken into account. The fact that the vertical drift $b^{\bar{i}}$ is orthogonal to y^i, that is, $y_k b^{\bar{i}}(x,y) = 0$, makes it possible to adapt the proof of Theorem 4.2 to the case at hand. It follows that $\eta(s)$ cannot have explosions. Because the frame $z(t)$ is orthonormal, it cannot have explosions either. To rule out the possibility that $\xi(s)$ may explode, we assume that $\phi(x)$ in (6.2) vanishes for $|x| > R$, where R is a large enough number. Biologically, this means that the model is no longer applicable if the biomass exceeds a certain maximum value (cf. [2]). For $|x| > R$, the SDEs (6.11) read

$$\begin{cases} d\xi^i &= z^i_j \circ dw^j - 2\eta^i ds, \\ d\eta^i &= z^i_j \circ dv^j + C^i(\eta)ds, \\ dz^i_j &= -C^i_{k\ell}(\eta)z^\ell_j \circ d\eta^k, \end{cases}$$

where $C^i = g^{ij}C^k_{jk}$. Since the right-hand sides of these SDEs are independent of ξ and neither $\eta(s)$ nor $z(s)$ can have explosions, it follows that $\xi(s)$ cannot explode either.

We denote by $\xi_{x,y}(s)$, $\eta_{x,y}(s)$ the solution of (6.11) with initial conditions $\xi(0) = x$, $\eta(0) = y$, $z(0) = z$ (the initial condition for the frame $z(s)$ can be omitted, since $(\xi(s), \eta(s))$ is a Markov process on its own; cf. Theorem 3.1). In addition, we denote by $\sigma_{x,y}$ the first time of hitting the surface $y^1y^2 = 0$ by $\eta_{x,y}(s)$. Then the solution to (6.7) can be written as

$$\rho(s,x,y) = E\Big\{\chi_{\{\sigma_{x,y}>s\}}\rho_0\big(\xi_{x,y}(s), \eta_{x,y}(s)\big) \exp\Big[\int_0^s V\big(\xi_{x,y}(u), \eta_{x,y}(u)\big)du\Big]\Big\}. \tag{6.12}$$

There are a number of conclusions concerning the qualitative behaviour of $\rho(s,x,y)$ one can draw from the above Feynman-Kac formula. This is, however, beyond the scope of this article and will be addressed in detail in a forthcoming paper devoted to the biological implications of the present example. For the time being, we just note that, in view of the Feynman-Kac formula (6.12) and the expression (6.10) for the 'potential' V in terms of Berwald's scalar curvature, it follows that, as far as the short-time asymptotics of $\rho(s, x, y)$ is concerned, large positive curvature will cause ρ to increase locally, while large negative curvature will have the opposite effect. This is in agreement with the well-known results on the stability of solutions of a deterministic system depending on the sign of the curvature (see [21], Chapt. 4.4).

References

[1] Antonelli, P.L. (ed.) (1985) *Mathematical Essays on Growth and Emergence of Form*, University of Alberta Press, Edmonton.

[2] Antonelli, P.L., Elliott, R.J. and Seymour, R.M. (1987) Nonlinear Filtering and Riemannian Scalar Curvature ***R***, *Adv. in Appl. Math.*, **8**, 237-253.

[3] Antonelli, P.L. (1988) A Curved Space Forward Cauchy Problem for Chemical Ecology on the Great Barrier Reef, *Stochastic Anal. Appli.*, **6**(3), 231-246.

[4] Antonelli, P.L. (1991) Finsler Volterra-Hamilton System in Ecology, *Tensor (N.S.)*, **50**, 22-31.

[5] Antonellli, P.L. (1992) A Mathematical Theory of Evolution by Heterotrony in Colonial Invertebrates, *Open Systems and Information Dynamics*, **1**, 57-74.

[6] Antonelli, P.L., Ingarden, R.S. and Matsumoto, M. (1993) *The Theory of Sprays and Finsler Spaces with Applications in Physics and Biology*, D. Reidel and Kluwer Acad. Press, Dordrecht, Netherlands.

[7] Antonelli, P.L. and Zastawniak, T.J. (1993) Diffusions on Finsler Manifolds, *Rep. Math. Phy.*, **33**, 303-315.

[8] Cartan, E. (1934) Les Espaces de Finsler, *Actualités*, **79**, Paris, 2nd ed. 1971.

[9] Doob, L. (1953) *Stochastic Processes*, John Wiley and Sons, New York.

[10] Elworthy, K.D. (1982) *Stochastic Differential Equations on Manifolds*, Cambridge University Press, Cambridge.

[11] Emery, M. (1989) *Stochastic Calculus in Manifolds*, Springer-Verlag, Berlin, Heidelberg.

[12] Hashiguchi, M. (1976) On Conformal Transformations of Finsler Metrics, *J. Math. Kyoto Univ.*, **16**(1), 25-50.

[13] Ikeda, N. and Watanabe, S. (1981) *Stochastic Differential Equations and Diffusion Processes*, North Holland, Amsterdam, Kodansha, Tokyo, 2nd ed. 1989.

[14] Ingarden, R.S. (1976) Differential Geometry and Physics, *Tensor (N.S.)*, **30**, 201-209.

[15] Itô, K. (1950) Stochastic Differential Equations in a Differentiable Manifold, *Nagoya Math. J.*, **1**, 35-47.

[16] Itô, K. (1950) Stochastic Difffential Equations in a Differentiable Manifold (2), *Mem. Coll. Sci. Univ. Kyoto Math.*, **28**, 81-85.

[17] Itô, K. (1963) The Brownian Motion and Tensor Fields on a Riemannian Manifold, *Proc. Intern. Congr. Math.*, Stockholm, 536-539.

[18] Itô, K. (1975) Stochastic Parallel Displacement, Probabilistic Methods in Differential Equations, *Lecture Notes in Math.*, **451**, Springer-Verlag, Berlin, 1-7.

[19] Kunita, H. (1990) *Stochastic Flows and Stochastic Differential Equations*, Cambridge University Press, Cambridge.

[20] Matsumoto, M. (1986) *Foundations of Finsler Geometry and Special Finsler Spaces*, Kasheisha Press, Saikawa 3-23-2, Otsushi, Shigaken.

[21] Rund, H. (1959) *The Differential Geometry of Finsler Spaces*, Springer-Verlag, Berlin.

[22] Szabó, Z.I. (1981) Positive Definite Berwald Spaces, (Structure Theorems on Berwald Spaces), *Tensor (N.S.)*, **35**, 25-39.

[23] Waddington, C.H. (1957) *The Strategy of the Genes*, Allen and Unwin Ltd., London.

[24] Wilson, E.O. (1975) *Sociobiology*, Belknap Press, Harward Univ.

[25] Yano, K. and Ishihara, S. (1973) *Tangent and Cotangent Bundles, Differential Geometry*, Marcel Dekker, New York.

DIFFUSION ON THE TANGENT AND INDICATRIX BUNDLES OF A FINSLER MANIFOLD

P.L. ANTONELLI * and T.J. ZASTAWNIAK

1 Introduction

The theory of diffusion processes on Riemannian manifolds, which goes back to the pioneering articles [7], [8], [9], [10] by Itô, has now become a classical branch of stochastic calculus (see, for example, [4], [5], [6]). Recently, the theory has been extended by the present authors [1], [2] to the case of diffusions on Finsler manifolds, the extension being motivated by certain models in developmental and population biology involving systems in a noisy environment. The goal of this article is to represent and study Finslerian diffusions as processes on the slit tangent bundle $\overset{\circ}{T}M$ and the indicatrix bundle IM of a Finsler manifold M.

In Sect. 2 it is demonstrated that the notion of a Finslerian hv-Brownian motion introduced in [1] and [2] is equivalent to a Riemannian Brownian motion with drift on the slit tangent bundle $\overset{\circ}{T}M$ equipped with a suitable Riemannian metric, the so-called diagonal lift of the Finsler metric. In Sect. 3 an alternative approach to Finslerian diffusions is developed with $\overset{\circ}{T}M$ replaced by the indicatrix bundle IM, in line with the view that Finslerian objects depend only on the direction of a tangent vector to M. This leads to the definition of an indicatrized Finslerian $hv-$Brownian motion. Similarly as in the case of processes on $\overset{\circ}{T}M$, in Sect. 4 it is demonstrated that the latter notion is equivalent to a Riemannian Brownian motion with drift on the indicatrix bundle IM equipped with a suitable Riemannian metric. The drift on $\overset{\circ}{T}M$ and the drift on IM are both expressed in terms of the torsion tensors of the Cartan connection.

Let M be an n-dimensional Finsler manifold with *metric function*

*Partially supported by NSERC A-7667. This article appeared in *Tensor, N.S.* **56**, 1995, 233-247.

$L : \mathring{T}M \to (0, \infty)$ defined and smooth on the slit tangent bundle $\mathring{T}M = TM \backslash \{0_x : x \in M\}$, where 0_x denotes the null vector in TM_x. It is assumed that $L(x, ky) = kL(x, y)$ for every $x \in M$, $0 \neq y \in TM_x$, and $k > 0$, that is, $L(x, y)$ is positively homogeneous of degree one in y. The *indicatrix bundle* $IM \subset \mathring{T}M$ is defined by $IM = \{(x, y) \in \mathring{T}M : L(x, y) = 1\}$. Given a local coordinate system (x^i) on an open set $U \subset M$, one can represent any vector $y \in TM_x$, where $x \in U$, as $y = y^i \partial/\partial x^i$ and use (x^i, y^i) as local coordinates defined on $\mathring{T}U \subset \mathring{T}M$. We shall refer to (x^i, y^i) as the *induced coordinates* on $\mathring{T}M$. We set $\partial_i = \partial/\partial x^i$ and $\partial_{(i)} = \partial/\partial y^i$. The *Finsler metric tensor* $g : \mathring{T}M \to T_2M$ such that $g(x, y) = g_{ij}(x, y)dx^i \otimes dx^j \in T_2M_x$ with

$$g_{ij}(x, y) = 2^{-1}\partial_{(i)}\partial_{(j)}L^2(x, y),$$

for any $x \in M$ and $0 \neq y \in TM_x$, will be assumed to be positive definite. (The latter assumption can be relaxed by restricting the diffusions considered throughout this article to a region in which $g_{ij}(x, y)$ is positive definite.)

Let M be equipped with the *Cartan connection* $(N^i_j, F^i_{jk}, C^i_{jk})$, N^i_j being the *nonlinear*, F^i_{jk} the *horizontal*, and C^i_{jk} the *vertical connection coefficients* defined by

$$\begin{aligned} N^i_j &= \partial_{(j)}(\gamma^i_{kl}y^k y^l), \\ F^i_{jk} &= 2^{-1}g^{il}(\delta_j g_{lk} + \delta_k g_{jl} - \delta_l g_{jk}), \\ C^i_{jk} &= 2^{-1}g^{il}(\partial_{(j)}g_{lk} + \partial_{(k)}g_{jl} - \partial_{(l)}g_{jk}) = 2^{-1}g^{il}\partial_{(j)}g_{lk}, \end{aligned}$$

where $\delta_i = \partial_i - N^j_i\partial_{(j)}$, g^{ij} is the inverse matrix to g_{ij}, and $\gamma^i_{jk} = 2^{-1}g^{il}(\partial_j g_{lk} + \partial_k g_{jl} - \partial_l g_{jk})$, by analogy with the expression for the Levi-Cività connection coefficients. Here g_{ij}, g^{ij}, γ^i_{jk}, N^i_j, F^i_{jk}, and C^i_{jk} are functions of both $x \in M$ and $0 \neq y \in TM_x$. In the above formulae and throughout the article the standard summation convention on upper and lower indices applies. The Cartan connection defined above is both $h-$ *metrical* and $v-$*metrical*, that is,

$$\delta_i g_{jk} = F^l_{ij}g_{lk} + F^l_{ik}g_{jl}, \qquad \partial_{(i)}g_{jk} = C^l_{ij}g_{lk} + C^l_{ik}g_{jl},$$

and *deflection-free*, that is,

$$N^i_j = F^i_{jk}y^k,$$

the latter three properties being essential for the construction of Finslerian Brownian motion.

We say that $A : \mathring{T}M \to TM$ is a Finsler vector field if $A(x,y) \in TM_x$ for every $x \in M$ and $0 \neq y \in TM_x$. The *horizontal* and *vertical covariant derivatives* of such a field will be denoted by

$$\nabla^h_j A^i = \delta_j A^i + F^i_{jk} A^k, \qquad \nabla^v_j A^i = \partial_{(j)} A^i + C^i_{jk} A^k.$$

These formulae can be extended to arbitrary Finsler tensor fields in the usual way.

For a systematic presentation of Finsler geometry the reader is referred to the monographs [13] and [15], as well as Cartan's original work [3].

Throughout the article we use the standard probabilistic setup (see, for example, [5], [11]). We fix a complete *probability space* $(\Omega, \mathcal{F}, P)$ equipped with a right-continuous *filtration* $(\mathcal{F}_t)_{t \geq 0}$ such that each $\mathcal{F}_t$ contains all the events of probability zero from the σ-algebra $\mathcal{F}$. We denote by $Ef = \int_\Omega f\, dP$ the *expectation* of a *random variable* $f : \Omega \to \mathbb{R}$. The *conditional expectation* of f given a sub-σ-field $\mathcal{G} \subset \mathcal{F}$ will be denoted by $E(f|\mathcal{G})$. A *stochastic process* $X : [0,\infty) \times \Omega \to \mathbb{R}$ is said to be *continuous* if its sample paths $t \mapsto X(t,\omega)$ are continuous for P-almost all $\omega \in \Omega$. A stochastic process X is *adapted* if the random variable $X(t)$ is $\mathcal{F}_t$-measurable for every $t \geq 0$. A *martingale* is an adapted process X such that $E|X(t)| < \infty$ for every $t \geq 0$ and $E(X(t)|\mathcal{F}_s) = X(s)$ for every $t > s \geq 0$. A *local martingale* is an adapted process X such that the process $(t,\omega) \mapsto X(\tau_n \wedge t, \omega)\chi_{\{0<\tau_n\}}(\omega)$ is a martingale for every n, where $\tau_n = \inf\{t \geq 0 : |X(t)| \geq n\}$ (as usual, we set $\inf \emptyset = \infty$) and where χ_A is the indicator function of $A \subset \Omega$. A *semimartingale* is the sum of a continuous local martingale and a continuous adapted process whose sample paths have P-almost surely (P-a.s.) finite variation on each compact interval in $[0,\infty)$.

By a *standard Brownian motion* on $\mathbb{R}^n$ (also known as the *Wiener process*) we understand a semimartingale $W : [0,\infty) \times \Omega \to \mathbb{R}^n$ such that $(t,\omega) \mapsto W^i(t,\omega)$ and $(t,\omega) \mapsto W^i(t,\omega)W^j(t,\omega) - t\delta^{ij}$ are martingales for all $i,j \in \{1,\dots,n\}$.

For a real-valued semimartingale X and a continuous adapted process Y, we denote by $\int Y\,dX$ the *Itô stochastic integral* of Y with respect to X. Given two real-valued semimartingales X and Y, the process $[X,Y] = XY - X(0)Y(0) - \int X\,dY - \int Y\,dX$ is called the *joint quadratic variation* of X and Y, and the process $\int Y \circ dX = \int Y\,dX + 2^{-1}[X,Y]$ is called the *Stratonovich integral* of Y with respect to X. We shall use the differential notation: if $Z = \int Y\,dX$, we shall write $dZ = Y\,dX$; if $Z = [X,Y]$, we shall write $dZ = dXdY$; and if $Z = \int Y \circ dX$, then we shall write $dZ = Y \circ dX$. Thus, in particular, $Y \circ dX = Y\,dX + 2^{-1}dXdY$. We shall also need the formula $Z \circ (Y \circ dX) = (ZY) \circ dX$ valid for arbitrary real-valued semimartingales X, Y, Z. It is well known that a standard Brownian motion

W on $\mathbb{R}^n$ satisfies the *Itô rules*

$$dW^i dW^j = \delta^{ij} dt, \qquad dW^i dt = 0, \qquad dtdt = 0.$$

Here dt stands for dX, where X is the deterministic process $X(t, \omega) = t$.

If $X^1, \dots, X^n$ are real-valued semimartingales and $f : \mathbb{R}^n \to \mathbb{R}$ is a function of class C^2, then we have the Itô change-of-variables formula

$$df(X) = \partial_i f(X)\, dX^i + 2^{-1} \partial_i \partial_j f(X)\, dX^i dX^j,$$

where $X = (X^1, \dots, X^n)$. If f is of class C^3, then the Itô formula can be written as

$$df(X) = \partial_i f(X) \circ dX^i.$$

A continuous stochastic process $X : [0, \infty) \times \Omega \to M$ with values on a manifold M is called an M-valued *semimartingale* if, for every smooth function $f : M \to \mathbb{R}$, the process $(t, \omega) \mapsto f(X(t, \omega))$ is a real-valued semimartingale. We shall also admit M-valued semimartingales defined on a *random time interval* $[0, \tau)$, where $\tau : \Omega \to [0, \infty]$ is a *stopping time*, i.e., a random variable such that $\{t < \tau\} \in \mathcal{F}_t$ for every $t \geq 0$.

Given two manifolds M and N, we shall consider *Stratonovich stochastic differential equations* (Stratonovich SDEs) of the form $dY = e(X, Y) \circ dX$, where $e(x, y) : TM_x \to TN_y$ are linear maps depending smoothly on $x \in M$ and $y \in N$, and where X is an M-valued semimartingale defined on a random time interval $[0, \tau)$. In local coordinates the above SDE reads $dY^i = e^i_j(X, Y) \circ dX^j$. The solution Y is an N-valued semimartingale defined on a random time interval $[0, \sigma)$. It is called a *maximal solution* if $P\{\sigma = \tau \text{ or } Y(t) \to \infty \text{ as } t \nearrow \sigma\} = 1$, where ∞ is the infinity of the one-point compactification of M. (If M is a compact manifold, then the condition reduces to $P\{\sigma = \tau\} = 1$.) If $\sigma < \tau$ with non-zero probability, then we say that the maximal solution explodes. From now on, by a solution of an SDE we shall always understand a maximal solution.

Definition 1.1. Let D be an elliptic second-order operator on a manifold M. An M-valued semimartingale X is called a *diffusion* on M with *generator* D if the process $(t, \omega) \mapsto f(X(t, \omega)) - f(X(0, \omega)) - \int_0^t Df(X(s, \omega))ds$ is a local martingale for every compactly supported smooth function $f : M \to \mathbb{R}$.

The simplest example of a diffusion is a standard Brownian motion on $\mathbb{R}^n$. In this case $D = 2^{-1}\Delta$, where Δ is the Laplace operator on $\mathbb{R}^n$. By analogy, Brownian motion on a Riemannian manifold can be defined as follows.

Definition 1.2. Let M be a Riemannian manifold with metric tensor g_{ij} and Levi-Cività connection ∇ with coefficients γ^i_{jk}. By an M−valued *Riemannian Brownian motion* X we mean a diffusion X on M with generator $D = 2^{-1}\Delta$, where

$$\Delta = g^{ij}\nabla_i\nabla_j = g^{ij}(\partial_i\partial_j - \gamma^k_{ij}\partial_k)$$

is the Laplace-Beltrami operator on M. Given a vector field B on M, an M−valued diffusion X with generator $D = 2^{-1}\Delta + B$ is called a Riemannian Brownian motion with *drift* B.

The standard construction of a Riemannian Brownian motion rests on the following well-known result (cf, for example, [4], [5], or [6]).

Theorem 1.3. *Let M be an n−dimensional Riemannian manifold with metric tensor g_{ij} and let Γ^i_{jk} be the coefficients of a metric connection (possibly other than the Levi-Cività connection γ^i_{jk}) on M. Let W be a standard Brownian motion on $\mathbb{R}^n$. Then the solution (X, Z) of the system*

$$\begin{cases} dX^i = Z^i_j \circ dW^j \\ dZ^i_j = -\Gamma^i_{kl}(X)Z^l_j \circ dX^k \end{cases} \tag{1.1}$$

of SDEs *with initial conditions $X(0) = x_0 \in M$ and $Z(0) = z_0$ an orthonormal frame in TM_{x_0} is a semimartingale on the orthonormal frame bundle OM over M, and X is an M−valued diffusion with generator*

$$D = 2^{-1}g^{ij}(\partial_i\partial_j - \Gamma^k_{ij}\partial_k) = 2^{-1}\Delta + B, \tag{1.2}$$

where Δ is the Laplace-Beltrami operator on M and

$$B = 2^{-1}g^{ik}T^j_{jk}\partial_i, \tag{1.3}$$

$T^i_{jk} = \Gamma^i_{jk} - \Gamma^i_{kj}$ being the torsion tensor of the connection at hand. This means that X is a Riemannian Brownian motion on M with drift B given by (1.3). *In particular, if Γ^i_{jk} is the Levi-Cività connection, then X is a Riemannian Brownian motion on M without drift.*

Definition 1.4. The process (X, Z) on OM described in Theorem 1.3 is called the *Riemannian stochastic development*.

For details on stochastic calculus, the theory of diffusions on Riemannian manifolds, and, in particular, the proof of Theorem 1.3 we refer to [4], [5], and [6].

2 Finslerian Diffusions on the Tangent Bundle

We assume that M is an n-dimensional Finsler manifold with metric function L, positive definite metric tensor g_{ij}, and the Cartan connection $(N^i_j, F^i_{jk}, C^i_{jk})$.

In [1] and [2] the system (1.1) of SDEs for stochastic development was generalized from the Riemannian to Finslerian case using a stochastic version of Cartan's concept of parallel transport and moving frames, or, equivalently, the concept of rolling a Finsler manifold along a curve. In the Finslerian case the system of SDEs for stochastic development reads (see [1], [2])

$$\begin{cases} dX^i = Z^i_j \circ dW^j \\ \delta Y^i = Z^i_j \circ dV^j \\ dZ^i_j = -F^i_{kl}(X,Y)Z^l_j \circ dX^k - C^i_{kl}(X,Y)Z^l_j \circ \delta Y^k, \end{cases} \tag{2.1}$$

where

$$\delta Y^i = dY^i + N^i_j(X,Y) \circ dX^j$$

and where W and V are independent standard Brownian motions on $\mathbb{R}^n$.

The following theorem, which is a generalization of Theorem 1.3 to the Finslerian case, has been stated and proved in [1] and [2]. To make the present article as self-contained as possible, we quote the theorem complete with the proof from [1] and [2].

Theorem 2.1. *Let (X,Y,Z) be the solution to* (2.1) *with initial conditions $X(0) = x_0 \in M$, $0 \neq Y(0) = y_0 \in TM_{x_0}$, and $Z(0) = z_0$ a frame in TM_{x_0} orthonormal with respect to the quadratic form $g(x_0, y_0)$. Then (X,Y,Z) is a semimartingale on the so-called orthonormal Finsler bundle $F\mathring{T}M$ over $\mathring{T}M$ defined below, and (X,Y) is a diffusion on $\mathring{T}M$ with generator*

$$\begin{aligned} D &= 2^{-1} g^{ij} \nabla^h_i \nabla^h_j + 2^{-1} g^{ij} \nabla^v_i \nabla^v_j \\ &= 2^{-1} g^{ij} \left(\delta_i \delta_j - F^k_{ij} \delta_k\right) + 2^{-1} g^{ij} \left(\partial_{(i)} \partial_{(j)} - C^k_{ij} \partial_{(k)}\right). \end{aligned} \tag{2.2}$$

Remark 2.2. The *orthonormal Finsler bundle* $F\mathring{T}M$ is by definition a principal bundle over $\mathring{T}M$ whose total space consists of elements (x,y,z), where $x \in M$, $0 \neq y \in TM_x$, and $z = (z_1, \ldots, z_n)$ is an orthonormal frame in TM_x relative to the quadratic form $g(x,y) \in T^0_2 M_x$, i.e.,

$$g_{ij}(x,y) z^i_k z^j_l = \delta_{kl},$$

with projection $\pi : F\mathring{T}M \ni (x, y, z) \mapsto (x, y) \in \mathring{T}M$. The differential structure of $F\mathring{T}M$ can be obtained from that of M. Namely, if (x^i) are local coordinates on M, then we can write $y = y^i\partial_i$ for any $y \in TM_x$ and $z_j = z^i_j\partial_i$ for each vector z_j of any linear frame $z = (z_1, \ldots, z_n)$ in TM_x, where $x \in M$, and use (x^i, y^i, z^i_j) as local coordinates on $F\mathring{T}M$. The structure group of $F\mathring{T}M$ is the group $O(n)$ of orthogonal transformations of $\mathbb{R}^n$ with the right action $R_h(x, y, z) = (x, y, zh)$, where $(zh)_i = h^j_i z_j$, for any $h \in O(n)$ and $(x, y, z) \in F\mathring{T}M$.

Proof of Theorem 2.1.

We claim that

$$g_{ij}(X, Y)Z^i_k Z^j_l = \delta_{kl} \quad P\text{-almost surely.} \tag{2.3}$$

Indeed, since the Cartan connection $(N^i_j, F^i_{jk}, C^i_{jk})$ is both $h-$ and $v-$metrical, using the Itô formula we find that

$$\begin{aligned} d[g_{ij}(X,Y)Z^i_k Z^j_l] &= Z^i_k Z^j_l \circ dg_{ij}(X,Y) + g_{ij}(X,Y)Z^i_k \circ dZ^j_l \\ &\quad + g_{ij}(X,Y)Z^j_l \circ dZ^i_k \\ &= Z^i_k Z^j_l (\delta_h g_{ij}(X,Y) - F^m_{hi}(X,Y)g_{mj}(X,Y) \\ &\quad - F^m_{hj}(X,Y)g_{im}(X,Y)) \circ dX^h \\ &\quad + Z^i_k Z^j_l (\partial_{(h)} g_{ij}(X,Y) - C^m_{hi}(X,Y)g_{mj}(X,Y) \\ &\quad - C^m_{hj}(X,Y)g_{im}(X,Y)) \circ \delta Y^h \\ &= 0. \end{aligned}$$

Because, be assumption, $g_{ij}(X(0), Y(0))Z^i_k(0)Z^j_l(0) = \delta_{kl}$, it follows by uniqueness that (2.3) holds true. As a consequence of (2.3), (X, Y, Z) is a semimartingale on $F\mathring{T}M$, as required.

Now, we compute the stochastic differentials dX^i and δY^i from (2.1):

$$\begin{aligned} dX^i &= Z^i_j dW^j - 2^{-1}F^i_{kl}(X,Y)Z^k_m Z^l_j dW^m dW^j \\ &\quad - 2^{-1}C^i_{kl}(X,Y)Z^k_m Z^l_j dV^m dW^j \\ &= Z^i_j dW^j - 2^{-1}g^{kl}(X,Y)F^i_{kl}(X,Y)dt, \\ \delta Y^i &= Z^i_j dV^j - 2^{-1}F^i_{kl}(X,Y)Z^k_m Z^l_j dW^m dV^j \\ &\quad - 2^{-1}C^i_{kl}(X,Y)Z^k_m Z^l_j dV^m dV^j \\ &= Z^i_j dV^j - 2^{-1}g^{kl}(X,Y)C^i_{kl}(X,Y)dt, \end{aligned}$$

since $dW^m dW^j = \delta^{mj}dt$ and $dV^m dV^j = \delta^{mj}dt$ by the Itô rules, $dV^m dW^j = dW^m dV^j = 0$ because V and W are independent, and $Z^k_m Z^l_j \delta^{mj} = g^{lk}(X,Y)$ by (2.3). We also compute

$$\begin{aligned}
dX^i dX^j &= Z^i_k Z^j_l dW^k dW^l = Z^i_k Z^j_l \delta^{kl} dt = g^{ij}(X,Y)dt,\\
\delta Y^i \delta Y^j &= Z^i_k Z^j_l dV^k dV^l = Z^i_k Z^j_l \delta^{kl} dt = g^{ij}(X,Y)dt,\\
dX^i \delta Y^j &= Z^i_k Z^j_l dW^k dV^l = 0, \qquad \delta Y^i dX^j = Z^i_k Z^j_l dV^k dW^l = 0.
\end{aligned}$$

Thus, for any smooth function $f : \mathring{T}M \to \mathbb{R}$,

$$\begin{aligned}
df(X,Y) &= \delta_i f(X,Y) \circ dX^i + \partial_{(i)} f(X,Y) \circ \delta Y^i\\
&= \delta_i f(X,Y) dX^i + \partial_{(i)} f(X,Y) \delta Y^i\\
&\quad + 2^{-1}\delta_i \delta_j f(X,Y) dX^i dX^j + 2^{-1}\delta_i \partial_{(j)} f(X,Y) dX^i \delta Y^j\\
&\quad + 2^{-1}\partial_{(i)} \delta_j f(X,Y) \delta Y^i dX^j + 2^{-1}\partial_{(i)}\partial_{(j)} f(X,Y) \delta Y^i \delta Y^j\\
&= \delta_i f(X,Y) Z^i_j dW^j + \partial_{(i)} f(X,Y) Z^i_j dV^j\\
&\quad + \big[2^{-1} g^{ij}(X,Y)(\delta_i \delta_j f(X,Y) - F^k_{ij}(X,Y)\delta_k f(X,Y))\\
&\quad + 2^{-1} g^{ij}(X,Y)(\partial_{(i)}\partial_{(j)} f(X,Y) - C^k_{ij}(X,Y)\partial_{(k)} f(X,Y))\big]dt.
\end{aligned}$$

As a consequence, (X,Y) is a diffusion with generator D given by the expression in square brackets multiplying dt above, which completes the proof of Theorem 2.1. □

Following [1] and [2], we introduce the definitions below.

Definition 2.3. The solution (X, Y, Z) to (2.1) with the above initial conditions is called the *Finslerian hv-stochastic development.*

Definition 2.4. A diffusion (X,Y) on $\mathring{T}M$ with the generator D given by (2.2) is called a *Finslerian hv-Brownian motion.*

Remark 2.5. In [1] and [2] we also introduced the notion of a Finslerian h-Brownian motion defined by the system of SDEs

$$\begin{cases}
dX^i = Z^i_j \circ dW^j\\
\delta Y^i = 0\\
dZ^i_j = -F^i_{kl}(X,Y) Z^l_j \circ dX^k,
\end{cases}$$

where

$$\delta Y^i = dY^i + N^i_j(X,Y) \circ dX^j$$

and where W is a standard Brownian motion on $\mathbb{R}^n$. The resulting process (X,Y) is a degenerate diffusion on $\mathring{T}M$ and remains beyond the scope of the present paper.

As follows from Theorem 2.1, the system (2.1) of SDEs for the Finslerian hv-stochastic development can be used to construct a Finslerian hv-Brownian motion. The goal of the present section is to demonstrate that a Finslerian hv-Brownian motion can also be obtained as a Riemannian Brownian motion with a suitable drift on the slit tangent bundle $\mathring{T}M$ equipped with the diagonal lift metric defined below, which turns $\mathring{T}M$ into a Riemannian manifold. The drift will be expressed in terms of the torsion tensors P^i_{jk} and C^i_{jk} of the Cartan connection.

Given a local coordinate system (x^i) on M, we take the induced coordinates (x^i, y^i) on $\mathring{T}M$ and consider the adapted frame $\delta_i, \partial_{(i)}$, where $\delta_i = \partial_i - N^j_i(x,y)\partial_{(j)}$. We set $\delta y^i = dy^i + N^i_j(x,y)dx^j$. Then $dx^i, \delta y^i$ is the dual frame to $\delta_i, \partial_{(i)}$. The *diagonal lift metric* on $\mathring{T}M$ is defined by (see [17] and [18])

$$G(x,y) = g_{ij}(x,y)dx^i \otimes dx^j + g_{ij}(x,y)\delta y^i \otimes \delta y^j, \tag{2.4}$$

where g is the Finsler metric tensor. If g is positive definite, then so is G. Equipped with the metric tensor G, $\mathring{T}M$ becomes a Riemannian manifold. We shall also need the following connection ∇ on $\mathring{T}M$, which we call the *horizontal lift of the Cartan connection to* $\mathring{T}M$ (see, for example, [12] or [14]):

$$\begin{aligned}
\nabla_{\delta_i}\delta_j &= \Gamma^k_{ij}\delta_k + \Gamma^{(k)}_{ij}\partial_{(k)} = F^k_{ij}\delta_k,\\
\nabla_{\delta_i}\partial_{(j)} &= \Gamma^k_{i(j)}\delta_k + \Gamma^{(k)}_{i(j)}\partial_{(k)} = F^k_{ij}\partial_{(k)},\\
\nabla_{\partial_{(i)}}\delta_j &= \Gamma^k_{(i)j}\delta_k + \Gamma^{(k)}_{(i)j}\partial_{(k)} = C^k_{ij}\delta_k,\\
\nabla_{\partial_{(i)}}\partial_{(j)} &= \Gamma^k_{(i)(j)}\delta_k + \Gamma^{(k)}_{(i)(j)}\partial_{(k)} = C^k_{ij}\partial_{(k)}.
\end{aligned} \tag{2.5}$$

Here Γ^k_{ij}, $\Gamma^{(k)}_{ij}$, $\Gamma^k_{i(j)}$, etc. are the coefficients of ∇ in the adapted frame $\delta_i, \partial_{(i)}$. It is easily seen that $\nabla G = 0$, i.e., ∇ is a metrical connection with respect to the Riemannian metric G on $\mathring{T}M$. However, in general, ∇ is non-symmetric, i.e., differs from the Levi-Cività connection corresponding to G. In the adapted frame $\delta_i, \partial_{(i)}$ the torsion tensor T of ∇ can be expressed as follows:

$$\begin{aligned}
T(\delta_i,\delta_j) &= T^k_{ij}\delta_k + T^{(k)}_{ij}\partial_{(k)} = R^k_{ij}\partial_{(k)},\\
T(\delta_i,\partial_{(j)}) &= T^k_{i(j)}\delta_k + T^{(k)}_{i(j)}\partial_{(k)} = -\,C^k_{ji}\delta_k - P^k_{ji}\partial_{(k)},\\
T(\partial_{(i)},\delta_j) &= T^k_{(i)j}\delta_k + T^{(k)}_{(i)j}\partial_{(k)} = +\,C^k_{ij}\delta_k + P^k_{ij}\partial_{(k)},\\
T(\partial_{(i)},\partial_{(j)}) &= T^k_{(i)(j)}\delta_k + T^{(k)}_{(i)(j)}\partial_{(k)} = 0,
\end{aligned} \tag{2.6}$$

where $R^i_{jk} = \delta_j N^i_k - \delta_k N^i_j$, $P^i_{jk} = \partial_{(j)} N^i_k - F^i_{kj}$, and C^i_{jk} are the non-vanishing torsion tensors of the Cartan connection $(N^i_j, F^i_{jk}, C^i_{jk})$. (As is well known, C^i_{jk} plays the double role of a torsion tensor and the vertical part of the Cartan connection.)

Now, let us use the system (1.1) of SDEs for the Riemannian stochastic development in the case of the Riemannian manifold $\mathring{T}M$ with the diagonal lift metric G defined by (2.4) and the metric connection ∇ with coefficients Γ^k_{ij}, $\Gamma^{(k)}_{ij}$, $\Gamma^k_{i(j)}$, etc. defined by (2.5). According to Theorem 1.3, in this way we obtain a Riemannian Brownian motion (X, Y) on $\mathring{T}M$ with drift B, where, by (1.3) and (2.6),

$$\begin{aligned} B &= 2^{-1} g^{ik} \left(T^j_{jk} + T^{(j)}_{(j)k} \right) \delta_i + 2^{-1} g^{ik} \left(T^j_{j(k)} + T^{(j)}_{(j)(k)} \right) \partial_{(i)} \\ &= 2^{-1} g^{ik} P^j_{jk} \delta_i - 2^{-1} g^{ik} C^j_{kj} \partial_{(i)}. \end{aligned} \tag{2.7}$$

Writing the first equality in (1.2) in terms of the adapted frame and using the expressions (2.5) for the connection coefficients, we find that the generator $D = 2^{-1}\Delta + B$ of (X, Y) is equal to the generator (2.2) of a Finslerian hv-Brownian motion. Thus, we proved the following theorem

Theorem 2.6. *A semimartingale (X, Y) on $\mathring{T}M$ is a Finslerian hv-Brownian motion if and only if it is a Riemannian Brownian motion with drift B on the slit tangent bundle $\mathring{T}M$ equipped with the Riemannian metric G defined by (2.4), where B is the vector field on $\mathring{T}M$ defined by (2.7) in terms of the torsion tensors P^i_{jk} and C^i_{jk} of the Cartan connection and the adapted frame $\delta_i, \partial_{(i)}$.*

3 Indicatrized Finslerian Stochastic Development

In the present section we adopt the point of view that Finslerian objects should essentially depend only on the direction of a tangent vector y to the Finsler manifold M at hand, the dependence on the length of y being completely determined by the appropriate homogeneity conditions. In line with this point of view, we reformulate the theory of Finslerian diffusions using the indicatrix bundle IM in place of the slit tangent bundle $\mathring{T}M$.

As in Sect. 2, we assume that M is an n-dimensional Finsler manifold with metric function L, positive definite metric tensor g_{ij}, and the Cartan connection $(N^i_j, F^i_{jk}, C^i_{jk})$.

While the system (2.1) of SDEs for the Finslerian hv-stochastic development can be interpreted in terms of 'rolling' along a Brownian motion

(W, V) on the 'tangent bundle' $\mathbb{R}^n \times \mathbb{R}^n$ over $\mathbb{R}^n$ (see [2]), we shall now modify (2.1) to describe 'rolling' along a Brownian motion (W, U) on the 'indicatrix bundle' $\mathbb{R}^n \times S^{n-1}$ over $\mathbb{R}^n$, where $S^{n-1} = \{x \in \mathbb{R}^n : \delta_{ij}x^i x^j = 1\}$ is the unit sphere in $\mathbb{R}^n$. Thus, we now assume that W is a standard Brownian motion on $\mathbb{R}^n$ and U is a Riemannian Brownian motion on the unit sphere $S^{n-1} \subset \mathbb{R}^n$ such that U and W are independent. (See the Appendix for information on Riemannian Brownian motion on the unit sphere.) In addition to the initial conditions for (2.1), we now assume that $Y^i(0) = U^j(0)Z^i_j(0)$. This clearly ensures that $Y(0) \in IM_{X(0)}$, where $IM_x = \{y \in TM_x : L(x, y) = 1\}$ is the *indicatrix* at $x \in M$. The following theorem describes the key properties of the solution (X, Y, Z) to the system (2.1) modified as above.

Theorem 3.1. *Let (X, Y, Z) be the solution of the system*

$$\begin{cases} dX^i = Z^i_j \circ dW^j \\ \delta Y^i = Z^i_j \circ dU^j \\ dZ^i_j = -F^i_{kl}(X, Y)Z^l_j \circ dX^k - C^i_{kl}(X, Y)Z^l_j \circ \delta Y^k \end{cases} \tag{3.1}$$

of SDEs, *where*

$$\delta Y^i = dY^i + N^i_j(X, Y) \circ dX^j$$

and where W is a standard Brownian motion on $\mathbb{R}^n$ and U a Riemannian Brownian motion on the unit sphere $S^{n-1} \subset \mathbb{R}^n$ such that U and W are independent, with initial conditions $X(0) = x_0 \in M$, $Y(0) = y_0 \in IM_{x_0}$, and $Z(0) = z_0$ a frame in TM_{x_0} orthonormal with respect to the quadratic form $g(x_0, y_0)$ such that $Y^i(0) = U^j(0)Z^i_j(0)$. Then (X,Y,Z) is a semimartingale on the orthonormal Finsler bundle FIM over IM, and (X, Y) is a diffusion on IM with generator

$$\begin{aligned} D = {} & 2^{-1}g^{ij}(\delta_i\delta_j - F^k_{ij}\delta_k) + 2^{-1}(g^{ij} - y^i y^j)(\partial_{(i)}\partial_{(j)} - C^k_{ij}\partial_{(k)}) \\ & - 2^{-1}(n-1)y^k\partial_{(k)}. \end{aligned} \tag{3.2}$$

Remark 3.2. The orthonormal Finsler bundle $FIM \subset F\mathring{T}M$ over $IM \subset \mathring{T}M$ is, by definition, a subbundle of $F\mathring{T}M$ induced by the projection $\pi : F\mathring{T}M \to \mathring{T}M$, that is, $FIM = \pi^{-1}(IM)$.

Remark 3.3. Even though the generator D given by (3.2) is expressed in terms of the adapted frame $\delta_i, \partial_{(i)}$ on $\mathring{T}M$, it can be regarded as an operator over IM in the sense that if $f_1, f_2 : \mathring{T}M \to \mathbb{R}$ are smooth functions such that $f_1 = f_2$ on IM, then $Df_1 = Df_2$ on IM. This property of D can be verified by direct computation, but it also follows from the proof of Theorem 3.1 as well as from assertion b) of Theorem 4.4.

Lemma 3.4. *Under the assumptions of Theorem* 3.1,
a) $Y^i = U^j Z^i_j$ *P-almost surely,*
b) $g_{ij}(X,Y)Z^i_k Z^j_l = \delta_{kl}$ *P-almost surely,*
c) $L(X,Y) = 1$ *P-almost surely.*

Proof of Lemma 3.4. a) Given the solution (X,Y,Z) to (3.1) with the initial conditions described in Theorem 3.1, we consider the SDE

$$dA^i = -F^i_{kl}(X,Y)A^l \circ dX^k - C^i_{kl}(X,Y)A^l \circ \delta Y^k + Z^i_j \circ dU^j. \tag{3.3}$$

Since $C^i_{jk}(x,y)y^k = 0$ and the Cartan connection is deflection-free, i.e., $F^i_{jk}(x,y)y^k = N^i_j(x,y)$, it follows from the second equation in (3.1) that $A^i = Y^i$ is a solution to (3.3). By the Itô formula, $d(U^j Z^i_j) = U^j \circ dZ^i_j + Z^i_j \circ dU^j$. Substituting for dZ^i_j from the third equation in (3.1), we find that $A^i = U^j Z^i_j$ is also a solution to (3.3). But, by assumption, $Y^i(0) = U^j(0)Z^i_j(0)$, which, by uniqueness, means that $Y^i = U^j Z^i_j$, as required.

b) The proof of assertion b) is exactly the same as that of (2.3) above.

c) By the Itô formula,

$$dL^2(X,Y) = \delta_i L^2(X,Y) \circ dX^i + \partial_{(i)} L^2(X,Y) \circ \delta Y^i.$$

Since the Cartan connection is deflection-free, it follows that $\delta_i L^2(x,y) = 0$. We also have $\partial_{(i)} L^2(x,y) = 2g_{ij}(x,y)y^j$. Thus, by the second equation in (3.1) and assertions a) and b) of the present lemma,

$$\begin{aligned} dL^2(X,Y) &= 2g_{ij}(X,Y)Y^i \circ \delta Y^j = 2g_{ij}(X,Y)U^k Z^i_k Z^j_l \circ dU^l \\ &= 2\delta_{kl}U^k \circ dU^l = 2d(\delta_{kl}U^k U^l) = 0, \end{aligned}$$

for U is a Brownian motion on the unit sphere, i.e., $\delta_{kl}U^k U^l = 1$. Because, by assumption, $L(X(0),Y(0)) = 1$, it follows that $L(X,Y) = 1$, as required. □

Proof of Theorem 3.1. Assertions b) and c) of Lemma 3.4 imply that (X,Y,Z) is a semimartingale on the orthonormal Finsler bundle FIM over IM with projection (X,Y) onto IM.

Let us now compute the stochastic differentials dX^i and δY^i from (3.1):

$$\begin{aligned} dX^i &= Z^i_j dW^j - 2^{-1}F^i_{kl}(X,Y)Z^k_m Z^l_j dW^m dW^j - 2^{-1}C^i_{kl}(X,Y)Z^k_m Z^l_j dU^m dW^j, \\ \delta Y^i &= Z^i_j dU^j - 2^{-1}F^i_{kl}(X,Y)Z^k_m Z^l_j dW^m dU^j - 2^{-1}C^i_{kl}(X,Y)Z^k_m Z^l_j dU^m dU^j. \end{aligned}$$

But $dW^m dW^j = \delta^{mj}dt$ by the Itô rules, and $dU^m dW^j = dW^m dU^j = 0$ because U and W are independent. Without loss of generality, we can assume that U is the representation of Brownian motion on the unit sphere $S^{n-1} \subset \mathbb{R}^n$ defined in the Appendix, which means that we can use formulae (A.2) and (A.3) for dU^j and $dU^m dU^j$. It follows that

$$\begin{aligned}
dX^i &= Z^i_j dW^j - 2^{-1}F^i_{kl}(X,Y)Z^k_m Z^l_j \delta^{mj}dt \\
&= Z^i_j dW^j - 2^{-1}g^{kl}(X,Y)F^i_{kl}(X,Y)dt, \\
\delta Y^i &= Z^i_j(\delta^j_k - \delta_{kl}U^j U^l)dV^k - 2^{-1}(n-1)U^j Z^i_j dt \\
&\quad - 2^{-1}C^i_{kl}(X,Y)Z^k_m Z^l_j(\delta^{mj} - U^m U^j)dt \\
&= Z^i_j(\delta^j_k - \delta_{kl}U^j U^l)dV^k - 2^{-1}(n-1)Y^i dt - 2^{-1}g^{kl}(X,Y)C^i_{kl}(X,Y)dt,
\end{aligned}$$

since $Z^k_m Z^l_j \delta^{mj} = g^{kl}(X,Y)$ by assertion b) of Lemma 3.4, $Y^i = U^j Z^i_j$ by assertion a) of Lemma 3.4, and $C^i_{kl}(x,y)y^l = 0$. It also follows that

$$\begin{aligned}
dX^i dX^j &= Z^i_k Z^j_l dW^k dW^l = Z^i_k Z^j_l \delta^{kl}dt = g^{ij}(X,Y)dt, \\
\delta Y^i \delta Y^j &= Z^i_k Z^j_l dU^k dU^l = Z^i_k Z^j_l(\delta^{kl} - U^k U^l)dt = \big(g^{ij}(X,Y) - Y^i Y^j\big)dt, \\
dX^i \delta Y^j &= Z^i_k Z^j_l dW^k dU^l = 0, \\
\delta Y^i dX^j &= Z^i_k Z^j_l dU^k dW^l = 0.
\end{aligned}$$

Thus, for any smooth function $f : \mathring{T}M \to \mathbb{R}$,

$$\begin{aligned}
df(X,Y) &= \delta_i f(X,Y) \circ dX^i + \partial_{(i)} f(X,Y) \circ \delta Y^i \\
&= \delta_i f(X,Y)dX^i + \partial_{(i)} f(X,Y)\delta Y^i \\
&\quad + 2^{-1}\delta_i\delta_j f(X,Y)dX^i dX^j + 2^{-1}\delta_i\partial_{(j)} f(X,Y)dX^i\delta Y^j \\
&\quad + 2^{-1}\partial_{(i)}\delta_j f(X,Y)\delta Y^i dX^j + 2^{-1}\partial_{(i)}\partial_{(j)} f(X,Y)\delta Y^i \delta Y^j \\
&= \delta_i f(X,Y)Z^i_j dW^j + \partial_{(i)} f(X,Y)Z^i_j(\delta^j_k - \delta_{kl}U^j U^l)dV^k \\
&\quad + \big[2^{-1}g^{ij}(X,Y)\big(\delta_i\delta_j f(X,Y) - F^k_{ij}(X,Y)\delta_k f(X,Y)\big) \\
&\quad + 2^{-1}\big(g^{ij}(X,Y) - Y^i Y^j\big)\big(\partial_{(i)}\partial_{(j)} f(X,Y) \\
&\quad - C^k_{ij}(X,Y)\partial_{(k)} f(X,Y)\big) - 2^{-1}(n-1)Y^k\partial_{(k)} f(X,Y)\big]dt.
\end{aligned}$$

As a consequence, (X,Y) is a diffusion with generator D given by the expression in square brackets multiplying dt above. Since, by assertion c) of Lemma 3.4, $L(X,Y) = 1$, that is, $(X,Y) \in IM$, it follows from the above calculation that D is an operator on IM in the sense of Remark 3.3. □

Definition 3.5. The solution (X,Y,Z) to (3.1) with the initial conditions specified in Theorem 3.1 will be called an *indicatrized Finslerian hv-stochastic development*.

Definition 3.6. A diffusion (X, Y) on IM with generator D given by (3.2) will be called an *indicatrized Finslerian hv-Brownian motion.*

We have just proved that if (X, Y, Z) is an indicatrized hv-stochastic development, then (X, Y) is an indicatrized hv-Brownian motion. In the next section we shall demonstrate that an indicatrized hv-Brownian motion can also be represented as a Riemannian Brownian motion with drift on the indicatrix bundle IM equipped with the Riemannian metric induced from $\mathring{T}M$.

4 Indicatrized hv-Brownian Motion as Riemannian Brownian Motion with Drift on the Indicatrix Bundle

As in the previous sections, we assume that M is an n-dimensional Finsler manifold with metric function L, positive definite metric tensor g_{ij}, and the Cartan connection $(N^i_j, F^i_{jk}, C^i_{jk})$. We shall use the standard summation convention on upper and lower indices supplemented with the rule that there is no summation with respect to $n = \dim M$. We also assume that Latin indices i, j, k, l, m run from 1 to n, while Greek indices κ, λ, μ run form 1 to $n-1$.

We consider the indicatrix bundle $IM \subset \mathring{T}M$ as a Riemannian manifold equipped with the metric $\hat{G}$ induced by the diagonal lift metric G defined on $\mathring{T}M$ by (2.4). We shall prove the following theorem.

Theorem 4.1. *A semimartingale (X, Y) on IM is an indicatrized Finslerian hv-Brownian motion if and only if it is a Riemannian Brownian motion with drift B on the indicatrix bundle IM equipped with the Riemannian metric $\hat{G}$ induced by the diagonal lift metric G defined by* (2.4), *the drift being expressed as*

$$B = 2^{-1} g^{ik} P^j_{jk} \delta_i - 2^{-1} g^{ik} C^j_{kj} \partial_{(i)} \tag{4.1}$$

in terms of the torsion tensors P^i_{jk} and C^i_{jk} of the Cartan connection and the adapted frame $\delta_i, \partial_{(i)}$.

Remark 4.2. The vector field B on $\mathring{T}M$ defined by (4.1) is tangent to IM, which means that it can also be regarded as a vector field on IM. This can be verified by direct computation, but it also follows from the proof of assertion a) of Theorem 4.4 below.

Theorem 4.1 will be proved at the end of the present section. Right now, let us introduce the notation and state some facts concerning the indicatrix bundle IM to be used in the remainder of this section. For more information on the indicatrix bundle the reader is referred to [12], [19], and [20].

Let (x^k) be local coordinates on M and let (x^k, y^k) be the induced coordinates on $\mathring{T}M$. The vector fields $\partial_k = \partial/\partial x^k$, $\partial_{(k)} = \partial/\partial y^k$, and $\delta_k = \partial_k - N_k^j(x,y)\partial_{(j)}$ on $\mathring{T}M$ are defined as before. We also choose another system $(\tilde{x}^k, \tilde{y}^k)$ of local coordinates on $\mathring{T}M$ defined by

$$\tilde{x}^k(x,y) = x^k, \quad \tilde{y}^\kappa(x,y) = y^\kappa/L(x,y), \quad \tilde{y}^n(x,y) = L(x,y), \tag{4.2}$$

subject to the condition $\partial_{(n)}L(x,y) \neq 0$. Since $y \neq 0$ and $\partial_{(k)}L(x,y) = y_k/L$ for any k, reshuffling the indices if necessary, one can always ensure that the latter condition is satisfied in a neighbourhood of any given point in $\mathring{T}M$. Then, by the implicit function theorem, $(\tilde{x}^k, \tilde{y}^k)$ are well defined local coordinates in a neighbourhood of the given point. Also, the indicatrix bundle IM is given by the equation $\tilde{y}^n(x,y) = 1$ and $(\tilde{x}^k, \tilde{y}^\kappa)$ are local coordinates on IM.

Using the identities $\partial_{(k)}L(x,y) = y_k/L(x,y)$ and $\partial_k L(x,y) = \gamma_{0k}^0(x,y)/L(x,y)$, where $\gamma_{0k}^0(x,y) = \gamma_{jk}^i(x,y)y_iy^j$, one can find from (4.2) that

$$\begin{gathered}\frac{\partial x^k}{\partial \tilde{x}^l} = \delta_l^k, \quad \frac{\partial x^k}{\partial \tilde{y}^l} = 0, \quad \frac{\partial y^\kappa}{\partial \tilde{x}^l} = 0, \quad \frac{\partial y^n}{\partial \tilde{x}^l} = -\frac{\gamma_{0l}^0}{y_n}, \\ \frac{\partial y^\kappa}{\partial \tilde{y}^\lambda} = \delta_\lambda^\kappa L, \quad \frac{\partial y^n}{\partial \tilde{y}^\lambda} = -\frac{Ly_\lambda}{y_n}, \quad \frac{\partial y^k}{\partial \tilde{y}^n} = \frac{y^k}{L}.\end{gathered} \tag{4.3}$$

We define the vector fields $\tilde{\partial}_k = \partial/\partial\tilde{x}^k$ and $\tilde{\partial}_{(k)} = \partial/\partial\tilde{y}^k$ on $\mathring{T}M$. From (4.3) it follows that

$$\tilde{\partial}_k = \partial_k - (\gamma_{0k}^0/y_n)\partial_{(n)}, \quad \tilde{\partial}_{(\kappa)} = L\partial_{(\kappa)} - (Ly_\kappa/y_n)\partial_{(n)}, \quad \tilde{\partial}_{(n)} = (y^i/L)\partial_{(i)}. \tag{4.4}$$

Clearly, the vector fields $\tilde{\partial}_k$ and $\tilde{\partial}_{(\kappa)}$ are tangent to IM.

Proposition 4.3. a) *The vector fields* $\delta_k = \partial_k - N_k^j(x,y)\partial_{(j)}$ *are tangent to* IM;
b) *The vector field* $\tilde{\partial}_{(n)} = \partial/\partial\tilde{y}^n$ *is perpendicular to* IM *in the diagonal lift metric* G *defined by* (2.4).

Proof of Proposition 4.3. The proof depends on the condition $N_k^j(x,y)y_j = \gamma_{0k}^0(x,y)$ satisfied by the nonlinear part N_j^i of the Cartan connection.

To prove assertion a), it suffices to express δ_k as a linear combination of $\tilde{\partial}_k$ and $\tilde{\partial}_{(\kappa)}$. With the aid of (4.4), we get

$$\begin{aligned}\delta_k &= \partial_k - N^l_k \partial_{(l)} = \tilde{\partial}_k - (N^\lambda_k/L)\tilde{\partial}_{(\lambda)} + [(\gamma^0_{0k}/y_n) - (N^\lambda_k y_\lambda/y_n) - N^n_k]\partial_{(n)} \\ &= \tilde{\partial}_k - (N^\lambda_k/L)\tilde{\partial}_{(\lambda)}.\end{aligned}$$

To prove assertion b), we shall demonstrate that $\tilde{\partial}_{(n)}$ is perpendicular to each of the vectors $\delta_k, \tilde{\partial}_{(\kappa)}$ tangent to IM, which clearly form a basis in TIM. By (2.4) and (4.4),

$$\begin{aligned}G(\delta_k, \tilde{\partial}_{(n)}) &= G(\delta_k, (y^i/L)\partial_{(i)}) = 0, \\ G(\tilde{\partial}_{(\kappa)}, \tilde{\partial}_{(n)}) &= G(L\partial_{(\kappa)} - (Ly_\kappa/y_n)\partial_{(n)}, (y^i/L)\partial_{(i)}) \\ &= y^i g_{\kappa i} - (y_\kappa y^i/y_n) g_{ni} = y_\kappa - y_\kappa = 0,\end{aligned}$$

as required. □

We set $d\tilde{x}^k = dx^k$ and $\delta\tilde{y}^k = (\partial\tilde{y}^k/\partial y^l)\delta y^l$, where $\delta y^l = dy^l + N^l_m(x,y)dx^m$. Then $d\tilde{x}^k, \delta\tilde{y}^k$ is the dual frame to $\delta_k, \tilde{\partial}_{(k)}$. Moreover, $d\tilde{x}^k, \delta\tilde{y}^\kappa$ is the dual frame to the frame $\delta_k, \tilde{\partial}_{(\kappa)}$ on IM.

Let G be the diagonal lift metric on $\mathring{T}M$ defined by (2.4). We set

$$g_{(k)(l)} = G(\tilde{\partial}_{(k)}, \tilde{\partial}_{(l)}) = \frac{\partial y^i}{\partial \tilde{y}^k}\frac{\partial y^j}{\partial \tilde{y}^l} g_{ij}.$$

Then, by Proposition 4.3 b), the metric tensor G on $\mathring{T}M$ defined by (2.4) and the induced metric tensor $\hat{G}$ on IM can be written as

$$G = g_{kl} d\tilde{x}^k \otimes d\tilde{x}^l + g_{(\kappa)(\lambda)} \delta\tilde{y}^\kappa \otimes \delta\tilde{y}^\lambda + g_{(n)(n)} \delta\tilde{y}^n \otimes \delta\tilde{y}^n, \tag{4.5}$$

$$\hat{G} = g_{kl} d\tilde{x}^k \otimes d\tilde{x}^l + g_{(\kappa)(\lambda)} \delta\tilde{y}^\kappa \otimes \delta\tilde{y}^\lambda. \tag{4.6}$$

From (4.5) and (4.6) it follows that the dual tensors $G^\sharp$ and $\hat{G}^\sharp$ to G and $\hat{G}$ can be expressed as

$$\begin{aligned}G^\sharp &= g^{kl}\delta_k \otimes \delta_l + g^{(\kappa)(\lambda)}\tilde{\partial}_{(\kappa)} \otimes \tilde{\partial}_{(\lambda)} + g^{(n)(n)}\tilde{\partial}_{(n)} \otimes \tilde{\partial}_{(n)}, \\ \hat{G}^\sharp &= g^{kl}\delta_k \otimes \delta_l + g^{(\kappa)(\lambda)}\tilde{\partial}_{(\kappa)} \otimes \tilde{\partial}_{(\lambda)},\end{aligned} \tag{4.7}$$

where g^{kl} and $g^{(k)(l)}$ are, respectively, the inverse matrices to g_{kl} and $g_{(k)(l)}$.

By transforming the formulae (2.5) for the horizontal lift connection ∇ on $\mathring{T}M$ into the frame $\delta_k, \tilde{\partial}_{(k)}$ and projecting the resulting expressions orthogonally onto TIM, we obtain the following formulae for the induced connection $\hat{\nabla}$ on IM (cf. [20], where $\hat{\nabla}$ is called the K-connection):

$$\begin{aligned}
\hat{\nabla}_{\delta_k}\delta_l &= \hat{\Gamma}^m_{kl}\delta_m &&+ \hat{\Gamma}^{(\mu)}_{kl}\tilde{\partial}_{(\mu)} &&= F^m_{kl}\delta_m,\\
\hat{\nabla}_{\delta_k}\tilde{\partial}_{(\lambda)} &= \hat{\Gamma}^m_{k(\lambda)}\partial_m &&+ \hat{\Gamma}^{(\mu)}_{k(\lambda)}\tilde{\partial}_{(\mu)} &&= \frac{\partial \bar{y}^\mu}{\partial y^m}\Big(\delta_k \frac{\partial y^m}{\partial \bar{y}^\lambda} + \frac{\partial y^l}{\partial \bar{y}^\lambda} F^m_{kl}\Big)\tilde{\partial}_{(\mu)},\\
\hat{\nabla}_{\tilde{\partial}_{(\kappa)}}\delta_l &= \hat{\Gamma}^m_{(\kappa)l}\delta_m &&+ \hat{\Gamma}^{(\mu)}_{(\kappa)l}\tilde{\partial}_{(\mu)} &&= \frac{\partial y^k}{\partial \bar{y}^\kappa} C^m_{kl}\delta_m,\\
\hat{\nabla}_{\tilde{\partial}_{(\kappa)}}\tilde{\partial}_{(\lambda)} &= \hat{\Gamma}^m_{(\kappa)(\lambda)}\delta_m &&+ \hat{\Gamma}^{(\mu)}_{(\kappa)(\lambda)}\tilde{\partial}_{(\mu)} &&= \frac{\partial \bar{y}^\mu}{\partial y^m}\Big(\frac{\partial^2 y^m}{\partial \bar{y}^\kappa \partial \bar{y}^\lambda} + \frac{\partial y^k}{\partial \bar{y}^\kappa}\frac{\partial y^l}{\partial \bar{y}^\lambda} C^m_{kl}\Big)\tilde{\partial}_{(\mu)}.
\end{aligned} \tag{4.8}$$

Since ∇ is a metrical connection on $\mathring{T}M$ with respect to the metric tensor G, it follows that $\hat{\nabla}$ is a metric connection on IM with respect to the induced metric tensor $\hat{G}$. In general, the induced connection $\hat{\nabla}$ has a non-vanishing torsion tensor $\hat{T}$, which can be obtained by transforming the formulae (2.6) for the torsion tensor T of ∇ into the frame $\delta_k, \tilde{\partial}_{(k)}$ and projecting the resulting expressions orthogonally onto TIM. This procedure yields

$$\begin{aligned}
\hat{T}(\delta_k,\delta_l) &= \hat{T}^m_{kl}\delta_m &&+ \hat{T}^{(\mu)}_{kl}\tilde{\partial}_{(\mu)} = \frac{\partial \bar{y}^\mu}{\partial y^m} R^m_{kl}\tilde{\partial}_{(\mu)},\\
\hat{T}(\delta_k,\tilde{\partial}_{(\lambda)}) &= \hat{T}^m_{k(\lambda)}\delta_m &&+ \hat{T}^{(\mu)}_{k(\lambda)}\tilde{\partial}_{(\mu)} = -\frac{\partial y^l}{\partial \bar{y}^\lambda} C^m_{lk}\delta_m - \frac{\partial y^l}{\partial \bar{y}^\lambda}\frac{\partial \bar{y}^\mu}{\partial y^m} P^m_{lk}\tilde{\partial}_{(\mu)},\\
\hat{T}(\tilde{\partial}_{(\kappa)},\delta_l) &= \hat{T}^m_{(\kappa)l}\delta_m &&+ \hat{T}^{(\mu)}_{(\kappa)l}\tilde{\partial}_{(\mu)} = +\frac{\partial y^k}{\partial \bar{y}^\kappa} C^m_{kl}\delta_m + \frac{\partial y^k}{\partial \bar{y}^\kappa}\frac{\partial \bar{y}^\mu}{\partial y^m} P^m_{kl}\tilde{\partial}_{(\mu)},\\
\hat{T}(\tilde{\partial}_{(\kappa)},\tilde{\partial}_{(\lambda)}) &= \hat{T}^m_{(\kappa)(\lambda)}\delta_m &&+ \hat{T}^{(\mu)}_{(\kappa)(\lambda)}\tilde{\partial}_{(\mu)} = 0.
\end{aligned} \tag{4.9}$$

We are now in a position to consider the Riemannian stochastic development defined by (1.1) in the case of IM equipped with the Riemannian metric $\hat{G}$ and metric connection $\hat{\nabla}$. By Theorem 1.3, the projection (X, Y) of this Riemannian stochastic development from the orthonormal frame bundle OIM onto IM is a Riemannian Brownian motion with drift B on IM.

Theorem 4.4. *Let (X, Y) be the above Riemannian Brownian motion with drift B on IM. Then*

a) *B can be expressed by formula* (4.1),

b) *(X, Y) is an indicatrized Finslerian hv-Brownian motion.*

Proof of Theorem 4.4. To prove assertion a), we compute the drift B using formula (1.3) from Theorem 1.3 in the case of the indicatrix bundle IM

with metric tensor $\hat{G}$ and metric connection $\hat{\nabla}$. By (4.7) and (4.9), we have

$$\begin{aligned} B &= 2^{-1}g^{km}\left(\hat{T}^l_{lm} + \hat{T}^{(\lambda)}_{(\lambda)m}\right)\delta_k + 2^{-1}g^{(\kappa)(\mu)}\left(\hat{T}^l_{l(\mu)} + \hat{T}^{(\lambda)}_{(\lambda)(\mu)}\right)\tilde{\partial}_{(\kappa)} \\ &= 2^{-1}g^{km}\frac{\partial y^l}{\partial \tilde{y}^\lambda}\frac{\partial \tilde{y}^\lambda}{\partial y^j}P^j_{lm}\delta_k - 2^{-1}g^{(\kappa)(\mu)}\frac{\partial y^m}{\partial \tilde{y}^\mu}\frac{\partial y^k}{\partial \tilde{y}^\kappa}C^l_{ml}\partial_{(k)} \\ &= 2^{-1}g^{km}\left(\delta^l_j - y^l y_j/L^2\right)P^j_{lm}\delta_k - 2^{-1}\left(g^{km} - y^k y^m/L^2\right)C^l_{ml}\partial_{(k)}. \end{aligned} \tag{4.10}$$

The last equality holds true, since

$$\frac{\partial y^i}{\partial \tilde{y}^\kappa}\frac{\partial \tilde{y}^\kappa}{\partial y_j} = \frac{\partial y^i}{\partial \tilde{y}^k}\frac{\partial \tilde{y}^k}{\partial y_j} - \frac{\partial y^i}{\partial \tilde{y}^n}\frac{\partial \tilde{y}^n}{\partial y_j} = \delta^i_j - y^i y_j/L^2, \tag{4.11}$$

$$g^{(\kappa)(\mu)}\frac{\partial y^i}{\partial \tilde{y}^\kappa}\frac{\partial y^j}{\partial \tilde{y}^\mu} = g^{(k)(m)}\frac{\partial y^i}{\partial \tilde{y}^k}\frac{\partial y^j}{\partial \tilde{y}^m} - g^{(n)(n)}\frac{\partial y^i}{\partial \tilde{y}^n}\frac{\partial y^j}{\partial \tilde{y}^n} = g^{ij} - y^i y^j/L^2. \tag{4.12}$$

Because $N^j_m(x,y)$ is homogeneous of degree one in y, it follows by the Euler theorem that $y^l\partial_{(l)}N^j_m(x,y) = N^j_m(x,y)$, which implies that

$$y^l y_j P^j_{lm} = y^l y_j\left(\partial_{(l)}N^j_m - F^j_{ml}\right) = y_j N^j_m - y_j N^j_m = 0.$$

We also have $y^m C^l_{ml}(x,y) = 0$. Since $L(x,y) = 1$ on IM, it therefore follows from (4.10) that B is indeed given by (4.1).

To prove assertion b), we compute the generator D of (X,Y) using formula (1.2) from Theorem 1.3 in the case of the indicatrix bundle IM with metric tensor $\hat{G}$ and metric connection $\hat{\nabla}$. Writing the first equality in (1.2) in terms of the frame $\delta_k, \tilde{\partial}_{(k)}$, by (4.7), (4.8), (4.11), (4.12), and the identity $y_i C^i_{kl}(x,y) = 0$, we have

$$\begin{aligned} D &= 2^{-1}g^{kl}\left(\delta_k\delta_l - \hat{\Gamma}^m_{kl}\delta_m - \hat{\Gamma}^{(\mu)}_{kl}\tilde{\partial}_{(\mu)}\right) \\ &\quad + 2^{-1}g^{(\kappa)(\lambda)}\left(\tilde{\partial}_{(\kappa)}\tilde{\partial}_{(\lambda)} - \hat{\Gamma}^m_{(\kappa)(\lambda)}\delta_m - \hat{\Gamma}^{(\mu)}_{(\kappa)(\lambda)}\tilde{\partial}_{(\mu)}\right) \\ &= 2^{-1}g^{kl}\left(\delta_k\delta_l - F^m_{kl}\delta_m\right) + 2^{-1}g^{(\kappa)(\lambda)}\Bigg(\frac{\partial y^k}{\partial \tilde{y}^\kappa}\frac{\partial y^l}{\partial \tilde{y}^\lambda}\partial_{(k)}\partial_{(l)} + \frac{\partial^2 y^m}{\partial \tilde{y}^\kappa \partial \tilde{y}^\lambda}\partial_{(m)} \\ &\quad - \frac{\partial^2 y^i}{\partial \tilde{y}^\kappa \partial \tilde{y}^\lambda}\frac{\partial \tilde{y}^\mu}{\partial y^i}\frac{\partial y^m}{\partial \tilde{y}^\mu}\partial_{(m)} - \frac{\partial y^k}{\partial \tilde{y}^\kappa}\frac{\partial y^l}{\partial \tilde{y}^\lambda}\frac{\partial \tilde{y}^\mu}{\partial y^i}\frac{\partial y^m}{\partial \tilde{y}^\mu}C^i_{kl}\partial_{(m)}\Bigg) \\ &= 2^{-1}g^{kl}\left(\delta_k\delta_l - F^m_{kl}\delta_m\right) + 2^{-1}\left(g^{kl} - y^k y^l/L^2\right)\left(\partial_{(k)}\partial_{(l)} - C^m_{kl}\partial_{(m)}\right) \\ &\quad + 2^{-1}g^{(\kappa)(\lambda)}\frac{\partial^2 y^i}{\partial \tilde{y}^\kappa \partial \tilde{y}^\lambda}y_i\left(y^m/L^2\right)\partial_{(m)}. \end{aligned}$$

To evaluate the last term in the above formula, we compute the following second-order partial derivative, the result being equal to zero, since $L(x, y) = \tilde{y}^n$ is independent of $\tilde{y}^\lambda$:

$$0 = \frac{\partial^2 L^2(x,y)}{\partial\tilde{y}^\kappa \partial\tilde{y}^\lambda} = 2\frac{\partial}{\partial\tilde{y}^\kappa}\left(\frac{\partial y^i}{\partial\tilde{y}^\lambda} y_i\right) = 2\frac{\partial^2 y^i}{\partial\tilde{y}^\kappa \partial\tilde{y}^\lambda} y_i + 2\frac{\partial y^i}{\partial\tilde{y}^\lambda}\frac{\partial y^j}{\partial\tilde{y}^\kappa} g_{ij}.$$

Thus

$$g^{(\kappa)(\lambda)}\frac{\partial^2 y^i}{\partial\tilde{y}^\kappa \partial\tilde{y}^\lambda} y_i = -g^{(\kappa)(\lambda)}\frac{\partial y^i}{\partial\tilde{y}^\lambda}\frac{\partial y^j}{\partial\tilde{y}^\kappa} g_{ij} = -(g^{ij} - y^i y^j / L^2) g_{ij} = -(n-1).$$

Since $L(x, y) = 1$ on IM, it follows that D is given by formula (3.2), i.e., (X, Y) is an indicatrized Finslerian hv-Brownian motion. □

Finally, we shall demonstrate that Theorem 4.1 is a simple consequence of Theorem 4.4.

Proof of Theorem 4.1. It follows from Theorem 4.4 that the generator D of an indicatrized hv-Brownian motion given by (3.2) is equal to $2^{-1}\Delta + B$, where Δ is the Laplace-Beltrami operator on the indicatrix bundle IM equipped with the Riemannian metric $\hat{G}$ and where B is the vector field on IM defined by (4.1). This proves Theorem 4.1. □

5 Conclusions

The main result of the present article, i.e., the representation of Finslerian Brownian motion as Riemannian Brownian motion with drift is not only interesting in its own right, but is expected to facilitate further study of Finslerian diffusions with the aid of the well-established methods of the theory of diffusions on Riemannian manifolds.

Two approaches to Finslerian diffusions have been followed. One, which was developed in [1] and [2], is based on the slit tangent bundle $\mathring{T}M$, on which the diffusions and, in fact, all the relevant Finslerian objects are defined. The other one, introduced in the present article, uses the indicatrix bundle IM in place of $\mathring{T}M$ to define and study Finslerian diffusions. The two approaches lead to slightly different results in that the generators (2.2) and (3.2) of a Finslerian Brownian motion on $\mathring{T}M$ and an indicatrized Finslerian Brownian motion on IM differ by the term $-2^{-1}(n-1)y^k\partial_{(k)}$. This is, however, a non-Finslerian effect, since, as follows from the proof of Theorem 3.1, it can be traced back to a similar term in the generator of a Riemannian Brownian motion on the unit sphere S^{n-1} (cf. formula

(A.1) in the Appendix). Modulo the non-Finslerian term in question, the generators (2.2) and (3.2) are the same, and also the drift fields (2.7) and (4.1) are the same, as one would expect of direction-dependent Finslerian objects.

6 Appendix

6.1 Brownian motion on the unit sphere S^{n-1}

We consider the unit sphere $S^{n-1} \subset \mathbf{R}^n$ as a Riemannian manifold with the metric induced from $\mathbf{R}^n$. By definition (see Sect. 1), a Riemannian Brownian motion U on S^{n-1} is an S^{n-1}-valued diffusion with generator $D = 2^{-1}\Delta$, where Δ is the Laplace-Beltrami operator on S^{n-1}, which, as is well known, can be written as

$$\Delta = (\delta^{ij} - x^i x^j)\partial_i\partial_j - (n-1)x^k\partial_k, \tag{A.1}$$

where x^i are the canonical Cartesian coordinates on $\mathbf{R}^n$ and $\partial_i = \partial/\partial x^i$. Even though expressed in terms of the operators ∂_i on $\mathbf{R}^n$, the above expression defines an operator on S^{n-1}, since, for any smooth functions $f_1, f_2 : \mathbf{R}^n \to \mathbf{R}$, if $f_1 = f_2$ on S^{n-1}, then $\Delta f_1 = \Delta f_2$ on S^{n-1}.

To construct a concrete specimen of a Riemannian Brownian motion U on S^{n-1}, one can employ the system (1.1) of SDEs for Riemannian stochastic development. However, for our purposes, it proves more convenient to use another representation of Brownian motion on S^{n-1} due to Stroock [16] (see also [6], Sect. III.2). Namely, let $A^i_j(x) = \delta^i_j - \delta_{jk}x^i x^k/|x|^2$ for any $0 \neq x \in \mathbf{R}^n$ and let V be a standard Brownian motion on $\mathbf{R}^n$. Then the solution U to the SDE

$$dU^i = A^i_j(U) \circ dV^j$$

with initial condition $U(0) \in S^{n-1}$ is a Brownian motion on S^{n-1}. It is easily verified that

$$dU^i = (\delta^i_j - \delta_{jk}U^i U^k)dV^j - 2^{-1}(n-1)U^i dt, \tag{A.2}$$

$$dU^i dU^j = (\delta^{ij} - U^i U^j)dt. \tag{A.3}$$

References

[1] Antonelli, P.L. and Zastawniak, T.J. (1993) Diffusions on Finsler Manifolds, *Rep. Math. Phys.*, **33**, 303-315.

[2] Antonelli, P.L. and Zastawniak, T.J. (1993) Stochastic Calculus on Finsler Manifolds and An Application in Biology, *(this Proceedings)*.

[3] Cartan, E. (1934), 2nd ed. (1971) Les Espaces de Finsler, *Actualités*, **79**, Paris.

[4] Elworthy, K.D. (1982) Stochastic Differential Equations on Manifolds, *Cambridge University Press, Cambridge*.

[5] Emery, M. (1989) *Stochastic Calculus in Manifolds*, Springer-Verlag, Berlin, Heidelberg.

[6] Ikeda, N. and Watanabe, S. (1981) 2nd ed. (1989) *Stochastic Differential Equations and Diffusion Processes*, North Holland, Amsterdam, Kodansha, Tokyo.

[7] Itô, K. (1950) Stochastic Differential Equations in a Differentiable Manifold, *Nagoya Math. J.*, **1**, 35-47.

[8] Itô, K. (1953) Stochastic Differential Equations in a Differentiable Manifold (2), *Mem. Coll. Sci. Univ. Kyoto Math.*, **28**, 81-85.

[9] Itô, K. (1962) The Brownian Motion and Tensor Fields on a Riemannian Manifold, *Proc. Intern. Congr. Math.*, Stockholm, Inst. Mittag-Leffler, Djursholm, 536-539.

[10] Itô, K. (1975) Stochastic Parallel Displacement, in: M. A. Pinsky (ed.), Probabilistic Methods in Differential Equations, *Lect. Notes in Math.*, **451**, Springer-Verlag, Berlin-Heidelberg-New York, 1-7.

[11] Kunita, H. (1990) *Stochastic Flows and Stochastic Differential Equations*, Cambridge University Press, Cambridge.

[12] Matsumoto, M. (1981) Differential-Geometric Properties of Indicatrix Bundle Over Finsler Space, *Publ. Math. Debrecen*, **28**, 281-293.

[13] Matsumoto, M. (1986) *Foundations of Finsler Geometry and Special Finsler Spaces*, Kasheisha Press, Saikawa 3–23–2, Otsushi, Shigaken.

[14] Miron, R. (1987) On the Finslerian Theory of Relativity, *Tensor, N.S.*, **44**, 63-81.

[15] Rund, H. (1959) *The Differential Geometry of Finsler Spaces*, Springer-Verlag, Berlin.

[16] Stroock, D.W. (1971) On the Growth of Stochastic Integrals, *Z. Wahr. verw. Geb.*, **18**, 340-344.

[17] Yano, K. and Davies, E.T. (1963) On the Tangent Bundles of Finsler and Riemannian Manifolds, *Rend. Circ. Mat. Palermo*, **12**, 211-218.

[18] Yano, K. and Ishihara, S. (1973) *Tangent and Cotangent Bundles. Differential Geometry*, Marcel Dekker Inc., New York.

[19] Yasuda, H. (1979) On the Indicatrix Bundle Endowed with the K-Connection Over a Finsler Space, *Ann. Rep. Asahikawa Med. Coll.*, **1**, 117-124.

[20] Yasuda, H. and Fukui, M. (1980) On the Curvature of the Indicatrix Bundle Over a Finsler Space, *Ann. Rep. Asahikawa Med. Coll.*, **2**, 1-21.

DIFFUSION ON THE TOTAL SPACE OF A VECTOR BUNDLE *

DRAGOŞ HRIMIUC

Introduction

In recent years, the study of the differential geometry of the total space E, of a vector bundle $\pi : E \rightarrow M$, initiated by R. Miron [11], [12] has been developed by many people (see [13] and the references therein). If we take a horizontal complement of the vertical subbundle VE, we can express the geometrical objects defined on E in a more simplified form and new geometric objects can be obtained.

Recently, P.L. Antonelli and T. Zastawniak in a series of papers [2], [3], [4] extended the Riemannian theory of diffusion processes and stochastic development to the case of Finsler manifolds, the extension being motivated by important problems in Biology [3], [5].

In this paper, we extend their formalism to study some geometric problems of the theory of the diffusion processes and the stochastic development on E, related to these new geometric objects on E. We thereby obtain further generalization and geometric meaning for certain results of [2], [3]. But few probabilistic calculations are given here, for they are given in [2], [3], [4]. In a forthcoming publication, as a particular case, the theory of diffusion and stochastic development on Lagrange manifolds will be discussed [9]. (See our article on m-homogeneous case in this issue.)

1 Preliminaries

Let $\pi : E \rightarrow M$ a smooth vector bundle over M. Suppose that M is a real $n-$dimensional differentiable manifold and the dimension of each fibre of E is m. Local coordinates on E are (x^i, y^a), where (x^i) are local coordinates on the base manifold. Always in this paper, the range for the indices

*This paper first appeared in *Balkan J. Geom. Appl.*, 1, (1996), 53-62.

$i, j, k, h, \ldots$ is $\{1, 2, \ldots, n\}$, for the indices $a, b, c, d, \ldots$ is $\{1, 2, \ldots, m\}$ and the summation convention is used.

A nonlinear connection on E is defined by a distribution HE, complementary to the vertical distribution of TE, i.e.

$$TE = HE \oplus VE. \tag{1.1}$$

A local frame for $TE_{(x,y)}$, $(x, y) \in E$, adapted to the splitting (1.1) is (δ_i, δ_a), where

$$\delta_i = \partial_i - N_i^a(x, y)\partial_a, \tag{1.2}$$

is a basis in $HE_{(x,y)}$ and $\partial_i := \frac{\partial}{\partial x^i}$, $\quad \partial_a := \frac{\partial}{\partial y^a}$.

We denote by $(dx^i, \delta y^a)$ the dual basis of (δ_i, ∂_a), where

$$\delta y^a = dy^a + N_i^a(x, y)dx^i. \tag{1.3}$$

Now we can introduce the algebra of $d-$tensor fields on E with respect to the horizontal and vertical distributions. This algebra is locally spanned by $(1, \delta_i, \partial_a)$.

A d-connection on E is a linear connection on E which preserves by parallelism the horizontal and the vertical distribution (see [13]).

The local coordinate expression for a $d-$connection ∇ on E is given by:

$$\begin{aligned} \nabla_{\delta_i}\delta_j = L^k_{ij}\delta_k; \quad \nabla_{\partial_a}\delta_i = C^j_{ai}\delta_j, \\ \nabla_{\delta_i}\partial_a = L^b_{ia}\partial_b; \quad \nabla_{\partial_a}\partial_b = C^c_{ab}\partial_c. \end{aligned} \tag{1.4}$$

A d-connection on E determines an $h-$ and $v-$algorithm of covariant derivation. For example, if

$$t = t^{ia}_{jb}\delta_i \otimes \delta_a \otimes dx^j \otimes \delta y^b$$

is a d-tensor field on E, the horizontal and vertical covariant derivative of t is given by

$$t^{ia}_{jb|k} = \delta_k t^{ia}_{jb} + L^i_{kh}t^{ha}_{jb} + L^a_{kc}t^{ic}_{jb} - L^h_{kj}t^{ia}_{hb} - L^c_{kb}t^{ia}_{jc}$$

and

$$t^{ia}_{jb}|_c = \partial_c t^{ia}_{jb} + C^i_{ch}t^{ha}_{jb} + C^a_{cd}t^{id}_{jb} - C^h_{cj}t^{ia}_{hb} - C^d_{cb}t^{ia}_{jd}.$$

If

$$G = g_{ij}dx^i \otimes dx^j + h_{ab}\delta y^a \otimes \delta y^b$$

is an (h, v)-metric on E, there exists $d-$connections compatible with G (see [13], Ch. III).

A systematic presentation of the geometry of E is given in the monograph of Miron and Anastasiei [13].

Throughout this paper, we shall use the usual set up for the general theory of stochastic calculus. We follow closely [3], [5]. For an introduction see [6], [7]. [10].

Let $(\Omega, \mathcal{F}, P)$ be a probability space endowed with a right continuous filtration $(\mathcal{F}_t)_{t>0}$ such that each $\mathcal{F}_t$ contains all negligible events in $\mathcal{F}$. If $f : \Omega \to \mathbb{R}$ is an integrable random variable we denote by $E(f) = \int_\Omega f dP$ its expectation and by $E(f/G)$ the conditional expectation of f given G. (G is a sub $\sigma-$field of $\mathcal{F}$.)

A stochastic process is a measurable function $x : [0, \infty) \times \Omega \to \mathbb{R}$. One says that a process is continuous if all its sample path, $t \to x(t, \omega)$, are continuous functions for almost all $\omega \in \Omega$.

A stochastic process is adapted if for each $t \geq 0$ the random variable $x(t)$ is $\mathcal{F}_t-$measurable.

A martingale is an adapted process such that for each $t \geq 0$ $x(t)$ is integrable and $x(s) = E\big(x(t)/\mathcal{F}_s\big)$ for every $t > s \geq 0$.

A continuous local martingale is a continuous adapted process x such that the process

$$(t, \omega) \to x(\tau_n \wedge t, \omega)\chi_{\{0<\tau_n\}}(\omega)$$

is a martingale for every n, where $\tau_n = \inf \{t \geq 0;\ |x(t)| \geq n\}$ and χ_A is the indicator function of $A \subset \Omega$.

A continuous $\mathcal{F}_t-$adapted process is called a semimartingale if it is written as the sum of a local martingale and a process of bounded variation. For a real semimartingale x and a continuous adapted process y we denote by $ydx := \int ydx$ the Itô stochastic integral of y with respect to x, $dydx := d(xy) - xdy - ydx$ the joint quadratic variation of x and y and $y \circ dx := \int y \circ dx$ the Stratonovich stochastic integral, where $y \circ dx = ydx + \frac{1}{2}\, dxdy$.

If M is a differentiable manifold, a $M-$valued semimartingale is a continuous process $x : [0, \infty) \times \Omega \to M$ such that $(t, \omega) \to f\big(x(t, \omega)\big)$ is a real-valued semimartingale for every smooth function $f : M \to \mathbb{R}$.

If D is an elliptic second-order operator on M, an $M-$valued semimartingale X is called a diffusion on M with generator D, if the process

$$(t, \omega) \to f\big(x(t, \omega)\big) - f\big(x, (0, \omega)\big) - \int_0^t Df\big(x(s, \omega)\big)ds,$$

is a local martingale for every smooth function $f : M \to \mathbb{R}$ with compact support.

2 Stochastic Parallelism on E

We consider an arbitrary smooth curve

$$c:\ [0,T]\to E,\quad c(t)=\big(x(t),y(t)\big),\quad t\in[0,T] \tag{2.1}$$

locally expressed by the equations:

$$x^i=x^i(t),\quad y^a=y^a(t),\quad t\in[0,T].$$

The tangent vector field $\dot{c}$ of c is given by

$$\dot{c}=\frac{dx^i}{dt}\,\delta_i+\frac{\delta y^a}{dt}\,\partial_a. \tag{2.2}$$

If $X=X^i\delta_i+X^a\partial_a$ then it is parallel along c if and only if

$$X^j_{|i}\,\frac{dx^i}{dt}+X^j|_a\,\frac{\delta y^a}{dt}=0 \tag{2.3}$$

and

$$X^b_{|i}\,\frac{dx^i}{dt}+X^b|_a\,\frac{\delta y^a}{dt}=0. \tag{2.4}$$

Let $X_0\in H_{c(0)}E$ be a horizontal tangent vector at $c(0)$. Since ∇ preserves by parallelism the horizontal distribution we can transport X_0 by parallelism along $c(t)$ that is, we can find a horizontal vector field $X=X^i\delta_i$ along c, solving of the following system of differential equations:

$$\frac{dX^j}{dt}+L^j_{ki}X^k\,\frac{dx^i}{dt}+C^j_{ka}X^k\,\frac{\delta y^a}{dt}=0 \tag{2.5}$$

with the initial condition

$$X(0)=X_0. \tag{2.6}$$

The solution of (2.5), (2.6) will be called the $h-$parallel transport of X_0 along C. Analogously, if $\overline{X}_0$ is a vertical vector at $c(0)$ the $v-$parallel transport of $\overline{X}_0$ along c is defined similarly as the solution of the system of differential equations:

$$\frac{d\overline{X}^b}{dt}+L^b_{ci}\overline{X}^c\,\frac{dx^i}{dt}+C^b_{ca}X^c\,\frac{\delta y^a}{dt}=0 \tag{2.7}$$

with the initial condition:

$$\overline{X}(0)=X_0. \tag{2.8}$$

From the assumption that the connection is $g-$metrical, if $X(t)$ and $Y(t)$ are two solutions of (2.5) we have

$$g_{ij}\big(x(t),y(t)\big)X^i(t)Y^j(t)=\text{ const.} \tag{2.9}$$

Also if $\overline{X}(t)$, $\overline{Y}(t)$ are two solutions of (2.7) we have

$$h_{ab}\big(x(t), y(t)\big)\overline{X}^a(t)\overline{Y}^b(t) = \text{ const.} \tag{2.10}$$

In this section we extend the concept of stochastic parallel transport along a trajectory of a diffusion process on E. For this, similar to that in [2], [6], [7], we shall approximate the diffusion by piecewise smooth sample path for which the parallel transport is defined by the equations (2.5), (2.6). The theorems below are those of [2], [3], but are here given in a more general geometric setting.

Let $c(t) = \big(x(t), y(t)\big)$ be a diffusion on E starting from (x_0, y_0). If $\pi : 0 = t_0 < t_1 < \cdots < t_n = T$ is a division of the time interval $[0, T]$ we can take a piecewise smooth approximation $c_\pi(t) = \big(x_\pi(t), y_\pi(t)\big)$ on E with bounded first second and third order derivative such that $c_\pi(t_\alpha) = c(t_\alpha)$, $\alpha \in \{0, 1, 2, \dots, n\}$. If $X_0 \in H_{(x_0,y_0)}E$ the $h-$parallel transport $X_\pi(t)$ of X_0 along $c_\pi(t)$ can be defined as a piecewise smooth function, solution of the following system of differential equations:

$$\frac{dX^j_\pi}{dt} + L^j_{ki}\big(x_\pi(t), y_\pi(t)\big)X^k_\pi \frac{dx^i_\pi}{dt} + C^j_{ka}\big(x_\pi(t), y_\pi(t)\big)X^k_\pi \frac{\delta y^a_\pi}{dt} = 0, \tag{2.11}$$

$$X_\pi(0) = X_0. \tag{2.12}$$

We have:

Theorem 1. *The solution of the family of ordinary differential equations (2.11), (2.12) converges in probability as mesh $\pi \to 0$, to the solution $X(t)$ of the Stratonovich stochastic differential equation*

$$dX^j + L^j_{ki}X^k \circ dx^i + C^j_{ka}X^k \circ \delta y^a = 0, \tag{2.13}$$

$$X(0) = X_0, \tag{2.14}$$

where

$$\delta y^a = dy^a + N^a_i \circ dx^i. \tag{2.15}$$

Definition 2.2. The solution of (2.13), (2.14) is called the stochastic $h-$parallel transport of X_0 along the diffusion c.

Similarly we shall be able to define the $v-$stochastic parallel transport of $\overline{X}_0 \in VE_{c(0)}$ along a diffusion c as the solution of the Stratonovich stochastic differential equation

$$d\overline{X}^b + L^b_{ci}\overline{X}^c \circ dx^i + C^b_{ca}\overline{X}^c \circ \delta y^a = 0, \tag{2.16}$$

$$\overline{X}(0) = \overline{X}_0. \tag{2.17}$$

Remark 2.3. Let $c(t) = \big(x(t), y(t)\big)$ be a diffusion on E starting at (x_0, y_0). If $X_0 = X^i \frac{\partial}{\partial x^i} \in TM_{x_0}$ and $A_0 = A^a s_a \in E_{x_0}$ we can take the horizontal and vertical lift of X_a and A_0

$$X_0^v = X^i \frac{\delta}{\delta x^i} \quad \text{and} \quad A_0^v = A^a \frac{\partial}{\partial y^a}\,. \tag{2.18}$$

Now, we can define the stochastic $h-$ and $v-$parallel transport for X_0 and A_0 along the diffusion c on E as the stochastic $h-$ and $v-$parallel transport for X_0^v and A_0^v.

Theorem 2. (i) *If $X(t)$ and $Y(t)$ are any two solutions of* (2.13) *then*

$$g_{ij}\big(x(t), y(t)\big) X^i(t) X^j(t) = \text{ const. } a.s. \tag{2.19}$$

(ii) *If $\overline{X}(t)$ and $\overline{Y}(t)$ are any two solutions of* (2.16) *then*

$$h_{ab}\big(x(t), y(t)\big) \overline{X}^a(t) \overline{Y}^b(t) = \text{ const. } a.s. \tag{2.20}$$

Let $c : [0, T] \to E$ be a diffusion on E, $c(t) = \big(x(t), y(t)\big)$. We say that c is a horizontal diffusion if $y(t)$ is a solution of the following stochastic differential equation

$$dy^a + N_i^a(x, y) \circ dx^i = 0, \quad y(0) = y_0. \tag{2.21}$$

Also we say that c is a vertical diffusion if $x(t) = x_0$ a.s. The system (2.21) has, generally a local solution. We cannot, in general, extend the solution for almost all $t \in [0, T]$ but there are some important situations in which we can do it [2], [3], [9]. We can give a geometric description of the solution of (2.21).

Let $x : [0, T] \to M$ be a diffusion on M starting from x_0 and x_π a piecewise smooth approximation with bounded first, second and third order derivative, associated to the division $\pi : 0 = t_0 < t_1 < \cdots < t_n = T$.

We define the nonlinear transport by parallelism of y_π along x_π as the solution of the following (nonlinear) system of ordinary differential equations

$$\begin{gathered} \frac{dy_\pi^a}{dt} + N_i^a(x_\pi, y_\pi)\, \frac{dx_\pi^i}{dt} = 0, \\ y_\pi(0) = y_0. \end{gathered} \tag{2.22}$$

Theorem 3. *The solution of* (2.22) *converges in probability as mesh* $\pi \to 0$ *to the solution of* $y(t)$ *of the Stratonovich stochastic differential equation* (2.21).

Remark 2.6. If the solution of system (2.22) is defined on $[0, T]$, the same is valid for the solution of the stochastic differential system (2.21).

Remark 2.7. If $c : [0, T] \to E, \quad c(t) = \big(x(t), y(t)\big)$ is a horizontal diffusion then the equations (2.13), (2.14) and (2.16), (2.17) are written:

$$dX^j + L^j_{ki} X^k \circ dx^i = 0, \quad X(0) = X_0$$

and

$$d\overline{X}^b + L^b_{ci} \overline{X}^c \circ dx^i = 0, \quad \overline{X}(0) = \overline{X}_0.$$

If $c(t) = \big(x(t), y(t)\big)$ is a vertical diffusion the equations (2.13), (2.14) and (2.16), (2.17) become

$$dX^j + C^j_{ka} X^k \circ dy^a = 0, \quad X(0) = X_0$$

and

$$d\overline{X}^b + C^b_{ca} \overline{X}^c \circ dy^a = 0, \quad \overline{X}(0) = \overline{X}_0.$$

3 Stochastic Development on E

Let $O'(E)$ be the principal bundle of frames on E defined as follows. The total space of $O'(E)$ consists of elements (x, y, z), where $x \in M$, $y \in E_x$ and $z = (e_1, \dots, e_n, \bar{e}_1, \dots, \bar{e}_m)$ is frame of $TE_{(x,y)}$ such that $(e_1, \dots, e_n)$ is an orthonormal frame of $HE_{(x,y)}$ relative to the metric structure g and $(\bar{e}_1, \dots, \bar{e}_m)$ is an orthonormal frame in $VE_{(x,y)}$ with respect to the metric structure h.

The differential structure of $O'(E)$ can be obtained from that of E as follows.

Let us consider (U_α, Φ_α) be a coordinate system of E and $\widetilde{U}_\alpha = \{(x, y, z) \in O'(E);\ (x, y) \in U_\alpha$ and z is a frame as above$\}$.

We define the mapping:

$$\widetilde{\Phi}_\alpha : \widetilde{U}_\alpha \to \Phi_\alpha(\widetilde{U}_\alpha) \times O(n) \times O(m) \subset \mathbb{R}^{h+m} \times \mathbb{R}^{n^2+m^2}$$

by

$$\widetilde{\Phi}(x, y, z) = \big(\Phi_\alpha(x, y), e^i_j, \bar{e}^a_b\big),$$

where $e_j = e_j^i \left(\frac{\delta}{\delta x^i}\right)_{(x,y)}$, $\bar{e}_b = \bar{e}_b^a \left(\frac{\partial}{\partial y^a}\right)_{(x,y)}$ and $O(n)$ is the group of orthogonal transformations in $\mathbb{R}^n$. The projection $\pi : O'(E) \to E$ is defined as usual by $\pi(x,y,z) = (x,y)$ and the right action by $R_v(x,y,z) = (x,y,z\cdot v)$, where $z \cdot v = (e_i u, \bar{e}_a \bar{u})$, $e_i u = u_i^j e_j$, $\bar{e}_a \bar{u} = \bar{u}_a^b \bar{e}_b$ for any $v = (u,\bar{u}) \in O(n) \times O(m)$ and $(x,y,z) \in O'(E)$.

If $\alpha(t) = \left(\gamma^i(t), \bar{\gamma}^a(t)\right)$, $t \in [0,T]$ is a smooth curve in $\mathbb{R}^{n+m}$ using the $d-$connection ∇ we can roll E along $\alpha(t)$ to obtain a curve $c(t)$ on E as a trace of $\alpha(t)$. In fact, if $(x_0, y_0, z_0) \in O'(E)$ we must find a smooth curve $\tilde{c}(t) = \left(x(t), y(t), z(t)\right)$ on $O'(E)$ such that

$$\begin{gathered}\frac{dx^i}{dt} = \frac{d\gamma^j}{dt}(t)e_j^i(t), \quad \frac{\delta y^a}{dt} = \frac{d\bar{\gamma}^b}{dt}\bar{e}_b^a(t),\\ \frac{de_h^j}{dt} + L_{ki}^j e_h^k \frac{dx^i}{dt} + C_{ka}^j e_h^k \frac{\delta y^a}{dt} = 0,\\ \frac{d\bar{e}_d^b}{dt} + L_{ci}^b \bar{e}_d^c \frac{dx^i}{dt} + C_{ca}^b \bar{e}_d^c \frac{\delta y^a}{dt} = 0,\\ x^i(0) = x_0^i, \quad y^a(0) = y_0^a, \quad e_i(0) = e_{0i}, \quad \bar{e}_a(0) = \bar{e}_{0a}.\end{gathered} \tag{3.1}$$

For the curve $c(t) = c(t, x_0, y_0, z_0, \alpha) = \pi\left(\tilde{c}(t)\right)$ it follows:

$$c(t, x_0, y_0, z_0 \cdot v, \alpha) = c(t, x_0, y_0, z_0, v \cdot \alpha), \quad t \in [0,T], \tag{3.2}$$

where $v = (u, \bar{u}) \in O(n) \times O(m)$ and $v \cdot \alpha$ is the curve in $\mathbb{R}^{n+m}$ defined by

$$(v \cdot \alpha)(t) = \left(u_j^i \gamma^j(t), \bar{u}_b^a \bar{\gamma}^b(t)\right). \tag{3.3}$$

Let $c(t) = \left(x(t), y(t)\right)$, $t \in [0,T]$ be a diffusion on E starting at (x_0, y_0) and z_0 an orthonormal frame in $TE_{(x_0,y_0)}$ as above.

We use the stochastic parallel transport to move this orthonormal frame along $c(t)$ and we shall obtain the moving frame $z(t) = \left(e_1(t), \ldots, e_n(t), \bar{e}_1(t), \ldots, \bar{e}_m(t)\right)$ such that the following stochastic differential equations are satisfied:

$$\begin{gathered}de_h^j + L_{ki}^j e_h^k \circ dx^i + C_{ka}^j e_h^k \circ \delta y^a = 0,\\ d\bar{e}_d^b + L_{ci}^b \bar{e}_d^c \circ dx^i + C_{ca}^b \bar{e}_d^c \circ y^a = 0\end{gathered} \tag{3.4}$$

and almost surely on $[0,T]$ we have:

$$\begin{gathered}g_{ij}\left(x(t), y(t)\right)e_h^i(t)e_k^j(t) = \delta_{hk},\\ h_{ab}\left(x(t), y(t)\right)e_c^a(t)e_d^b(t) = \delta_{cd}.\end{gathered} \tag{3.5}$$

We regard $\left(x(t), y(t), z(t)\right)$ as a stochastic process on the orthogonal bundle $O'(E)$.

Now, we can study the concept of stochastic development on E or rolling the total space E along a standard Brownian motion in $\mathbb{R}^{n+m}$, extending the Riemannian and Finslerian stochastic development of [2], [3].

Let $w(t), v(t)$ be two independent standard Brownian motion in $\mathbb{R}^n$ and $\mathbb{R}^m$, thus $(w(t), v(t))$ is a standard Brownian motion in $\mathbb{R}^{n+m}$.

The system of stochastic differential equations for the stochastic development on E can be written as:

$$\begin{aligned} dx^i &= e^i_j \circ dw^j; \quad \delta y^a = \bar{e}^a_b \circ dv^b, \\ de^j_h &+ L^j_{ki} e^k_h \circ dx^i + C^j_{ka} e^k_h \circ \delta y^a = 0, \\ d\bar{e}^b_d &+ L^b_{ci} \bar{e}^c_d \circ dx^i + C^b_{ca} \bar{e}^c_d \circ \delta y^a = 0, \\ x^i(0) &= x^i_0, \quad y^a(0) = y^a_0, \quad e_i(0) = e_{0i}, \quad \bar{e}_a(0) = \bar{e}_{0a}, \end{aligned} \tag{3.6}$$

with $(x_0, y_0, z_0) \in O'(E)$.

From (3.5) it follows that the solution of (3.6) is a process on $O'(E)$.

Definition 3.1. The solution $(x(t), y(t), z(t))$ of (3.6) is called the stochastic development on E.

Theorem 4. *The solution of stochastic differential equation* (3.6) *defines a flow of diffeomorphisms on* $O'(E)$ $(x(t), y(t), z(t))$, *whose projection* $(x(t), y(t))$, *from* $O'(E)$ *to* E *is a diffusion on* E *starting at* (x_0, y_0) *which have the probability law independent of the choice of the initial orthonormal frame* z_0 *in* $E_{(x_0,y_0)}$ *and whose generator is*

$$D = \tfrac{1}{2}\, g^{ij}(\delta_i\delta_j - L^k_{ij}\delta_k) + \tfrac{1}{2}\, h^{ab}(\partial_a\partial_b - C^c_{ab}\partial_c). \tag{3.7}$$

This result is that of [2], [3] but is here given a more general geometric setup. The reader should consult these papers for proof of (3.7).

4 Examples

We consider the tangent bundle $\pi : TM \to M$ with the (h, v)–metric

$$G = g_{ij} dx^i \otimes dx^j + h_{ab}\delta y^a \otimes \delta y^b$$

G is called h–Riemannian (v–Riemannian) if its horizontal (vertical) part $g_{ij}(x, y)$ $(h_{ab}(x, y))$ are functions of position only, i.e. depend on x alone. If G is h– and v–Riemannian we shall say that G is (h, v)–Riemannian. Also, we say that G is locally-Minkowski if locally $h_{ab}(x, y) = h_{ab}(y)$. The

$v-$metric h_{ab} is called weakly regular if the $d-$tensor field $\tilde{h}_{ab} = \frac{1}{2}\,\partial_a\partial_b\varepsilon$ is nondegenerate, where $\varepsilon = h_{ab}(x,y)y^a y^b$ (see [13]).

1. We suppose G is Riemannian-locally Minkowski metric on TM and $h_{ab}(y)$ is weakly regular. Then we can take $N_i^a = 0$ (see [13], page 126) and the coefficients of the canonical $d-$connection compatible with G are given by

$$L^i_{jk} = \gamma^i_{jk}(x) = \frac{1}{2}\, g^{ih}(\partial_k g_{hj} + \partial_j g_{hk} - \partial_h g_{jk}), \quad L^a_{ib} = 0$$

$$C^i_{cj} = 0; \quad C^a_{bc} = \frac{1}{2}\, h^{ad}(\partial_b h_{dc} + \partial_c h_{db} - \partial_d h_{bc}).$$

The system (3.6) for the stochastic development takes the form

$$dx^i = e^i_j \circ dw^j; \quad de^i_h + \gamma^j_{ki} e^k_h \circ dx^i = 0$$
$$dy^a = \bar{e}^a_b \circ dv^b; \quad d\bar{e}^b_d + C^b_{ca}\bar{e}^c_d \circ dy^a = 0.$$

2. Let G be a $(h,v)-$Riemannian metric. Then we can take $N^a_i = \gamma^a_{ib}y^b$ as a nonlinear connection on TM (γ^i_{jk} are the Christoffel symbols formed with g_{ij}). The coefficients of the canonical $d-$connection, compatible with G are given by

$$L^i_{jk} = \gamma^i_{jk}; \quad L^a_{bk} = \gamma^a_{bk} + \frac{1}{2}\, h^{ac}h_{bc\|k}; \quad C^i_{jk} = 0; \quad C^a_{bc} = 0$$

where $h_{ab\|k}$ denote the covariant derivative of h_{ab} with respect to γ^i_{jk}. The system (3.6) for the stochastic development reads:

$$dx^i = e^i_j \circ dw^j$$

$$de^j_h + \gamma^i_{ki} e^k_h \circ dx^i = 0$$

$$\delta y^a = \bar{e}^a_b \circ dv^b$$

$$d\bar{e}^b_d + (\gamma^b_{ai} + \frac{1}{2}\, h^{bc}h_{ac\|i})\bar{e}^a_d \circ dx^i = 0.$$

References

[1] Anastasiei, M. (1986) Vector Bundles. Einstein Equations, *An. Şt. Univ. Iaşi*, **32**, s I a Mat, 17-24.

[2] Antonelli, P.L. and Zastawniak, T.J. (1993) Diffusions on Finsler Manifolds, *Rep. on Math. Phys.*, **33**, 303-315 (also in this issue).

[3] Antonelli, P.L. and Zastawniak, T.J. (1994) Stochastic Calculus on Finsler Manifolds and An Application in Biology, *Nonlinear World*, **1**, 149-171 (also in this issue).

[4] Antonelli, P.L. and Zastawniak, T.J. (1995) Diffusion on the Tangent and Indicatrix Bundle on a Finsler Manifold, *Tensor N.S.*, **56**, 233-247 (also in this issue).

[5] Antonelli, P.L. and Zastawniak, T.J. (editors) (1994) *Lagrange Differential Geometry, Finsler Spaces and Diffusions Applied in Biology and Physics*, Pergamon Press, **20**, Math. and Compt. Mod.

[6] Elworthy, K.D. (1982) *Stochastic Differential Equations on Manifolds*, Cambridge University Press, Cambridge.

[7] Emery, M. (1989) *Stochastic Calculus in Manifolds*, Springer-Verlag, Berlin, Heidelberg.

[8] Hrimiuc, D. (1989) On the Geometry of the Total Space of an Infinite Dimensional Vector Bundle, *An. Şt. Univ. "Al.I. Cuza", Iaşi*, **35**, 77-87.

[9] Hrimiuc, D. (1996) Diffusions on Lagrange Manifolds, *An. St. Univ. "Al.I. Cuza", Iaşi*, **XLI**, 247-252.

[10] Ikeda, H. and Watanabe, S. (1989) *Stochastic Differential Equations and Diffusion Processes*, North Holland, Amsterdam, Kodansha, Tokyo.

[11] Miron, R. (1980) Introduction to the Theory of Finsler Spaces, *Proc. of the Nat. Seminar on Finsler spaces*, Braşov, 131-183.

[12] Miron, R. (1982) Vector Bundles Finsler Geometry, *Proc. of the Nat. Seminar on Finsler spaces*, Braşov, 147-188.

[13] Miron, R. and Anastasiei, M. (1994) *The Geometry of Lagrange Spaces, Theory and Applications*, Kluwer Academic Press, FTPH, No. 59.

DIFFUSIONS AND LAPLACIANS ON LAGRANGE MANIFOLDS

P.L. ANTONELLI * and D. HRIMIUC

1 Introduction

The theory of Brownian motion and stochastic development on Finsler spaces cannot be extended to all classes of Lagrange manifolds. Nevertheless, the hv-stochastic development of Antonelli and Zastawniak can be presented in the more general setting of vector bundles, [3], while the related h-stochastic parallel transport cannot. Here, we construct the h-stochastic parallel transport for some special classes of Lagrange spaces and describe a process for generating diffusions on $\overset{\circ}{T}M$. It will be seen that an essential part of this construction depends on the existence of a certain complete nonlinear connection for these particular Lagrange manifolds.

2 Nonlinear Parallelism

Let M be a C^∞-differentiable manifold, n-dimensional, and $\pi : TM \to M$ its tangent bundle. If (x^i) is a local system of coordinates on a domain U of a chart on M, the induced system of coordinates on $\pi^{-1}(U)$ in TM is (x^i, y^i). We set

$$\partial_i := \partial/\partial x^i, \quad \dot{\partial}_i := \partial/\partial y^i, \quad i \in \{1, \ldots, n\}. \tag{1.1}$$

A connection (nonlinear) on TM is defined by a distribution HTM over TM, supplementary to the vertical distribution VTM. A connection having local coefficients $N = (N^i_j)$ generates an adapted frame $(\delta_i, \dot{\partial}_i)$ with its dual $(dx^i, \delta y^i)$ where

$$\delta_i := \partial_i - N^j_i \dot{\partial}_j, \quad \delta y^i = dx^i + N^i_j dx^j \tag{1.2}$$

*Partially supported by NSA Grant No. 144-FK181

A regular Lagrangian is a real valued continuous function, $L : TM \to \mathbb{R}$, smooth on the slit tangent bundle $\mathring{T}M = TM \backslash \{0_x,\ x \in M\}$ such that the matrix with entries

$$g_{ij} := \dot{\partial}_i \dot{\partial}_j L \tag{1.3}$$

is everywhere nondegenerate on $\mathring{T}M$.

A *Lagrange manifold* is a pair (M, L) where M is a smooth manifold and L is a regular Lagrangian. Throughout this paper we suppose that the metric tensor (1.3) is positive definite.

If $N = (N_i^j)$ is a fixed connection on the Lagrange manifold (M, L) we can associate to N and L some important geometrical structures:

a metric tensor on $\mathring{T}M$

$$G(x,y) = g_{ij} dx^i \otimes dx^j + g_{ij} \delta y^i \otimes \delta y^j,$$

an almost product structure

$$P = \delta_i \otimes dx^i - \dot{\partial}_i \otimes \delta y^i,$$

and an almost complex structure

$$F = -\dot{\partial}_i \otimes dx^i - \delta_i \otimes \delta y^i.$$

It can be proved [6], [9] that there exists a linear connection ∇ on $\mathring{T}M$ such that

$$\nabla P = 0, \quad \nabla F = 0, \quad \nabla G = 0.$$

Locally, ∇ is given by a pair of coefficients $C\Gamma = (L^i_{jk}, C^i_{jk})$ where

$$\begin{aligned} L^i_{jk} &= \frac{1}{2} g^{ih}(\delta_j g_{hk} + \delta_k g_{jh} - \delta_h g_{jk}) \quad \text{and} \\ C^i_{jk} &= \frac{1}{2} g^{ih} \dot{\partial}_j g_{hk}. \end{aligned} \tag{1.4}$$

It should be noted that ∇ preserves by parallelism the horizontal and vertical distributions.

Let $x : [0, T] \to M$ be a smooth curve on M and $0 \neq y_0 \in T_{x_0}M$.

1.1. Definition: $y(t) \in TM_{x(t)}$ is called the *nonlinear parallel transport* of y_0 along $x(t)$ if

$$\frac{dy^i}{dt} + N^i_j(x(t), y(t))\, \frac{dx^j}{dt} = 0 \tag{1.5}$$

$$y(0) = y_0. \tag{1.6}$$

In general, this system has only a local solution. We are interested to find a connection such that the solutions of (1.5), (1.6) are defined on the whole interval $[0, T]$, that is, the *connection we seek should be complete.*

For a regular Lagrangian on M we have the semispray associated to L (the Euler-Lagrange vector field)

$$\xi = y^i \partial_i + \xi^i \dot{\partial}_i, \quad \xi^i = g^{ij}(\partial_j L - y^k \dot{\partial}_j \partial_k L). \tag{1.7}$$

For Lagrange manifolds the geodesics are extremals of the action integral of L and coincide with the integral curves of ξ, that is, the solutions of the equations

$$\frac{d^2 x^i}{dt^2} = \xi^i\left(x(t), \frac{dx^i}{dt}\right). \tag{1.8}$$

The semispray given by (1.7) generates the so called *canonical connection*, whose coefficients are given by

$$\overset{c}{N}{}^i_j = -\frac{1}{2}\dot{\partial}_j \xi^i. \tag{1.9}$$

Generally, the paths of $\overset{c}{N} = (\overset{c}{N}{}^i_j)$, that is the solutions of the system

$$\frac{d^2 x^i}{dt^2} + \overset{c}{N}{}^i_j\left(x(t), \frac{dx}{dt}\right)\frac{dx^j}{dt} = 0, \tag{1.10}$$

do *not* coincide with the integral curves of ξ.

Now let

$$E = y^i \dot{\partial}_i L - L \tag{1.11}$$

be the *energy function* associated to L.

1.2. Definition: We say that $N = (N^i_j)$ is *conservative with respect to L*, if

$$\delta_i E = \partial_i E - N^j_i \dot{\partial}_j E = 0, \quad i \in \{1, \ldots, n\}.$$

This is equivalent to the fact that E is constant along any *horizontal curve* (that is, a curve $c(t) = \big(x(t), y(t)\big)$ on $\overset{\circ}{T}M$ such that (1.5) is true).

Please note that in general, the canonical connection $\overset{c}{N} = (\overset{c}{N}{}^i_j)$ is *not* conservative. Let us now consider another connection on $\overset{\circ}{T}M$, $N = (N^i_j)$, such that

$$N^i_j = \overset{c}{N}{}^i_j + T^i_j,$$

where T^i_j are the components of $(1,1)$-*tensor field* (see [9]).

Theorem 1. *If* $T^i_j g_{ik} = T^i_k g_{ij}$, *the following conditions are equivalent:*

(a) *The paths of* N *coincide with the integral curves of* ξ.
(b) *The nonlinear connection* N, *is conservative.*

Proof. From $\delta_i E = 0$ we get

$$y^h \partial_i \dot{\partial}_h L - \partial_i L - N_i^k(\dot{\partial}_k L + y^\ell g_{k\ell} - \dot{\partial}_k L) = 0,$$

that is,

$$y^h \partial_i \dot{\partial}_h L - \partial_i L - \overset{c}{N}{}_i^k y^\ell g_{k\ell} - T_i^k g_{k\ell} y^\ell = 0. \qquad (*)$$

The paths of N coincide with the integral curves of ξ if and only if

$$\xi^i + N_j^i y^j = 0.$$

This is equivalent to

$$\partial_j L - y^k \dot{\partial}_j \partial_k L + \overset{c}{N}{}_k^i g_{ij} y^k + T_k^i g_{ij} y^k = 0.$$

Using the identity

$$\overset{c}{N}{}_k^i g_{ij} = \overset{c}{N}{}_j^i g_{ik} + \dot{\partial}_j \partial_k L - \dot{\partial}_k \partial_j L, \quad \text{(see [6])}$$

This becomes $(*)$. □

1.4. Remark: $N = (N_j^i)$ is a conservative connection, if T_j^i are solutions of the system of equations:

$$T_j^i g_{ik} y^j = y^h \dot{\partial}_k \partial_h L - \partial_k L - \overset{c}{N}{}_h^i g_{ik} y^h,$$
$$T_j^i g_{ik} = T_k^i g_{ij}.$$

A solution of this system is given by

$$T_j^i = \frac{B}{2A} y_j y^i - \frac{1}{2A}(\xi^{*i} y_j + \xi_j^* y^i), \qquad (1.12)$$

where $A = y_i y^i = g_{ij} y^i y^j$, $B = \xi_i^* y^i$ and $\xi^{*i} = y^k \dot{\partial}_k \xi^i - 2\xi^i$ (see also [4]).

Assume that the local expression of the energy function has the following properties

(P_1) $\lim\limits_{\|y\| \to \infty} |E(x,y)| = \infty$

(P_2) for every x, $E(x,y) = 0$ iff $y = 0$.

1.5. Remark: The properties above are true for some classes of Lagrangians:

Example 1.

$$L(x,y) = \frac{1}{2} F^2(x,y) + b_i(x) y^i$$

where F is a Finsler function and b_i are the components of a covector. In this case, $E(x,y) = \frac{1}{2}F^2(x,y)$ and both (P_1) and (P_2) are verified.

Example 2. If L is m-homogeneous $m \in \mathbb{R}$, $m \neq 0$, $m \neq 1$, (see [1]) then $E(x,y) = (m-1)L(x,y) = m(m-1)^2 g_{ij}(x,y)y^i y^j$ and (P_1), (P_2) are true.

Example 3. If L is a φ-Lagrangian, that is $L = \varphi(F^2)$, where F is a Finsler function and $\varphi : \mathbb{R}_+ \to \mathbb{R}$ a c^3-differentiable function (see [1]), it is easy to find conditions on φ such that (P_1) and (P_2) are true.

We are now ready to state and prove

Theorem 2. *If $N = (N^i_j)$ is a conservative connection and the energy function satisfies (P_1) and (P_2), then N is complete.*

Proof. The system (1.5), (1.6) has a local solution $y(t)$ defined on $[0,\tau)$. This solution can not explode when $t \to T$ because of (P_1) and $E(x(t),y(t))$ is a nonzero constant, $\forall t \in [0,\tau)$. On the other hand, the solution can not hit the zero section when t approaches τ because of (P_2) (in (1.6) $y_0 \neq 0$). □

From now on we assume that E satisfies (P_1) and (P_2). Let $x : [0,\tau] \to M$ be a smooth curve on M and $0 \neq u_0 \in T_{x(0)}M$. Consider the horizontal lift $(x(t),y(t))$ of $x(t)$ along u_0 (that is $y(t)$ is a solution of (1.5), (1.6), where $y_0 = u_0$) we say that $u(t) \in T_{x(t)}M$ is the *parallel transport* of u_0 along $x(t)$ if

$$\frac{du^i}{dt} + L^i_{jk}(x(t),y(t))u^j\frac{dx^k}{dt} = 0, \quad \text{and} \tag{1.13}$$

$$u(0) = u_0. \tag{1.14}$$

Because the linear connection (1.4) preserves the horizontal distribution $u(t)$ is horizontal, $t \in [0,T]$. If $u(t), v(t)$ are two solutions of (1.13), then

$$g_{ij}(x(t),y(t))u^i v^j = \text{const.}, \quad t \in [0,T]. \tag{1.15}$$

The differential equation (1.13) is linear and because of (1.15) the solution of (1.13), (1.14) can be extended to $[0,T]$ and $u(t) \neq 0$, $t \in [0,T]$.

3 h-Stochastic Parallel Transport

In this section, we define the concept of h-stochastic parallel transport along a trajectory of a horizontal diffusion on M, for a class of Lagrange manifolds. So we extend some results of [7], [8] which hold for Riemannian manifolds, and those of [2], [3] which are for Finsler manifolds.

Let $x(t)$ be a diffusion on M defined for all $t \in [0, T]$, where T is a stopping time. We shall approximate the diffusion by piecewise smooth sample paths on M.

Given a partition

$$\pi : 0 = t_0 < t_1 < \cdots < t_n = T,$$

we take a piecewise smooth curve $x_\pi(t)$, with bounded first, second and third order derivative such that $x_\pi(t_i) = x(t_i)$, $i \in \{1, 2, \ldots, n\}$. We can lift each of these to $\mathring{T}M$ and thus we obtain a process $y_\pi(t) \in T_{x_\pi(t)}M$ whose sample paths satisfy the ordinary differential equation:

$$\frac{dy^i_\pi(t)}{dt} + N^i_j(x_\pi(t), y_\pi(t)) \frac{dx^j}{dt} = 0, \tag{2.1}$$

with the initial condition

$$y_\pi(0) = y_0 \neq 0. \tag{2.2}$$

The following result can be proved as in [2].

Theorem 3. *The solution of* (2.1), (2.2) *converges in probability as* mesh $\pi \to 0$ *to the solution of the Stratonovich stochastic differential equation:*

$$dy^j + N^j_i(x(t), y(t)) \circ dx^i = 0, \tag{2.3}$$

$$y(0) = y_0. \tag{2.4}$$

2.2. Definition: The solution of (2.3), (2.4) is called the *stochastic non-linear parallel transport* of y_0 along the diffusion $x(t)$ and the diffusion $(x(t), y(t))$ will be called the *horizontal lift* of the diffusion $x(t)$.

By using the Itô formula, it is easy to prove

Proposition 4. *If* $y(t)$ *is a solution of* (2.3) *then*

$$E(x(t), y(t)) = \text{const. a.s. on } [0, T]. \tag{2.5}$$

We can also define stochastic parallelism along a horizontal diffusion on $\mathring{T}M$ as in [2], [3]. Let $c(t) = (x(t), y(t))$ be a horizontal diffusion on $\mathring{T}M$ starting at (x_0, y_0) and $c_\pi(t) = (x_\pi(t), y_\pi(t))$ a piecewise smooth, bounded first, second and third derivative such that $c_\pi(t_i) = c(t_i)$, $i \in \{1, \ldots, n\}$.

Let $u_\pi(t) \in T_{x_\pi(t)}M$ be a process whose sample paths satisfy the ordinary differential equations of parallel transport:

$$\frac{du^i_\pi}{dt} + L^i_{jk} u^j_\pi \frac{dx^k_\pi}{dt} = 0, \tag{2.6}$$

$$u_\pi(0) = u_0 \neq 0, \tag{2.7}$$

where $u_0 \in T_{x_0}M$ is $\mathcal{F}_0$-measurable random vector and π is an arbitrary partition of $[0,T]$.

The following result can be proved as in [2].

Theorem 5. *The solution of the family of differential equations* (2.6), (2.7) *converges in probability when* mesh $\pi \longrightarrow 0$ *to the solution* $u(t)$ *of the stochastic differential equation*

$$du^i + L^i_{jk} u^j \circ dx^k = 0 \tag{2.8}$$

$$u(0) = u_0. \tag{2.9}$$

One consequence of the Itô formula is

Proposition 6. *If* $u(t), v(t)$ *are two solutions of* (2.8) *then*

$$g_{ij}(x(t), y(t)) u^i u^j = \text{ const. a.s. on } \quad [0,T]. \tag{2.10}$$

4 Stochastic Development

Let $O(M)$ be the principal bundle of frames on M. The total space of $O(M)$ consists of elements (x,y,z), $x \in M$, $y \in T_xM$ and $z = (e_1, e_2, \ldots, e_n)$ is an orthonormal frame of $HTM_{(x,y)}$ relative to the metric g_{ij}.

If $\gamma(t) = (\gamma^i(t))$, $t \in [0,T]$ is a smooth curve in $\mathbb{R}^n$ using the linear connection (1.4), we can roll M along γ to obtain a curve on M (see [8], [3]) as a trace of γ.

In fact, if $(x_0, y_0, z_0) \in O(M)$ we must find a smooth curve $c(t) = (x(t), y(t), z(t))$ on OM such that

$$\begin{aligned} &\frac{dx^i}{dt} = \frac{d\gamma^j}{dt} e^i_j(t), \\ &\frac{\delta y^i}{dt} = 0, \\ &\frac{de^j_h}{dt} + L^j_{ki} e^k_h \frac{dx^i}{dt} = 0, \\ &x^i(0) = x^i_0, \quad y^i(0) = y^i_0, \quad e_i(0) = e_{0i}. \end{aligned} \tag{3.1}$$

Let $c(t) = \big(x(t), y(t)\big), \quad t \in [0,T]$ be a horizontal diffusion on $\mathring{T}M$ and z_0 an orthonormal frame in $HTM_{(x_0,y_0)}$ as above.

We use the stochastic parallel transport to move this orthonormal frame along $c(t)$ obtaining the moving frame $z(t) = \big(e_1(t), \ldots, e_h(t)\big)$ such that the following stochastic differential equations are satisfied:

$$de^j_h + L^j_{ki} e^k_h \circ dx^i = 0, \tag{3.2}$$

and a.s. on $[0,T]$

$$g_{ij}\big(x(t), y(t)\big) e^i_h(t) e^j_k(t) = \delta_{hk}. \tag{3.3}$$

Now let $w(t)$ be a standard Brownian motion on $\mathbb{R}^n$. The system of stochastic differential equations for the stochastic development on M can be written as:

$$\begin{aligned} dx^i &= e^i_j \circ dw^j, \\ \delta y^i &= 0, \\ de^j_h + L^j_{ki} e^k_h \circ dx^i &= 0, \\ x^i(0) = x^i_0, \quad y(0) &= y_0, \quad e_i(0) = e_{0i}, \end{aligned} \tag{3.4}$$

with $(x_0, y_0, z_0) \in O(M)$.

The solution of (3.4) is called the *stochastic h-development over M.*

In a similar way as in [2], [3] we can prove:

Theorem 7. *The solution of stochastic differential equations* (3.4) *define a flow of diffeomorphisms on* $O(M)$, $\big(x(t), y(t), z(t)\big)$ *whose projection* $\big(x(t), y(t)\big)$ *from* $O(M)$ *to* $\mathring{T}M$ *is a Markov diffusion on* $\mathring{T}M$ *starting at* (x_0, y_0), *which has probability law independent of the choice of the initial orthonormal frame* z_0 *in* $TM_{(x_0,y_0)}$, *and whose generator is*

$$D = \frac{1}{2}\, g^{ij}(\delta_i \delta_j - L^k_{ij} \delta_k). \tag{3.5}$$

Note: (3.5) can be regarded as a generalization of Laplace-Beltrami operator for some special Lagrange manifolds (see [2], [3] for Finsler manifolds) given in examples 1,2,3 above in Section 1.5.

References

[1] Antonelli, P.L. and Hrimiuc, D. (1996) On the Theory of φ -Lagrange Manifolds With Applications in Biology and Physics, *Nonlinear World*, **3**, 299-333.

[2] Antonelli, P.L. and Zastawniak, T.J. (1993) Diffusions on Finsler Manifolds, *Rep. Math. Phy.*, **33**, 303-315 (also, this issue).

[3] Antonelli, P.L. and Zastawniak, T.J. (1994) Stochastic Calculus on Finsler Manifolds and An Application in Biology, *Nonlinear World*, **1**, 149-171 (also, this issue).

[4] Grifone, J. (1972) Structure Presque-Tangente et Connections I, *Ann. Inst. Fourier*, Grenable, **22**, no. 1, 287-334.

[5] Hrimiuc, D. (1996) Diffusion on the Total Space of a Vector Bundle, *Balkan J. Geom. Appl.*, **1**, no. 2, 53-62 (also, this issue).

[6] Hrimiuc, D. and Shimada, H. (1996) On the $\mathcal{L}$ -Duality Between Lagrange and Hamilton Manifolds, *Nonlinear World*, **3**, 613-641.

[7] Itô, K. (1975) Stochastic Parallel Displacement, *Probabilistic Methods in Differential Equations*, L.N. Math. **451**, Springer, Berlin, 1-7.

[8] Ikeda, N. and Watanabe, S. (1981) *Diffusion Processes and Stochastic Differential Equations*, North Holland, Amsterdam.

[9] Miron, R. and Anastasiei, M. (1994) *The Geometry of Lagrange Spaces, Theory and Applications*, Kluwer Academic Press, FTPH, No. 59.

φ-LAGRANGE LAPLACIANS

P.L. ANTONELLI * and D. HRIMIUC

1 Introduction

A Lagrangian on n-dim C^∞ manifold M^n is a real-valued function

$$L : (x, y) \in TM^n \longrightarrow L(x, y) \in \mathbb{R}^1, \quad \forall (x, y) \in TM^n,$$

C^∞ on $\widetilde{TM^n} = TM^n - \{0\}$ and continuous on the zero section of π : $TM^n \to M$. Introduce a metric tensor by

$$g_{ij}(x, y) := \frac{1}{2} \cdot \dot{\partial}_i \dot{\partial}_j L(x, y).$$

If we require max rank on $\widetilde{TM^n}$, then L is *regular*

Definition. A *Lagrange manifold* is a pair (M^n, L) with L regular on $\widetilde{TM^n}$ and g_{ij} being of constant signature (eg., positive definite)

Examples. (1) $L = F^2$, $F(x, y) > 0$ on $\widetilde{TM^n}$, p-homogeneous of degree *one* in y. Then $(M^n, L) \equiv (M^n, F^2)$ is a *Finsler Manifold.*

Given (2) (M^n, F^2), a Finsler manifold, set $(M^n, \varphi(F^2)) = (M^n, L)$, where

$$\varphi'(t) \neq 0, \qquad \forall t \in \mathrm{Im}\,(F^2), \quad \text{and,}$$
$$\varphi'(t) + 2t \cdot \varphi''(t) \neq 0,$$

then (M^n, φ) is a *φ-Lagrange manifold,* [2].

Examples. $\varphi(t) = t^{\frac{m}{2}}$, $m \in Z_+$, then $(M^n, \varphi(F^2))$ is called an *m-homogeneous Lagrange manifold.* $L = \varphi(F^2)$ is *positive definite* and $L =$

*Invited lecturer at A.M.S. Annual Meeting, San Diego, California, Jan. 10, 1997. Research at the University of Alberta partially supported by NSERC-A-7667. This paper is dedicated to Prof. Radu Miron on the occasion of his 70th birthday.

$\frac{2}{m(m-1)} g_{ij} y^i y^j \iff F$ is a positive definite Finsler metric. *We assume positive definite from here on, so m is even,* [1], [2].

If $F(x,y) = \left(\sum_{i=1}(y^i)^m\right)^{\frac{1}{m}}$ is defined on $\widetilde{T\mathbb{R}^n}$, then $\varphi(F^2) = L = e^{m\phi} \cdot \sum_{i=1}^{n}(y^i)^m$ is a regular Lagrangian and positive definite. This L is called the *Ecological* m^{th}*-power Lagrangian*, while $e^{\phi} \cdot F$ is called the mth*-root Ecological Finsler metric function. See the appendix for an example, and the references.*

Theorem 1. *A φ-Lagrange manifold $(M, \varphi(F^2))$ is projective to its associated Finsler space (M, F^2). This means the geodesics are identical, in fact.*

Now employ R. Miron's Theory of d-connections on Lagrange spaces [7]. We have

Theorem 2. *Given a φ-Lagrange space $(M, \varphi(F^2))$ with $\varphi' + 5\varphi'' t + 2\varphi''' t \neq 0$ on $\operatorname{Im}(F^2)$, $\exists!$ d-connection $C\Gamma = (N^i_j, L^i_{jk}, C^i_{jk})$ such that,*

(1) $\delta_i g_{jk} = L^\ell_{ij} g_{\ell k} + L^\ell_{ik} g_{j\ell}$,

(2) $\dot{\partial}_i g_{jk} = C^\ell_{ij} g_{\ell k} + C^\ell_{ik} g_{j\ell}$,

(3) $N^i_j = L^i_{jk} y^k$

(4) $T^i_{jk} = 0$

(5) $S^i_{jk} = 0$.

In fact,

$$C^i_{jk} = \underbrace{\overset{\circ}{C}{}^i_{jk}}_{\text{Cartan's Tensor}} + \frac{A}{F}(\delta^i_j \ell_k + \delta^i_k \ell_j) + \frac{A}{(1+2A)F} \cdot h_{jk}\ell^i + \frac{BF^2 - A}{F} \ell_i \ell_j \ell^i$$

where $A = A(F^2)$, $B = B(F^2)$ *with*

$$A(t) := \frac{\varphi''}{\varphi'} t,$$

$$B(t) := 2 \cdot \frac{\varphi'\varphi'' + (\varphi'\varphi''' - (\varphi'')^2)t}{\varphi'(\varphi' + 2t\varphi'')}$$

$$\ell_k := \dot{\partial}_i F \quad \textit{and}$$

$$h_{j\ell} = \textit{angular metric of } (M^n, F^2).$$

In addition, N^i_j *is Cartan's nonlinear connection for* (M^n, F^2) *and* $L^i_{jk} = \Gamma^{*i}_{jk}$ *is Cartan's Finsler Connection for* (M^n, F^2).

Remark. All tensors have components relative to the basis for vector fields in $\widetilde{TM^n}$ provided by $\{1, \delta_i, \dot{\partial}_i\}$, where $\delta_i = \partial_i - N^j_i \cdot \dot{\partial}_j$, and its dual $\{1, \delta y^i, dx^i\}$ where, $\delta y^i = dy + N^i_j dx^j$. Such tensors are called d-tensors.

Theorem 3. *The equivalence problem for m-homogeneous φ-Lagrange space is solved by their structure equations:*

$$\omega^i_j = L^i_{jk} dx^k + C^i_{jk} \delta y^k \qquad \textit{Connection forms}$$

$$\left.\begin{aligned} d(dx^i) - dx^h \wedge \omega^i_h &= -\Omega^i \\ d(\delta y^i) - dy^h \wedge \omega^i_h &= -\widetilde{\Omega}^i \end{aligned}\right\} \qquad \textit{Torsion forms}$$

$$d\omega^i_j - \omega^h_j \wedge \omega^i_h = -\Omega^i_j \qquad \textit{Curvature forms}$$

$$\Omega^i = C^i_{jk} dx^j \wedge dy^k;$$

$$\widetilde{\Omega}^i = \frac{1}{2} \underset{\text{Finsler}}{\overset{\circ}{R}{}^i_{jk}} dx^j \wedge dx^k + \underset{\text{Finsler}}{\overset{\circ}{P}{}^i_{jk}} dx^j \wedge \delta y^k,$$

$$\Omega^i_j = \frac{1}{2} R^i_{j\ kh} dx^k \wedge dx^h + P^i_{j\ kh}\, dx^k \wedge \delta y^h + \frac{1}{2} S^i_{j\ kh} \delta y^k \wedge \delta y^h,$$

$$y^\ell R^r_{\ell jh} := R^r_{0jh} = \frac{m}{2} \overset{\circ}{R}{}^r_{jh},$$

$$P^r_{0jh} = \frac{m}{2} \overset{\circ}{P}{}^r_{jh}, \qquad S^r_{0jk} = 0.$$

Remark. For the special *Ecological m-homogeneous example,* $S^i_{j\ kh} = 0$, *so that* $\widetilde{T_x M^n}$ *is flat.* However, for $n \geq 3$, $\overset{\circ}{S}{}^i_{j\ h\ell} \neq 0$, [8]. Also, $\overset{\circ}{R}{}^i_{jk} \neq 0$, [3]. This refers to the *Ecological* m^{th}*-power Lagrangian.*

2 The Stochastic Laplacian

The geodesics for either manifold, (ecological) *m-homogeneous or* (ecological) *Finsler*, are the same from Theorem 1. How do we perturb to obtain a model with noise? Do we use m^{th}-root or m^{th}–power theory? One answer is in the *Lagrange frame bundle*, $\widetilde{LTM^n}$.

Theorem 4. *Let* $(X,Y,Z) \in \widetilde{LTM^n}$ *be a solution to*

$$\begin{aligned} dX^i &= Z^i_j \circ dW^j \\ \delta Y^i &= Z^i_j \circ dV^j \\ dZ^i_j &= -\Gamma^{*i}_{k\ell} Z^\ell_j \circ dX^k - C^i_{k\ell} Z^\ell_j \circ \delta Y^k, \end{aligned}$$

($\circ$ *means Stratonovich Stochastic Calculus)*

where, $\delta Y^i = dY^i + N^i_j \circ dX^j$, *with initial conditions* $X(0) = x_0 \in M$, $0 \neq Y(0) = y_0 \in \widetilde{T_{x_0}M^n}$, *and* $Z(0) = z_0$, *a frame in* $\widetilde{T_{x_0}M^n}$ *orthonormal with respect to* $g_{ij}(x_0, y_0)$. *Then* (X,Y,Z) *is a semi-martingale on the orthonormal m-homogeneous bundle of frames,* $\widetilde{LTM^n}$ *over* $\widetilde{TM^n}$, *and* (X,Y) *is a* **diffusion** *on* $\widetilde{TM^n}$ **with generator**

$$\mathbb{D} = \frac{1}{2} g^{ij}(\delta_i \delta_j - \Gamma^{*k}_{ij}\delta_k) + \frac{1}{2} g^{ij}(\dot{\partial}_i \dot{\partial}_j - C^k_{ij}\dot{\partial}_k),$$

with C^k_{ij} *given in Theorem 2 above, with* $A = \frac{m}{2}$, $B = 0$. *The probability law of this Markov diffusion is independent of the initial frame* $Z(0)$.

Remark. Theorem 4 generalizes the Finsler Theory of the Stochastic (Finsler) Laplacian due to Antonelli and Zastawniak (1993) and used successfully on problems in biology, [4], [5], [6]. Theorem 4 is due to the present authors, using also the paper [1].

Consider diagonal lift metric on $\widetilde{TM^n}$,

$$G(x,y) = g_{ij}(x,y)dx^i \otimes dx^j + g_{ij}(x,y)\delta y^i \otimes \delta y^j.$$

If g is *positive definite* so is G, Consider the *Miron lift* of a φ-Lagrange connection $C\Gamma$. This is a linear connection ∇ on $\widetilde{TM^n}$ such that

$$\nabla_{\delta_i}\delta_j = L^k_{ij}\delta_k, \qquad \nabla_{\delta_i}\dot{\partial}_j = L^k_{ij}\dot{\partial}_k$$
$$\nabla_{\dot{\partial}_i}\delta_j = C^k_{ij}\delta_k, \qquad \nabla_{\dot{\partial}_i}\dot{\partial}_j = C^k_{ij}\dot{\partial}_k,$$
$$T(\delta_i,\delta_j) = \overset{\circ}{R}{}^{k}_{ij}\dot{\partial}_k,$$
$$T(\delta_i,\dot{\partial}_j) = -C^k_{ji}\delta_k - \overset{\circ}{P}{}^{k}_{ji}\dot{\partial}_k,$$
$$T(\dot{\partial}_i,\delta_j) = C^k_{ij}\delta_k + \overset{\circ}{P}{}^{k}_{ij}\dot{\partial}_k, \qquad T(\dot{\partial}_i,\dot{\partial}_j) = 0.$$

Here, $R^b_{ij}k = \overset{\circ}{R}{}^{k}_{ij} = \delta_j N^i_k - \delta_k N^i_j$, $P^i_{jk} = \overset{\circ}{P}{}^{i}_{jk} = \dot{\partial}_j N^i_k - L^i_{jk}$ and C^i_{jk} is given as in Theorem 4 above.

Using the general result of [7, p. 179] we now have the

Theorem 5. *$\widetilde{TM^n}$, with metric G and Almost Complex structure induced by N^i_j, is Almost Kaehler with symplectic form $\theta = g_{ij}\delta y^i \wedge dx^j$, such that $\theta = d\omega$, $\omega = \frac{1}{2}\dot{\partial}_i\varphi(F^2)dx^i$ being the Hilbert 1-form, with $d\theta = 0$.*

3 Stochastic Imbedding into Riemannian Diffusion

Theorem 6. *The m-homogeneous Lagrange Laplacian $\mathbb{D}$ is a Riemannian Laplacian, $\frac{1}{2}\Delta_G$, plus a nonzero drift field on $\widetilde{TM^n}$,*

$$B = \frac{1}{2}g^{ik}\overset{\circ}{P}{}^{j}_{jk}\delta_i - \frac{1}{2}g^{ik}C^j_{jk}\dot{\partial}_i.$$

Remarks. (1) The m-homogeneous Brownian motion is a Riemannian Brownian motion plus drift B, given by Finsler torsion $\overset{\circ}{P}{}^{j}_{ik}$ of F and with C^j_{ik} as in Theorem 4.

(2) Also, $g_{ij}(x,y)$ is not a Finsler metric tensor, if $m \geq 3$. *It can be diagonal,* which could never happen in a non-Riemannian Finsler space (see appendix).

References

[1] Anastasiei, M. and Antonelli, P. L. (1996) Lagrangians Which Generate Sprays in *Lagrange and Finsler Geometry. Applications to Physics and Biology*, Antonelli, P. L. and Miron, R., eds., Kluwer Academic Press, Dordrecht, 279 pp.

[2] Antonelli, P. L. and Hrimiuc, D. (1996) On the Theory of φ-Lagrange Manifolds with Applications in Biology and Physics, *Nonlinear World*, **3**, pp. 299–333.

[3] Antonelli, P. L. and Shimada, H. (1991) On 1-Form Finsler Connections with Constant Coefficients, *Tensor, N.S.*, **50**, 263-275.

[4] Antonelli, P. L. and Zastawniak, T. J. (1993) Diffusion Theory on Finsler Manifolds, *Rep. Math. Phys.*, **33**, 303-315.

[5] Antonelli, P. L. and Zastawniak, T. J. (1996) Diffusion on the Tangent and Indicatrix Bundles of a Finsler Manifold *Tensor, N.S.*, **56**, 233–247.

[6] Antonelli, P. L. and Zastawniak, T. J. eds. (1994) Lagrange Geometry, Finsler Spaces and Noise Applied in Biology and Physics, *Math. and Compt. Mod.* special issue, **20**, 4/5.

[7] Miron, R. and Anastasiei, M. (1994) *The Geometry of Lagrange Spaces: Theory and Applications*, Kluwer Academic Press, Dordrecht, 285 pp.

[8] Shimada, H. (1979) On Finsler Spaces with the Metric $L = \sqrt[m]{a_{i_1 \dots i_m}(x) y^{i_1} \dots y^{1_m}}$, *Tensor, N.S.*, **33**, 365-372.

Appendix

Let us give an example. Let $L = e^{m\alpha_i x^i}[(y^1)^m + (y^2)^m]$ where α_1, α_2 are fixed constants. Then the geodesics of this associated m-homogeneous Lagrange space are given by

$$\begin{aligned}\frac{dy^1}{ds} &= -\alpha_1(y^1)^2 - \frac{m}{m-1}\,\alpha_2 \cdot y^1 y^2 \\ &\quad + \frac{\alpha_1}{m-1} \cdot (\frac{y^2}{y^1})^{m-2} \cdot (y^2)^2 \\ \frac{dy^2}{ds} &= -\alpha_2(y^2)^2 - \frac{m}{m-1}\,\alpha_1 \cdot y^1 y^2 \\ &\quad + \frac{\alpha_2}{m-1} \cdot (\frac{y^1}{y^2})^{m-2} \cdot (y^1)^2.\end{aligned}$$

These are precisely the geodesics of the ecological Finsler metric.

For $n = 3$,

$$L = e^{m\alpha_i \cdot x^i}[(y^1)^m + (y^2)^m + (y^3)^m] = e^{m\phi} \cdot \mathcal{L}^{(m)}$$

results in the geodesic equations

$$\begin{aligned}\frac{dy^1}{ds} &= -\alpha_1(y^1)^2 + \frac{\alpha_1}{m-1}\left[\left(\frac{y^2}{y^1}\right)^{m-2} \cdot (y^2)^2 + \left(\frac{y^3}{y^1}\right)^{m-2} \cdot (y^3)^2\right] \\ &\quad - \frac{m}{m-1}\, y^1(\alpha_2 y^2 + \alpha_3 y^3) \\ \frac{dy^2}{ds} &= -\alpha_2(y^2)^2 + \frac{\alpha_2}{m-1}\left[\left(\frac{y^3}{y^2}\right)^{m-2} \cdot (y^3)^2 + \left(\frac{y^1}{y^2}\right)^{m-2} \cdot (y^1)^2\right] \\ &\quad - \frac{m}{m-1}\, y^2(\alpha_1 y^1 + \alpha_3 y^3) \\ \frac{dy^3}{ds} &= -\alpha_3(y^3)^2 + \frac{\alpha_3}{m-1}\left[\left(\frac{y^1}{y^3}\right)^{m-2} \cdot (y^1)^2 + \left(\frac{y^2}{y^3}\right)^{m-2} \cdot (y^2)^2\right] \\ &\quad - \frac{m}{m-1}\, y^3(\alpha_1 y^1 + \alpha_2 y^2).\end{aligned}$$

Then $g_{ij} = \frac{1}{2}\,\dot{\partial}_i\dot{\partial}_j L = \frac{1}{2}\,\dot{\partial}_i\big(m(y^j)^{m-1} \cdot e^{m\alpha_i x^i}\big)$ so that

$$(g_{ij}) = \begin{pmatrix} (y^1)^{m-2} & 0 & 0 \\ 0 & (y^2)^{m-2} & 0 \\ 0 & 0 & (y^3)^{m-2} \end{pmatrix} \cdot \frac{m(m-1)}{2} \cdot e^{m\alpha_i x^i}.$$

DIFFUSION, LAPLACIAN AND HODGE DECOMPOSITION ON FINSLER SPACES

P.L. ANTONELLI and T.J. ZASTAWNIAK

1 Introduction

Throughout this paper M will be a compact Finsler space with positive definite metric. That is to say, M will be a boundaryless compact manifold equipped with a positive smooth $\mathbb{R}$-valued function F defined on the slit tangent bundle $\mathring{T}M$ (the tangent bundle TM with the zero section removed) such that $F(x, y)$ is a positive homogeneous function of degree one in $y \in \mathring{T}_x M$ for any $x \in M$ and the Finsler metric tensor defined by

$$g_{ij}(x,y) = \frac{1}{2}\partial_{y^i}\partial_{y^j}\left(F^2(x,y)\right)$$

is positive definite. The latter is a tensor in the Finslerian sense, that is, such that the diagram

$$\begin{array}{ccc} & g & \\ \mathring{T}M & \longrightarrow & T^2M \\ \pi_1 \searrow & & \swarrow \pi_2 \\ & M & \end{array}$$

commutes. (Here π_1 and π_2 are the projections of the bundles $\mathring{T}M$ and T^2M.)

In their ground-breaking 1996 paper [1] Bao and Lackey proposed an extension of the Laplace-Beltrami operator from Riemannian spaces and obtained a Hodge decomposition theorem on compact Finsler spaces with positive definite metric. We shall refer to this extension as the Bao-Lackey Laplacian or the BL-Laplacian for short. It is constructed by considering the indicatrix bundle IM with fibres $I_x M = \{y \in T_x M : F(y) = 1\}$ as a Riemannian manifold with the diagonal lift metric $\hat{g}$ (for definition see below), which gives rise to a Hodge star $*$ and thus an inner product

$$\langle \eta, \theta \rangle = \int_{IM} \frac{(-1)^{n-1}}{\mathrm{Vol}(I_x M)} \eta \wedge *\theta$$

of differential forms η, θ on IM. This can also be regarded as an inner product of differential forms η, θ on M, since any such form can be identified with a form on IM (with no y dependence). Taking d_M to be the exterior differential on M and d_M^* the adjoint to d_M in the sense that

$$\langle d_M^* \eta, \theta \rangle = \langle \eta, d_M \theta \rangle$$

for any differential forms η, θ on M, the BL-Laplacian Δ_{BL} on M can be defined by

$$\Delta_{\mathrm{BL}} = d_M^* d_M + d_M d_M^*.$$

Because an inner product has been introduced on the elliptic complex of differential forms on M, a Hodge decomposition theorem follows by a general result about elliptic complexes in Wells [5], Chap. 4, Theorem 5.2.

It is clear that Δ_{BL} reduces to the standard Laplacian if M is a Riemannian manifold. But are there any other natural extensions of the Laplace-Beltrami operator to Finsler spaces? After all, as far as the Hodge theorem is concerned, it only matters that there should be a well-defined inner product on differential forms, which leaves quite a lot of room for choice. In this paper we seek such an alternative from the viewpoint of diffusion theory and come up with a different candidate, which will be called the AZ-Laplacian and denoted by Δ_{AZ}. Is any of the two Laplacians Δ_{BL} and Δ_{AZ} more natural than the other? We are inclined to believe that the one emerging from diffusion theory is perhaps more suitable for defining harmonic forms as those which realize heat equilibrium. Further investigations may elucidate this point or reveal more candidates.

Our construction of the Laplacian will proceed as follows. First we shall introduce a random walk on a Finsler manifold M extending Pinsky's concept of the isotropic transport process on a Riemannian manifold [3], [4]. Then a limit will be taken under suitable scaling as in the Central Limit Theorem, producing a diffusion generator on M, which is obviously a second-order operator. The quadratic form of this operator will give rise to an inner product of differential forms on M, leading eventually to the definition of our Laplacian Δ_{AZ} in a similar way as for Δ_{BL} above. Harmonic forms are then introduced and a Hodge decomposition obtained again by the general result about elliptic complexes in [5].

We need to state some standard definitions and notation for Finsler spaces to be used in this paper. The *formal Christoffel symbols* are defined by

$$\gamma^i_{jk} = \frac{1}{2} g^{il} \left(\partial_{x^j} g_{lk} + \partial_{x^k} g_{jl} - \partial_{x^l} g_{jk} \right).$$

Here g^{ij} is the inverse matrix to g_{ij}. Note that the $\gamma^i_{jk}(x,y)$ pick up the y-dependence from $g_{jk}(x,y)$ and they *do not* transform like connection coefficients in contrast to the Levi-Cività connection coefficients in Riemannian geometry. We define the horizontal vector fields

$$\delta_i = \partial_{x^i} - \gamma^j_{ik}(x,y)y^k\partial_{y^j}$$

along with the basis $dx^i, \delta y^i$, $i = 1, \dots, n$, of one-forms on $\mathring{T}M$ dual to δ_i, ∂_{y^i}. Here $\delta y^i = dy^i + N^i_j dx^j$. Then the diagonal lift metric tensor $\hat{g}$ on $\mathring{T}M$ is given by

$$\hat{g} = g_{ij}dx^i \otimes dx^j + g_{ij}\delta y^i \otimes \delta y^j,$$

which is clearly a Riemannian metric on $\mathring{T}M$, since g is positive definite. We shall use the same symbol $\hat{g}$ to denote the Riemannian metric induced on the submanifold $IM \subset \mathring{T}M$. It will be called the *diagonal lift metric* on IM.

2 Finslerian Isotropic Transport Process

To define this process we first introduce the *geodesic flow* on a Finsler manifold. This is a deterministic motion $\xi_t(x,y)$ obtained by following the unit speed geodesic with initial position $\xi_0(x,y) = x \in M$ and initial velocity $\dot{\xi}_0(x,y) = y \in I_xM$. It satisfies the geodesic equation $\ddot{\xi}^i_t + \gamma^i_{jk}(\xi_t, \dot{\xi}_t)\dot{\xi}^j_t\dot{\xi}^k_t = 0$, where γ^i_{jk} are the Finslerian formal Christoffel symbols. If we put

$$(T^0_t f)(x,y) = f(\xi_t(x,y), \dot{\xi}_t(x,y))$$

for any function $f \in C^\infty(IM)$, then $T^0_{s+t} = T^0_s T^0_t$ and $u(t,x,y) = \left(T^0_t f\right)(x,y)$ satisfies the p.d.e. $\partial_t u = Zu$, where the first-order operator

$$Z = y^i\partial_{x^i} - \gamma^i_{jk}(x,y)y^jy^k\partial_{y^i} = y^i\delta_i$$

is known as the *geodesic flow field* or the *canonical horizontal vector field* on IM. It is the infinitesimal operator of the semigroup T^0_t, $t \geq 0$.

Next, for any $x \in M$ we consider the indicatrix I_xM as a submanifold of IM with the diagonal lift metric $\hat{g}$ induced from IM. This means simply that $g_{ij}(x,y)$ considered as a function of y with x fixed is taken as a metric tensor on $\mathring{T}_xM$ and this is then restricted to $I_xM \subset \mathring{T}_xM$. Equipped with this metric the indicatrix I_xM is a compact Riemannian manifold. We denote by ω_x the volume measure of I_xM in this metric and assume that it is normalized so that $\int_{I_xM} \omega_x(dy) = 1$. The *canonical projection operator* $\Pi : C^\infty(IM) \to C^\infty(M)$ is then defined by

$$(\Pi f)(x) = \int_{I_xM} f(x,y)\, d\omega_x(y).$$

Let us take a sequence of independent random variables $e_1, e_2, \ldots$ with exponential distribution

$$P\{e_n > t\} = e^{-t}, \quad t \geq 0,$$

on some probability space (Ω, Σ, P). Then $t_n = e_1 + \cdots + e_n$ are the jump times of a Poisson process with unit rate. We also put $t_0 = 0$ for convenience.

The isotropic transport process can be defined by induction. For a given starting point $x \in M$ and starting velocity $y \in I_xM$ we put $x_0 = x$ and $y_0 = y$. Assuming that $x_i \in M$ and $y_i \in I_{x_i}M$ have been constructed for $i = 0, \ldots, n-1$, we put

$$x_n = \xi_{e_n}(x_{n-1}, y_{n-1})$$

and require that the conditional distribution of $y_n \in I_{x_i}M$ given $e_1, \ldots, e_n$ and $y_1, \ldots, y_{n-1}$ should be uniform on $I_{x_n}M$ under the normalized volume measure ω_{x_n}, that is,

$$E\{f(x_n, y_n)|e_1, \ldots, e_n, y_1, \ldots, y_{n-1}\} = (\Pi f)(x_n)$$

for any $f \in C^\infty(I_{x_n}M)$. Finally, we put

$$X_t = \xi_{t-t_n}(x_n, y_n) \quad \text{and} \quad \mathrm{Y_t = \dot{X}_t} \quad \text{for } \mathrm{t \in [t_n, t_{n+1}),\ n = 0, 1, 2, \ldots}\ .$$

In this way we obtain a Markov process (X_t, Y_t) on the indicatrix bundle IM defined for all $t \geq 0$, called the *isotropic transport process*. The sample paths of X_t are continuous piecewise geodesic with jumps of the tangent vector Y_t at the jump times t_n of a Poisson process.

The infinitesimal operator L of (X_t, Y_t) can be computed by observing that on $\{t < e_1\}$, which has probability e^{-t}, the process moves along the original geodesic $\xi_t(x,y)$, and on $\{t \geq e_1\}$, which has probability $1 - e^{-t}$, a new direction y_1 is chosen according to the measure ω_{x_1} at time t_1. For any $f \in C^\infty(IM)$ we have

$$f(X_t, Y_t) = f(\xi_t(x,y), \dot{\xi}_t(x,y)) = f(x,y) + t(Zf)(x,y) + o(t)$$

on $\{t < e_1\}$ and

$$\begin{aligned} f(X_t, Y_t) &= f(x_1, y_1) + o(1) \\ (\Pi f)(x_1) &= (\Pi f)(x) + o(1) \end{aligned}$$

on $\{t \geq e_1\}$. Using the fact that $E\{f(x_1, y_1)|e_1\} = (\Pi f)(x_1)$, we obtain

$$\begin{aligned} Ef(X_t, Y_t) &= \int_{\{t<e_1\}} f(X_t, Y_t)dP + \int_{\{t\geq e_1\}} f(X_t, Y_t)dP \\ &= e^{-t}[f(x,y) + t(Zf)(x,y)] + \int_{\{t\geq e_1\}} f(x_1, y_1)dP + o(t) \\ &= e^{-t}[f(x,y) + t(Zf)(x,y)] + \int_{\{t\geq e_1\}} (\Pi f)(x_1)dP + o(t) \\ &= e^{-t}[f(x,y) + t(Zf)(x,y)] + (1-e^{-t})(\Pi f)(x) + o(t) \\ &= f(x,y) + t[(Zf)(x,y) + (\Pi f)(x) - f(x,y)] + o(t). \end{aligned}$$

It follows that

$$Lf = \lim_{t\downarrow 0} \frac{1}{t}(f \circ T_t - f) = Zf + \Pi f - f.$$

Here T_t is the semigroup of (X_t, Y_t),

$$(T_t f)(x, y) = Ef(X_t, Y_t)$$

for any $f \in C^\infty(IM)$.

3 Central Limit Theorem

We shall scale the isotropic transport process in such a way that the mean time and distance travelled between consecutive jumps of direction will be ε^2 and ε, respectively, for some $\varepsilon > 0$. In other words, the scaled process X_t^ε will be piecewise geodesic with speed $1/\varepsilon$ and jumps of direction at the jump times of a Poisson process with rate $1/\varepsilon^2$. The velocity part of the process will be defined by $Y_t^\varepsilon = \varepsilon \dot{X}_t^\varepsilon$ to ensure that the joint process $(X_t^\varepsilon, Y_t^\varepsilon)$ lives on the unit-radius indicatrix bundle IM. The infinitesimal operator of the scaled process $(X_t^\varepsilon, Y_t^\varepsilon)$ is then found to be

$$L^\varepsilon = (1/\varepsilon)Z + (1/\varepsilon^2)(\Pi - 1).$$

We claim that X_t^ε tends weakly to a diffusion process on M as $\varepsilon \downarrow 0$.

First we observe that $\Pi^2 = \Pi$ and $\Pi Z \Pi = 0$. The latter equality depends on the fact that any linear function of y on $I_x M$ has integral zero with respect to the normalized volume measure ω_x. This will be so if the indicatrix $I_x M$ is symmetric about the origin in the tangent space $T_x M$, which is the case if the Finsler metric function F is homogeneous in y, that is, $F(x, ky) = |k|F(x, y)$, and not just positive homogeneous. Using these two equalities, we obtain the identity

$$L^{\varepsilon}(1 + \varepsilon Z + \varepsilon^2 Z^2)\Pi = \Pi Z^2 \Pi + \varepsilon Z^3 \Pi.$$

This implies the following lemma.

Lemma 1. *For any $f \in C^\infty(M)$ put $f^\varepsilon = (1 + \varepsilon Z + \varepsilon^2 Z^2) f \in C^\infty(IM)$. Then*

$$\lim_{\varepsilon \downarrow 0} f^\varepsilon = f \quad \text{and} \quad \lim_{\varepsilon \downarrow 0} \mathrm{L}^\varepsilon \mathrm{f}^\varepsilon = \Pi \mathrm{Z}^2 \mathrm{f}.$$

The operator ΠZ^2 on the right-hand side is the generator of a diffusion X_t on M. The theory of convergence of semigroups, in particular, the limit theorem for semigroups in application to random evolution proved by Kurtz in [2] can now be applied to show that

$$\lim_{\varepsilon \downarrow 0} e^{tL^\varepsilon} f = e^{t\Pi Z^2} f$$

for all $f \in C^\infty(M)$, which, in turn, implies that X_t^ε converges weakly to X_t in the sense that

$$\lim_{\varepsilon \downarrow 0} E\{f(X_t^\varepsilon)\} = E\{f(X_t)\}$$

for every $f \in C(M)$. All technical assumptions leading to these results are trivially satisfied if M is compact.

The diffusion generator ΠZ^2 can be expressed as

$$(\Pi Z^2 f)(x) = \int_{I_x M} y^i y^j \left(\partial_{x^i} \partial_{x^j} - \gamma_{ij}^k(x, y)\partial_{x^k}\right) f(x)\, d\omega_x(y)$$

for any $f \in C^\infty(M)$.

This limiting argument is based Pinsky's approach to diffusion on Riemannian manifolds, see [3], [4], where further technical details can also be found. In particular, if M is a Riemannian manifold, then $g^{ij}(x) = (\dim M) \int_{I_x M} y^i y^j \, d\omega_x(y)$ and $\Delta = (\dim M)\Pi Z^2$, the Laplace-Beltrami operator, so $X_{t/\dim M}$ is a Brownian motion on M. This has been extended above to the case of Finsler manifolds with smooth compact indicatrix.

4 Laplacian, Harmonic Forms and Hodge Decomposition

The diffusion generator ΠZ^2 is a second-order differential operator with leading quadratic form

$$H^{ij}(x) = \int_{I_x M} y^i y^j \, d\omega_x(y),$$

which is clearly non-degenerate and positive definite. When M is Riemannian, it reduces to the metric tensor $g^{ij}(x) = (\dim M)\, H^{ij}(x)$. Similarly, we define

$$H^{i_1 j_1 \dots i_p j_p}(x) = \int_{I_x M} y^{i_1} y^{j_1} \dots y^{i_p} y^{j_p} \, d\omega_x(y),$$

which can be written briefly as $H^{IJ}(x) = \int_{I_x M} y^I y^J \, d\omega_x(y)$ for multi-indices I and J. Because

$$H^{IJ}(x) u_I u_J = \int_{I_x M} \left(y^I u_I\right)^2 d\omega_x(y) > 0 \tag{1}$$

for any p-form $u \neq 0$, it follows that $H^{IJ}(x)$ defines an inner product on the exterior algebra $\Lambda^p_x M$, namely,

$$(\eta|\theta) = \int_M \eta_I \theta_J H^{IJ} \sqrt{G} dx,$$

where

$$\sqrt{G} = \int_{I_x M} \sqrt{g} \, d\omega_x(y), \qquad \sqrt{g} = \sqrt{\det(g_{ij})}.$$

It is easily seen that $\sqrt{G}$ is a tensor density of weight 1 and H^{IJ} is a contravariant tensor of rank $2p$ on M, so the product $(\eta|\theta)$ is well defined. It follows from (1) that H^{IJ} is invertible as a map from p-forms to p-forms. The inverse will be denoted by $H_{IJ} = H_{i_1 j_1 \dots i_p j_p}$.

We denote by d the exterior differential on M and by δ the codifferential defined to be the adjoint to d in the above inner product,

$$(d\eta|\theta) = (\eta|\delta\theta)$$

for any $\eta \in \Lambda^{p-1} M$ and $\theta \in \Lambda^p M$. Our Laplacian is then defined by

$$\Delta_{\mathrm{AZ}} = \delta d + d\delta.$$

Proposition 2. *Let M be a boundaryless compact Finsler manifold whose metric tensor $g_{ij}(x,y)$ is positive definite and whose indicatrix is smooth and compact. Then*

$$\Delta_{AZ}\eta = 0 \Leftrightarrow d\eta = \delta\eta = 0 \tag{2}$$

for any $\eta \in \Lambda^p M$.

Indeed, if $\Delta_{AZ}\eta = 0$, then

$$0 = (\Delta_{AZ}\eta|\eta) = (\delta d\eta|\eta) + (d\delta\eta|\eta) = (d\eta|d\eta) + (\delta|\delta\eta),$$

so $d\eta = \delta\eta = 0$.

A form $\eta \in \Lambda^p M$ satisfying either side of the equivalence (2) will be called *harmonic* and the space of such forms will be denoted by $\mathcal{H}^p M$.

Finally, we are in a position to state the following Hodge decomposition theorem based on our Laplacian Δ_{AZ}.

Theorem 3. *Let M be a boundaryless compact Finsler manifold whose metric tensor $g_{ij}(x,y)$ is positive definite, and whose indicatrix is smooth and compact, with harmonic functions defined by means of the AZ-Laplacian Δ_{AZ}. Then*

(a) *Each cohomology class $H^p M$ contains a unique harmonic representative.*

(b) *$\mathcal{H}^p M$ is finite dimensional, its dimension being equal to the pth Betti number of M.*

(c) *$\Lambda^p M = \mathcal{H}^p M \oplus d\Lambda^{p-1} M \oplus \delta\Lambda^{p+1} M$, the three spaces on the right-hand side being mutually orthogonal in L_2.*

The proof is exactly the same as in Bao and Lackey's paper [1]. It follows directly from the general result about elliptic complexes in Wells [5], Chap. 4, Theorem 5.2, or can be proved by the elegant and simple argument presented in [1]. Rather than repeating this word-by-word, we refer the reader to the sources.

References

[1] Bao, David and Lackey, Brad (1996) A Hodge Decomposition Theorem for Finsler Spaces, *C.R. Acad. Sci. Paris*, **323**, (1), 51-56.

[2] Kurtz, T.G. (1973) A Limit Theorem for Perturbed Operator Semigroups with Applications to Random Evolutions, *J. Functional Analysis*, **12**, 55–67.

[3] Pinsky, Mark A. (1978) Stochastic Riemannian Geometry, in *Probabilistic Analysis and Related Topics*, 1, ed. A.T. Bharucha-Reid, Academic Press, Inc., New York-San Francisco-London, 199-235.

[4] Pinsky, Mark A. (1992) Inverse Questions in Stochastic Differential Geometry, in *Probability Theory*, eds. L. Chen, K. Choi, K. Hu and J.H. Lou, Walter de Gruyter & Co., Berlin-New York, 3-28.

[5] Wells, R.O. (1980) *Differential Analysis on Complex Manifolds*, Springer-Verlag.

A MEAN-VALUE LAPLACIAN FOR FINSLER SPACES *

PAUL CENTORE

Introduction

On a Riemannian space, the Laplace operator (both for forms and functions) is a natural and important operator. It leads to the Hodge Decomposition Theorem, which gives topological information about the space, and is essential to investigating the diffusion of heat. These considerations also make sense on the more general Finsler spaces, but so far it is not clear what we should use as a Laplacian on Finsler spaces. In this paper, we seek to generalize the Laplacian (first for functions and then for forms) on a Riemannian space to a Laplacian on a Finsler space. We do this by generalizing an important property of the Laplacian on Riemannian space, and that is that the Laplacian (at least infinitesimally) measures the average value of a function around a point.

1 Notation

We will use the following notation. M or M^n is a compact manifold of dimension n with local co-ordinates x^i around a point x. If $X \in T_xM$, then $X = X^i\frac{\partial}{\partial x^i}$, where the X^i are co-ordinates for the tangent bundle canonically induced from the x^i for the base manifold. F is a Finsler metric, i.e. a function $F : TM \longrightarrow \mathbb{R}$ where

1. F is positive-definite: $F(x, X) \geq 0$, with equality iff $X = \vec{0}$.

2. F is smooth except along the zero-section: $F|_{TM\setminus\{(x,\vec{0})|x\in M\}}$ is C^∞.

*This paper contains results obtained for a doctoral dissertation, under the supervision of Prof. John Bland, at the University of Toronto, and was presented at the University of Alberta in August, 1997.

3. F is strictly convex: at any (x, X), rank $\left[\frac{\partial^2 F}{\partial X^i \partial X^j}\right] = n - 1$.

4. F is homogeneous: $F(x, kX) = |k|F(x, X)$, for all $k \in \mathbb{R}$.

Some authors do not demand the full strength of condition 4., but only require *positive* homogeneity: $F(x, kX) = kF(x, X)$, for all $k > 0$. Condition 4, however, will be essential in what follows, because it guarantees that the unit ball

$$I = \{X \in T_xM | F(X) \leq 1\}$$

is centered at $\vec{0}$, i.e.

4.′ F is symmetric: $F(x, -X) = F(x, X)$.

In addition to the unit ball, we can consider a ball (in the tangent space T_xM) of any radius $\varepsilon > 0$, and we will define the indicatrix I_ε of radius ε to be this (solid) ball:

$$I_\varepsilon = \{X \in T_xM | F(X) \leq \varepsilon\}.$$

Apart from balls I_ε *in the tangent space*, we will see that we can talk about balls Σ_ε *on the manifold*, and we must be careful to distinguish the two.

Associated with any Finsler metric are several commonly used quantities. All these quantities appear also in Riemannian space, and generally have the same interpretation here, but there is an important difference—most important quantities in a Finsler space are not functions solely of M but in fact of TM. This difference is most obvious when we write the argument as (x, X) but in the sequel, to preserve space, we will often suppress arguments. Quantities, and relations they satisfy, which we will use are:

$$\begin{aligned}
g_{ij}(x, X) &:= \frac{1}{2}\frac{\partial^2 F^2(x, X)}{\partial X^i \partial X^j} \\
F^2(X^i \frac{\partial}{\partial x^i}) &= g_{ij}(x, X) X^i X^j \\
g^{jk} g_{ij} &= \delta_i^k \\
g(x, X) &:= \det[g_{ij}(x, X)] \\
\Gamma^i_{jk}(x, X) &:= \frac{1}{2} g^{ir}(x, X)\left(\frac{\partial g_{jr}}{\partial x^k}(x, X) - \frac{\partial g_{jk}}{\partial x^r}(x, X) + \frac{\partial g_{rk}}{\partial x^j}(x, X)\right)
\end{aligned}$$

$f : M \longrightarrow \mathbb{R}$ will be a smooth function with derivatives $f_i = \frac{\partial f}{\partial x^i}$, $f_{ij} = \frac{\partial^2 f}{\partial x^i \partial x^j}$, etc. We are looking for a Laplacian $\Delta f : M \longrightarrow \mathbb{R}$.

2 Geometric Motivation

Our general approach to extending the Laplacian to Finsler spaces will be to choose an operator which measures the mean value of a function at a point.

In the Euclidean space $\mathbb{R}^n$, we say a function f is *harmonic*, or has zero Laplacian, if

$$\begin{aligned} f(x) &= \text{average value of } f \text{ on a solid sphere (of any radius) around } \quad x \\ &= \frac{\int_{\Sigma_\varepsilon} f dV}{\int_{\Sigma_\varepsilon} dV} \end{aligned}$$

where

$$\begin{aligned} \Sigma_\varepsilon &: = \{y \in \mathbb{R}^n | d(x,y) \le \varepsilon\} \\ dV &: = dx^1 \dots dx^n. \end{aligned}$$

Thus

$$\frac{\int_{\Sigma_\varepsilon} f dV}{\int_{\Sigma_\varepsilon} dV} - f(x) = 0$$

on harmonic functions in Euclidean space. To make sense of the expression $\frac{\int_{\Sigma_\varepsilon} f dV}{\int_{\Sigma_\varepsilon} dV}$ for a general function, we must "infinitesimalize," i.e. consider the above quantity as $\varepsilon \to 0$. Furthermore, we know from the more usual expression, $\Delta f = f_{ii}$, that we are only interested in the 2nd-order term of this expression, so we should divide by ε^2 as $\varepsilon \to 0$. In addition, we need a normalizing factor of $2(n+2)$, and we obtain

$$\Delta f(x) = 2(n+2) \lim_{\varepsilon \to 0} \frac{1}{\varepsilon^2} \Big(\frac{\int_{\Sigma_\varepsilon} f dV}{\int_{\Sigma_\varepsilon} dV} - f(x) \Big).$$

We would like to extend this to Riemannian space. The extension should be easy, because our definition is based on spheres, which are defined in Riemannian space, and the correct choice of a volume form

$$dV = \sqrt{g}\, dx.$$

We define, for a Riemannian space,

$$\Delta f(x) = 2(n+2) \lim_{\varepsilon \to 0} \frac{1}{\varepsilon^2} \Big(\frac{\int_{\Sigma_\varepsilon} f \sqrt{g} dx}{\int_{\Sigma_\varepsilon} \sqrt{g} dx} - f(x) \Big).$$

(In Riemannian space, it is necessary to infinitesimalize even to define harmonic functions. On a Euclidean space, a harmonic function f is a function whose value at any point x *equals* the average of f on any ball around x. In a Riemannian space, this is generally not true. It holds only on so-called *harmonic* spaces, which are known to contain symmetric spaces and to be contained in Einstein spaces [RWW]. What is true is that the average value of f over a ball of radius ε *tends to* the Laplacian of f at x as $\varepsilon \to 0$.)

Now take the final step and extend to a Finsler space. Again, we can talk about spheres because these are well-defined in Finsler spaces, but now we have a difficulty in that we need a canonical volume form. We will see in the next section that there is a canonical volume form, which for now we will denote as $dV = \omega(x)dx$. With this volume form, we can define Δf on a Finsler space:

$$\Delta f(x) = 2(n+2)\lim_{\varepsilon\to 0}\frac{1}{\varepsilon^2}\Big(\frac{\int_{\Sigma_\varepsilon} f\omega(x)dx}{\int_{\Sigma_\varepsilon}\omega(x)dx} - f(x)\Big).$$

Let's examine this expression. We note first of all that Δf is invariant, that is, it doesn't depend on the choice of co-ordinates. We note secondly that Δf is linear. There is an obvious question: Is Δ a differential operator? Somewhat surprisingly, perhaps, the answer is yes. To see this, express $f(x)$ and $\omega(x)$ as Taylor series and use the inverse of the exponential map (to be defined precisely later):

$$x^i = X^i - \frac{1}{2}\Gamma^i_{jk}(X)X^jX^k - o(|X|^3),$$

where $|X| = \sqrt{\Sigma(X^i)^2}$, to express the integrals as integrals on the tangent space, instead of on the manifold. We get:

Theorem 1. *Δf is a linear, 2nd order differential operator given by*

$$\Delta f(x) = (n+2)\Bigg(\frac{\int_I X^iX^jdX}{\int_I dX} f_{ij} + f_i\Big(\frac{2w_j}{w}\frac{\int_I X^iX^jdX}{\int_I dX} - \frac{2\int_I \Gamma^q_{jq}X^iX^jdX}{\int_I dX} - \frac{\int_I \Gamma^i_{jk}X^jX^kdX}{\int_I dX}\Big)\Bigg).$$

Its two most important properties are:

Theorem 2. *Δf is given by*

$$\Delta f = \frac{1}{\omega(x)}\frac{\partial}{\partial x^i}\Big(f_jK^{ij}\omega(x)\Big).$$

where $\omega(x)$ is the volume form of (M, F), and K^{ij} is a canonically associated Riemannian metric (to be specified later). In addition, Δ is self-adjoint, i.e.

$$\int_M u(x)\Delta v(x)\omega(x)dx = \int_M v(x)\Delta u(x)\omega(x)dx.$$

The following sections will work out this expression in detail, show that it really does generalize the Riemannian case, and investigate some of its properties.

3 Concomitants of the Finsler Metric

1. Let M be a differentiable manifold with a Finsler metric F. Then we have a distance function between pairs of points, and this function satisfies certain properties. For $x \in M$, let $U \subset M$ be a co-ordinate chart with $x \in U$. Then define a function $\rho_x : U \longrightarrow \mathbb{R}$, where $\rho_x(y)$ is the Finsler distance from y to x. ρ_x satisfies:

1. $\rho_x(x) = 0$,
2. $\rho_x > 0$ on $U \setminus \{x\}$,
3. ρ_x is continuous on U,
4. ρ_x is differentiable on $U \setminus \{x\}$,

Define $\Sigma_\varepsilon = \{y \in M | \rho_x(y) \leq \varepsilon\}$ —think of Σ_ε as the ball of radius ε around x. These balls are defined intrinsically on the Finsler manifold, and will be essential to our Laplacian.

2. We can also use F to define a nowhere-zero volume form [Busemann, §6]. To see how, note that, as a metric space, (M, F) has defined on it a natural Hausdorff measure. Choose the volume form which generates the n-dimensional measure, i.e. take

$$dV = \left(\frac{\kappa_n}{\int_I dX}\right) dx,$$

where

$\kappa_n =$ volume of the unit sphere in $\mathbb{R}^n$

$I = \{X \in T_pM | F(X) \leq 1\}$

$=$ the solid unit ball *in the tangent space*

$dX = dX^1 dX^2 \ldots dX^n$

$=$ Euclidean volume form *in the tangent space* with the co-ordinates X^i

This generalizes the Riemannian volume form and is written in co-ordinates as

$$dV(p) := \kappa_n \Big(\int_{\{X \in T_pM | F(X) \leq 1\}} dX^1 \ldots dX^n \Big)^{-1} dx^1 \ldots dx^n.$$

(In the Riemannian case, $\kappa_n \left(\int_{\{X \in T_pM | F(X) \leq 1\}} dX^1 \ldots dX^n \right)^{-1} = \sqrt{\det g_{ij}}$, where $F(X) = \sqrt{g_{ij}X^iX^j}$; this is the standard Riemannian volume form. The factor κ_n is necessary because the standard Riemannian volume form $\sqrt{\det g_{ij}}dx^1 \ldots dx^n$ gives the volume of a *parallelipiped* (in the tangent space) with sides $X_1, X_2, \ldots, X_n$, whereas the expression $\int_{\{X \in T_pM | F(X) \leq 1\}} \times dX^1 \ldots dX^n$ gives the volume of a Finsler *sphere* (in the tangent space).) Then

$$dV(X_1, \ldots, X_n) = \kappa_n \det \left[X_i^j \right] \left(\int_{\{X \in T_pM | F(X) \leq 1\}} dX^1 \ldots dX^n \right)^{-1},$$

where

$$X_i = X_i^j \frac{\partial}{\partial x^j}.$$

Proposition 3. *dV is a well-defined n-form.*

Proof. dV is an n-form if, when we change co-ordinates to $\tilde{x}^i = \tilde{x}^i(x)$, generating a new expression

$$\tilde{dV} = \kappa_n \left(\int_{\{\tilde{X} \in T_pM | F(\tilde{X}) \leq 1\}} d\tilde{X}^1 \ldots d\tilde{X}^n \right)^{-1} d\tilde{x}^1 \ldots d\tilde{x}^n,$$

and a new co-ordinate expression $\tilde{X}_i$ for X_i, then

$$dV(X_1, \ldots, X_n) = \tilde{dV}(\tilde{X}_1, \ldots, \tilde{X}_n).$$

□

1. Write

$$\tilde{X}_i^k \frac{\partial}{\partial \tilde{x}^k} = X_i^j \frac{\partial}{\partial x^j}.$$

Then

$$\tilde{X}_i^k \frac{\partial x^j}{\partial \tilde{x}^k} \frac{\partial}{\partial x^j} = X_i^j \frac{\partial}{\partial x^j},$$

so

$$\tilde{X}_i^k \frac{\partial x^j}{\partial \tilde{x}^k} = X_i^j$$
$$\det\left[\tilde{X}_i^k\right] \cdot \det\left[\frac{\partial x^j}{\partial \tilde{x}^k}\right] = \det\left[X_i^j\right]. \tag{III.2-1}$$

2. Calculate

$$\begin{aligned} dV(X_1, \dots, X_n) &= \kappa_n \det\left[X_i^j\right] \left(\int_{\{X \in T_pM \mid F(X) \leq 1\}} dX^1 \dots dX^n\right)^{-1} \\ &= \kappa_n \det\left[\tilde{X}_i^k\right] \cdot \det\left[\frac{\partial x^j}{\partial \tilde{x}^k}\right] \\ &\quad \times \left(\int_{\{\tilde{X} \in T_pM \mid F(\tilde{X}) \leq 1\}} \det\left[\frac{\partial x^l}{\partial \tilde{x}^m}\right] d\tilde{X}^1 d\tilde{X}^2 \dots d\tilde{X}^n\right)^{-1} \\ &\quad \text{(using (1))} \\ &= \kappa_n \det\left[\tilde{X}_i^k\right] \left(\int_{\{\tilde{X} \in T_pM \mid F(\tilde{X}) \leq 1\}} d\tilde{X}^1 \dots d\tilde{X}^n\right)^{-1} \\ &= d\tilde{V}(\tilde{X}_1, \dots, \tilde{X}_n). \end{aligned}$$

3. Like a Riemannian manifold, a Finsler manifold also has an intrinsically defined exponential map, which sends one-dimensional subspaces of a tangent space isometrically onto geodesics. We'll express the exponential map exp in local co-ordinates, and compute some expressions to be used in later calculations. A geodesic in a Finsler manifold is a parametrized path $x(t) = (x^1(t), x^2(t), \dots x^n(t))$ which satisfies the differential equation

$$\ddot{x}^i = -\Gamma^i_{jk}(x, \dot{x})\dot{x}^j \dot{x}^k,$$

where a dot above the x indicates differentiation with respect to t. We note, in contradistinction to the Riemannian case, that the Christoffel symbols Γ^i_{jk} are functions on the tangent bundle rather than the manifold. This equation allows us to write exp in local co-ordinates around a point $p \in M$ as

$$\exp : T_pM \longrightarrow M$$

$$\begin{aligned} x^i(\exp(X)) = X^i &- \frac{1}{2}\Gamma^i_{jk}(X)X^j X^k - \frac{1}{6}\Gamma^i_{jk|l}X^j X^k X^l \\ &+ \frac{1}{3}\Gamma^i_{jk}\Gamma^k_{pq}X^j X^p X^q + o(|X|^4), \end{aligned} \tag{III.3-2}$$

where we are writing $X \in T_pM$ as $X = X^j \frac{\partial}{\partial x^j}$, and $\Gamma^i_{jk|l} = \frac{\partial}{\partial x^l}\Gamma^i_{jk}$.

In a Riemannian space, exp is C^∞ everywhere on T_pM; in a Finsler space, exp is C^∞ everywhere except the origin $(p, \vec{0})$, at which it is only C^1 [Rund, Chap. 3, §6]. This deficiency, however, will not cause us any trouble. We will use the exponential map to move integrals from the manifold to the tangent space, where the structure of F is exhibited more clearly. In particular, $\exp^{-1}$ takes the sphere Σ_ε on M to the indicatrix of "radius" ε on T_pM, and on T_pM indicatrices of different radii have a natural scaling property.

Indeed, T_pM can be thought of as a manifold on its own, and exp as a map between the manifolds T_pM and M. Since exp is invertible, we can pull back forms on M to forms on T_pM, and we'll calculate pullbacks (in local co-ordinates) for some special cases. To begin with, let X^i be the co-ordinate functions on T_pM, canonically induced by the co-ordinate functions x^i on M. Then forms on the manifold will be written as wedge-products of the 1-forms dx^i, and forms on the tangent space will be written as wedge-products of the 1-forms dX^i. Take the exterior derivative of both sides of (2). (This is allowed because exp is C^1 at p, although higher derivatives do not generally exist.)

$$\begin{aligned}
dx^i &= \frac{\partial}{\partial X^p}\Big(X^i - \frac{1}{2}\Gamma^i_{jk}(X)X^jX^k - \frac{1}{6}\Gamma^i_{jk|l}X^jX^kX^l \\
&\quad + \frac{1}{3}\Gamma^i_{jk}\Gamma^k_{pq}X^jX^pX^q + o(|X|^4)\Big)dX^p \\
&= dX^i - \frac{1}{2}\Gamma^i_{jk}\delta^j_pX^kdX^p - \frac{1}{2}\Gamma^i_{jk}X^j\delta^k_pdX^p \\
&\quad - \frac{1}{2}\Big(\frac{\partial}{\partial X^p}\Gamma^i_{jk}\Big)X^jX^kdX^p - \frac{1}{3}\Gamma^i_{pk|l}X^kX^ldX^p \\
&\quad - \frac{1}{6}\Gamma^i_{jk|p}X^jX^kdX^p + \frac{1}{3}\Gamma^i_{pk}\Gamma^k_{rq}X^qX^rdX^p \\
&\quad + \frac{2}{3}\Gamma^i_{jk}\Gamma^k_{pq}X^jX^qdX^p + o(|X|^3) \\
&= dX^i - \Gamma^i_{pk}X^kdX^p + c^i_{pjk}X^jX^kdX^p + o(|X|^3),
\end{aligned} \tag{III.3-3}$$

where we are writing

$$c^i_{pjk} = -\frac{1}{2}\frac{\partial}{\partial X^p}\Gamma^i_{jk} - \frac{1}{3}\Gamma^i_{pj|k} - \frac{1}{6}\Gamma^i_{jk|p} + \frac{1}{3}\Gamma^i_{pl}\Gamma^l_{jk} + \frac{2}{3}\Gamma^i_{jl}\Gamma^l_{pk}. \tag{III.3-4}$$

To work with n-forms, we'll also need to calculate $dx := dx^1 \ldots dx^n$ in terms of $dX := dX^1 dX^2 \dots dX^n$.

$$\begin{aligned} dx &= \left(dX^1 - \Gamma^1_{pk}X^k dX^p + c^1_{pjk}X^j X^k dX^p\right) \\ &\quad \times \left(dX^2 - \Gamma^2_{pk}X^k dX^p + c^2_{pjk}X^j X^k dX^p\right) \\ &\quad \dots \left(dX^n - \Gamma^n_{pk}X^k dX^p + c^n_{pjk}X^j X^k dX^p\right) \\ &= (1 - \Gamma^j_{jk}X^k + (\Gamma^a_{ak}\Gamma^b_{bj} - \Gamma^a_{bk}\Gamma^b_{aj} + c^r_{rjk})X^j X^k)dX \\ &= \Bigg(1 - \Gamma^j_{jk}X^k + \Big(\Gamma^a_{ak}\Gamma^b_{bj} - \frac{1}{3}\Gamma^a_{bk}\Gamma^b_{aj} - \frac{1}{3}\Gamma^i_{ik|j} - \frac{1}{6}\Gamma^i_{jk|i} \\ &\quad + \frac{1}{3}\Gamma^i_{il}\Gamma^l_{jk} - \frac{1}{2}\Big(\frac{\partial\Gamma^i_{jk}}{\partial X^i}\Big)\Big)X^j X^k\Bigg)dX, \end{aligned} \tag{III.3-5}$$

where we are neglecting terms of higher order than 2. Because exp is only C^1, expression (5) is only continuous and not necessarily differentiable at p.

4. The exponential map allows us to change co-ordinates so that the spheres Σ_ε on M become indicatrices of radius ε on T_pM. Thus any integral $\int_{\Sigma_\varepsilon}\omega$ (of an n-form ω) becomes an integral $\int_{I_\varepsilon}\exp^*\omega$ on the tangent space, and, if $\exp^*\omega(X)$ (where $\exp^*\omega = \exp^*\omega(X)dX$) is homogeneous we can use the scaling property of the I_ε's to write $\int_{I_\varepsilon}\exp^*\omega$ as $\varepsilon^k\int_I \exp^*\omega$ for some appropriate exponent k. In particular, when we come to integrate $\int_{\Sigma_\varepsilon} f\omega$ for a function f, we will replace f by its power series around p to get

$$\int_{\Sigma_\varepsilon} f\omega = f(p)\int_{\Sigma_\varepsilon}\omega + f_i(p)\int_{\Sigma_\varepsilon}x^i\omega + \frac{1}{2}f_{ij}(p)\int_{\Sigma_\varepsilon}x^i x^j\omega + \int_{\Sigma_\varepsilon}o(|x|^3)\omega, \tag{III.4-1}$$

where $|x| = \sqrt{\Sigma(x^i)^2}$. After moving all these new integrals to the tangent space, we will see that we do in fact have homogeneous functions, so we can write them all as integrals over I, which is the same set as I_1. We'll now determine the exponent k given above for integrals of the form $\int_{\Sigma_\varepsilon} x^{|\alpha|}dV$, where α is a multi-exponent for the x^i's.

Case A. $\int_{\Sigma_\varepsilon} dV$. This amounts to finding the volume of the sphere of radius ε in the Finsler manifold. We'll develop a power series in terms of ε for this volume.

$$\begin{aligned} \text{Vol}\,(\Sigma_\varepsilon) &= \int_{\Sigma_\varepsilon} dV \\ &= \int_{\Sigma_\varepsilon}\omega(x)dx, \end{aligned}$$

where $\omega(x) = \frac{\kappa_n}{\int_I dX}$ are the coefficients of the volume form as given in Section III.2. Replace $\omega(x)$ by its Taylor series around p:

$$= \int_{\Sigma_\varepsilon} \omega(p) + \omega_i(p)x^i + \frac{1}{2}\omega_{ij}(p)x^i x^j + o(|x|^3)dx.$$

Now use $\exp^{-1}$ to move to the tangent space, so that Σ_ε is mapped to I_ε, and dx becomes the expression given in Equation (III.3-3). Meanwhile, any x^i in the integrand is replaced by an expression in X^i using Equation (III.3-2).

$$= \int_{I_\varepsilon} \left(\omega(p) + \omega_i(p)\left(X^i - \frac{1}{2}\Gamma^i_{jk}(x, X)X^j X^k\right) + \frac{1}{2}\omega_{ij}(p)X^i X^j\right)$$
$$\times \left(1 - \Gamma^j_{jk}X^k + c_{jk}X^j X^k\right) dX.$$

(Here we are ignoring terms of order higher than 2, and abbreviating

$$c_{jk} := \Gamma^a_{ak}\Gamma^b_{bj} - \frac{1}{3}\Gamma^a_{bk}\Gamma^b_{aj} - \frac{1}{3}\Gamma^i_{ik|j} - \frac{1}{6}\Gamma^i_{jk|i} + \frac{1}{3}\Gamma^i_{il}\Gamma^l_{jk} - \frac{1}{2}\left(\frac{\partial}{\partial X^i}\Gamma^i_{jk}\right)$$

from Equation (III.3-5).)

$$= \int_{I_\varepsilon} \left(\omega(p) + \left(w_k(p) - \omega(p)\Gamma^j_{jk}\right)X^k\right.$$
$$\left. + \left(-\frac{1}{2}\Gamma^i_{jk}\omega_i(p) + \frac{1}{2}\omega_{jk}(p) - \omega_j(p)\Gamma^l_{lk} + \omega(p)c_{jk}\right)X^j X^k\right)dX. \tag{III.4-2}$$

Notice that I_ε is symmetric on T_pM. This is an advantage of working in the tangent space because symmetry ensures that $\int_{I_\varepsilon} X^k dX = 0$, or more generally $\int_{I_\varepsilon} h(X)dX = 0$ if h is any odd function. In the line above we see the integral $\int_{I_\varepsilon} \Gamma^j_{jk}(X)X^k dX$. Consider

$$\Gamma^j_{jk}(X) := \frac{1}{2}g^{jr}(x, X)\left(\frac{\partial g_{jr}}{\partial x^k}(x, X) - \frac{\partial g_{jk}}{\partial x^r}(x, X) + \frac{\partial g_{rk}}{\partial x^j}(x, X)\right). \tag{III.4-3}$$

Since $F^2(X)$ is an even function, and since $g_{ij} = \frac{1}{2}\frac{\partial^2 F^2(x,X)}{\partial X^i \partial X^j}$ is obtained by taking two derivatives of an even function, it must be that $g_{ij}(X)$ is also even. Then $g^{ij}(X)$ is even, and derivatives of the form $\frac{\partial g_{ij}}{\partial x^k}(X)$, with respect to co-ordinates x^k in the manifold are also even (as functions of the tangent space co-ordinates X). We see from Equation (3) that $\Gamma^j_{jk}(X)$ must then be even. The function X^k (on the tangent space) is odd, so the product $\Gamma^j_{jk}(X)X^k$ of an even function and an odd function must again be odd. Thus $\int_{I_\varepsilon} \Gamma^j_{jk}(X)X^k dX = 0$, and Equation (2) simplifies to

$$= \int_{I_\varepsilon} \omega(p)dX$$
$$+ \int_{I_\varepsilon} \left(-\frac{1}{2}\Gamma^i_{jk}(X)\omega_i(p) + \frac{1}{2}\omega_{jk}(p) - \omega_j(p)\Gamma^l_{lk}(X) + \omega(p)c_{jk} \right) X^j X^k dX. \tag{III.4-4}$$

Note that the first term of Equation (4) is the integral of a constant; because of the scaling property $I_{r\varepsilon} = rI_\varepsilon$, we can change our region of integration to I_1, via

$$\int_{I_\varepsilon} \omega(p)dX = \varepsilon^n \int_{I_1} \omega(p)dX$$
$$= \varepsilon^n \int_I \omega(p)dX,$$

where n is the dimension of M, and, of course $I = I_1$. Similarly, the second term of (4) is the integral of a function homogeneous of degree 2, so we can again change our region of integration to I, but this time with a factor of ε^{n+2}:

$$= \varepsilon^n \int_I \omega(p)dX + \varepsilon^{n+2}$$
$$\times \int_I \left(-\frac{1}{2}\Gamma^i_{jk}(X)\omega_i(p) + \frac{1}{2}\omega_{jk}(p) - \omega_j(p)\Gamma^l_{lk}(X) + \omega(p)c_{jk} \right) X^j X^k dX. \tag{III.4-5}$$

Recall that $\omega(p)$ is given by $\omega(p) = \frac{\kappa_n}{\int_I dX}$, where neither the numerator nor the denominator depends on X. Thus we can take $\omega(p)$ outside the integral:

$$= \varepsilon^n \frac{\kappa_n}{\int_I dX} \int_I dX + 0 \cdot \varepsilon^{n+1} + c_{n+2} \cdot \varepsilon^{n+2} + o(\varepsilon^{n+3}),$$

where c_{n+2} will be written explicitly later. In summary,

$$\text{Vol}\,(\Sigma_\varepsilon) = \kappa_n \varepsilon^n + c_{n+2}\varepsilon^{n+2} + o(\varepsilon^{n+3}). \tag{III.4-6}$$

Case B. $\int_{\Sigma_\varepsilon} x^i dV$. The function x^i is "odd" about the origin, which is the center of Σ_ε, so we would expect the integral to be near zero; in fact, this integral has order ε^{n+2} instead of the ε^{n+1} which seems natural. Again, we'll work this out by transferring $\int_{\Sigma_\varepsilon} x^i dV$ to T_pM, and using the symmetry of the indicatrix there.

$$\begin{aligned}
\int_{\Sigma_\varepsilon} x^i dV &= \int_{\Sigma_\varepsilon} x^i \omega(x) dx \\
&= \int_{\Sigma_\varepsilon} x^i \omega(p) + \omega_j(p) x^i x^j + \frac{1}{2}\omega_{jk}(p) x^i x^j x^k + o(|x|^4) dx \\
&= \int_{I_\varepsilon} \Big[\left(X^i - \frac{1}{2}\Gamma^i_{jk}(X) X^j X^k \right) \omega(p) \\
&\quad + \omega_j(p) \left(X^i - \frac{1}{2}\Gamma^i_{lk}(X) X^l X^k \right) \left(X^j - \frac{1}{2}\Gamma^j_{pq}(X) X^p X^q \right) \\
&\quad + \frac{1}{2}\omega_{jk}(p) \left(X^i - \frac{1}{2}\Gamma^i_{pq}(X) X^p X^q \right) \left(X^j - \frac{1}{2}\Gamma^j_{rs} X^r X^s \right) \\
&\quad \times \left(X^k - \frac{1}{2}\Gamma^k_{tu}(X) X^t X^u \right) \Big] \left(1 - \Gamma^j_{jk}(X) X^k + c_{jk} X^j X^k \right) dX \\
&= \int_{I_\varepsilon} \omega(p) X^i + d_{ij} X^i X^j + o(|X|^3) dX
\end{aligned}$$

for some coefficient d_{ij}. Again, oddness of X^i ensures that $\int_{I_\varepsilon} \omega(p) X^i dX$ vanishes, and homogeneity allows us to evaluate I_ε solely on I by adding an appropriate factor:

$$= d_{ij} \left(\int_I X^i X^j dX \right) \varepsilon^{n+2} + d_{n+3}\varepsilon^{n+3} + o(\varepsilon^{n+4}),$$

for some constant d_{n+3}. The important result here is that we don't need any terms involving ε^{n+1}, so

$$\left| \int_{\Sigma_\varepsilon} x^i dV \right| \leq d_{n+2}\varepsilon^{n+2} + o(\varepsilon^{n+4}).$$

With Cases A and B worked out, the method of finding expressions for $\int_{\Sigma_\varepsilon} x^\alpha dV$ is clear for any value of $|\alpha|$: move to the tangent space and transfer all integrals to the unit indicatrix. By homogeneity, the coefficients of ε^{n+k} will disappear if k is odd, or if $k < |\alpha|$. The first conclusion assures us that integrals of the form $\int_{\Sigma_\varepsilon} x^i dV$ are actually of the same order of magnitude as the integrals $\int_{\Sigma_\varepsilon} x^i x^j dV$, and the second conclusion allows us to ignore higher-order terms when they appear in integrands.

Furthermore, the expression (6) from Case A is interesting because of its parallels with the Riemannian case. The first term in both the Finsler and Riemannian case is $\kappa_n \varepsilon^n$, which is simply the Euclidean volume of a sphere of radius ε in $\mathbb{R}^n$, so even in a Finsler space, volume measurement doesn't differ dramatically from the Euclidean case. (This is actually a direct consequence of using the Hausdorff measure.) Also, in both cases the coefficient of ε^{n+1} vanishes. The most interesting coefficient is c_{n+2}, which measures how our space differs from Euclidean space. In the Riemannian case, this coefficient (up to a constant) is actually the scalar curvature [GV]. In the 2-dimensional case, this is clear from Puiseux's formula

$$\text{Gaussian Curvature} = \lim_{\varepsilon \to 0} \frac{12}{\pi} \frac{\pi\varepsilon^2 - \operatorname{Vol}(\Sigma_\varepsilon)}{\varepsilon^4}. \tag{III.4-7}$$

Since expression (6) is invariant by its very construction, the foregoing clearly suggests thinking of the second coefficient as a scalar curvature for Finsler spaces.

Definition 1. Define the *Puiseux curvature* P at x of the Finsler manifold (M, F) to be

$$P = \alpha_n c_{n+2},$$

t where $\alpha_n = \frac{6(n+2)\Gamma\left(\left(\frac{n+2}{2}\right)\right)}{\pi^{\left(\frac{n}{2}\right)}}$ is taken from [Hotelling, §5].

We can get an expression for the Puiseux curvature by starting with (III.4-5) and replacing $\omega(p)$ with $\omega(p) = \frac{\kappa_n}{\int_I dX}$:

$$\begin{aligned} P &= \alpha_n c_{n+2} \\ &= \alpha_n \int_I \Bigg[-\frac{1}{2}\Gamma^i_{jk}\frac{\partial}{\partial x^l}\left(\int_I dX\right)^{-1} + \frac{1}{2}\frac{\partial}{\partial x^i}\frac{\partial}{\partial x^k}\left(\int_I dX\right)^{-1} \\ &\quad - \frac{\partial}{\partial x^j}\left(\int_I dX\right)^{-1}\Gamma^l_{lk} + \left(\int_I dX\right)^{-1} \\ &\quad \times \left(\Gamma^a_{ak}\Gamma^b_{bj} - \frac{1}{3}\Gamma^a_{bk}\Gamma^b_{aj} - \frac{1}{3}\Gamma^i_{ik|j} - \frac{1}{6}\Gamma^i_{jk|i} + \frac{1}{3}\Gamma^i_{ik}\Gamma^l_{jk} - \frac{1}{2}\frac{\partial}{\partial X^i}\Gamma^i_{jk}\right)\Bigg] \\ &\quad \times X^j X^k dX, \end{aligned}$$

Lemma 4. $P : M \longrightarrow \mathbb{R}$ *given by* $x \mapsto$ *Puiseux curvature at* x *is* C^∞.

Proof. The expression for c_{n+2} involves integrals over I, but an integral over I is equal to an integral over $I \setminus \{\vec{0}\}$. If c_{n+2} is seen as an integral over $I \setminus \{\vec{0}\}$, then all limits of integration and integrands are smooth functions of x, so P is smooth. □

4 Existence of the Laplacian and a Co-ordinate Expression for It

Though the expression for the Laplacian given in Section II is well-motivated, it is not very usable as stated. Indeed, it is not even clear *a priori* that this Laplacian actually exists for any given function. In this section, we'll use the constructions of Section III to show Δ exists, and to derive a co-ordinate expression for Δ, which will make clear that Δ is a differential operator.

Firstly, we will state formally the definition of the Laplacian.

Definition 2. Let M be a differentiable manifold with a Finsler metric F, and let $f : M \longrightarrow \mathbb{R}$ be a smooth function. Let $dV = \omega(x)dx$, where $\omega(x) = \frac{\kappa_n}{\int_I dX}$ is the coefficient of the canonical volume form. Define the Laplacian Δ at a point p by

$$\Delta f(x) = 2(n+2)\lim_{\varepsilon\to 0}\frac{1}{\varepsilon^2}\left(\frac{\int_{\Sigma_\varepsilon} f dV}{\int_{\Sigma_\varepsilon} dV} - f(x)\right),$$

or, if we write $dV = \omega(x)dx$,

$$\Delta f(x) = 2(n+2)\lim_{\varepsilon\to 0}\frac{1}{\varepsilon^2}\left(\frac{\int_{\Sigma_\varepsilon} f\omega(x)dx}{\int_{\Sigma_\varepsilon}\omega(x)dx} - f(x)\right).$$

Theorem 5. *Let M be a differentiable manifold with a Finsler metric F, and let $f : M \longrightarrow \mathbb{R}$ be a smooth function. Let p be a point in M. Then $\Delta f(p)$ exists and is a linear, 2nd order differential operator with the co-ordinate expression:*

$$\Delta f = (n+2)\Bigg(\frac{\int_I X^i X^j dX}{\int_I dX} f_{ij} + f_i\Big(\frac{2w_j}{w}\frac{\int_I X^i X^j dX}{\int_I dX} - \frac{2\int_I \Gamma^q_{jq} X^i X^j dX}{\int_I dX} - \frac{\int_I \Gamma^i_{jk} X^j X^k dX}{\int_I dX}\Big)\Bigg). \tag{IV-1}$$

Proof. Write f as a Taylor series around p

$$f(x) = f(p) + f_i(p)x^i + \frac{1}{2}f_{ij}(p)x^i x^j + \frac{1}{6}f_{ijk}(p)x^i x^j x^k + o(|x|^4).$$

Substitute this expression into the definition:

$$2(n+2)\lim_{\varepsilon\to 0}\frac{1}{\varepsilon^2}\left(\frac{\int_{\Sigma_\varepsilon} f dV}{\int_{\Sigma_\varepsilon} dV} - f(p)\right)$$

$$= 2(n+2)\lim_{\varepsilon\to 0}\frac{1}{\varepsilon^2}$$

$$\times\left(\frac{\int_{\Sigma_\varepsilon} f(p) + f_i(p)x^i + \frac{1}{2}f_{ij}(p)x^ix^j + o(|x|^3)dV}{\int_{\Sigma_\varepsilon} dV} - \frac{\int_{\Sigma_\varepsilon} f(p)dV}{\int_{\Sigma_\varepsilon} dV}\right)$$

$$= 2(n+2)\lim_{\varepsilon\to 0}\frac{\int_{\Sigma_\varepsilon} f_i(p)x^i + \frac{1}{2}f_{ij}x^ix^j + \frac{1}{6}f_{ijk}(p)x^ix^jx^k + o(|x|^4)dV}{\varepsilon^2\int_{\Sigma_\varepsilon} dV}$$

$$= 2(n+2)\lim_{\varepsilon\to 0}$$

$$\times\frac{f_i(p)\int_{\Sigma_\varepsilon} x^i dV + \frac{1}{2}f_{ij}(p)\int_{\Sigma_\varepsilon} x^ix^j dV + \frac{1}{6}f_{ijk}(p)\int_{\Sigma_\varepsilon} x^ix^jx^k dV + \int_{\Sigma_\varepsilon} o(|x|^4)dV}{\varepsilon^2\int_{\Sigma_\varepsilon} dV}.$$

Now using the integral estimates from Section III:

$$= 2(n+2)\lim_{\varepsilon\to 0}\frac{c_1\varepsilon^{n+2} + c_2\varepsilon^{n+2} + c_3\varepsilon^{n+3} + o(\varepsilon^{n+4})}{\kappa_n\varepsilon^{n+2} + c_5\varepsilon^{n+3} + o(\varepsilon^{n+4})}$$

for some constants c_i; we have suppressed higher-order terms because they don't affect the limit. In the denominator, we see that the dominating term is of order ε^{n+2} and it isn't 0 (because κ_n is the volume of the unit sphere in $\mathbb{R}^n$). Since the numerator has no smaller terms, the limit exists.

Leaving aside the last two lines, we'll continue deriving the co-ordinate expression for Δ by writing $dV = \omega(x)dx$, replacing ω by its Taylor series around p, and moving to the tangent space.

$$= 2(n+2)\lim_{\varepsilon\to 0}\frac{1}{\varepsilon^2} \times \frac{\int_{\Sigma_\varepsilon}\left(f_i(p)x^i + \frac{1}{2}f_{ij}(p)x^ix^j + o(|x|^3)\right)\omega(x)dx}{\int_{\Sigma_\varepsilon}\omega(x)dx}$$

$$= 2(n+2)\lim_{\varepsilon\to 0}\frac{1}{\varepsilon^2} \times \frac{\begin{array}{c}\int_{I_\varepsilon}\left(f_i(p)\left(X^i - \frac{1}{2}\Gamma^i_{jk}X^jX^k\right) + \frac{1}{2}f_{ij}(p)\left(X^i - \frac{1}{2}\Gamma^i_{pq}X^pX^q\right)\left(X^j - \frac{1}{2}\Gamma^j_{rs}X^rX^s\right)\\ \left(\omega(p) + \omega_i(p)\left(X^i - \frac{1}{2}\Gamma^i_{pq}X^pX^q\right)\right)\left(1 - \Gamma^r_{rs}X^s\right)dX\end{array}}{\int_{I_\varepsilon}\left(\omega(p) + \omega_i(p)\left(X^i - \frac{1}{2}\Gamma^i_{pq}X^pX^q\right)\right)\left(1 - \Gamma^r_{rs}X^s\right)dX}$$

$$= 2(n+2)\lim_{\varepsilon\to 0} \times \frac{\begin{array}{c}\int_{I_\varepsilon} f_i(p)\omega(p)X^i + \frac{1}{2}f_{ij}(p)\omega(p)X^iX^j + f_i(p)\omega_j(p)X^iX^j\\ -f_i(p)\omega(p)\Gamma^r_{rs}X^iX^s - \frac{1}{2}f_i(p)\omega(p)\Gamma^i_{jk}X^jX^kdX\end{array}}{\varepsilon^2\int_{I_\varepsilon}\left(\omega(p) + \omega_i(p)X^i + o(|X|^2)\right)dX}.$$

Again, eliminating integrands homogeneous of degree 1, and ignoring terms of higher order than $n+2$, and moving to I, we have

$$= 2(n+2)\lim_{\varepsilon\to 0} \times \frac{\varepsilon^{n+2}\int_I\left(\frac{1}{2}f_{ij}(p)\omega(p) + f_i(p)\omega_j(p) - f_i(p)\omega(p)\Gamma^r_{rj} - \frac{1}{2}f_k(p)\omega(p)\Gamma^k_{ij}\right)X^iX^jdX}{\varepsilon^{n+2}\omega(p)\int_I dX}$$

Separating into coefficients of f_{ij} and f_i,

$$\Delta f = (n+2)\left(\frac{\int_I X^iX^jdX}{\int_I dX}f_{ij} + f_i\left(\frac{2\omega_j}{\omega}\frac{\int_I X^iX^jdX}{\int_I dX} - \frac{2\int_I\Gamma^q_{jq}X^iX^jdX}{\int_I dX} - \frac{\int_I\Gamma^i_{jk}X^jX^kdX}{\int_I dX}\right)\right) \tag{IV-1}$$

This expression clearly shows that Δ is a differential operator.

Remark 3. We could have deduced that Δ is a differential operator solely from the facts that Δ is linear and decreases support, but this explicit expression will be more direct to work with.

5 Our Laplacian Generalizes the Riemannian Laplacian

As a first application of the co-ordinate expression, we will prove directly that Δ generalizes the Riemannian Laplacian. To begin with, note the leading coefficients $K^{ij} = (n+2)\frac{\int_I X^i X^j dX}{\int_I dX}$ in the co-ordinate expression for Δ. These coefficients have a considerable amount of structure, and lead to a surprising definition.

Lemma 6. *Let M be a manifold of dimension n with Finsler metric F. Define*

$$K^{ij} = (n+2)\frac{\int_I X^i X^j dX}{\int_I dX}.$$

Then K^{ij} are the components of a symmetric, twice-contravariant, positive-definite tensor.

Proof. Clearly $K^{ij} = K^{ji}$, so these components are symmetric. To see that

$$K^{ij} = \frac{\int_I X^i X^j dX}{\int_I dX} \tag{V-1}$$

are the components of a tensor, make a co-ordinate change

$$\bar{x}^i = \bar{x}^i(x) \tag{V-2}$$

on the manifold. Then we have the following relations on the tangent space:

$$\overline{X}^i = \frac{\partial x^i}{\partial \bar{x}^j} X^j \tag{V-3}$$

$$d\bar{X} = \det\left[\frac{\partial x^i}{\partial \bar{x}^j}\right] dX. \tag{V-4}$$

In the barred co-ordinate system,

$$\begin{aligned}\overline{K}^{ij} &= \frac{\int_I \overline{X}^i \overline{X}^j d\overline{X}}{\int_I d\overline{X}} \\ &= \frac{\int_I \frac{\partial x^i}{\partial \bar{x}^k} \frac{\partial x^j}{\partial \bar{x}^l} X^k X^l \det\left[\frac{\partial x^i}{\partial \bar{x}^j}\right] dX}{\int_I \det\left[\frac{\partial x^i}{\partial \bar{x}^j}\right] dX}.\end{aligned}$$

Since none of the terms $\frac{\partial x^i}{\partial \bar{x}^k}$, $\frac{\partial x^j}{\partial \bar{x}^l}$, and $\det[\frac{\partial x^i}{\partial \bar{x}^j}]$ depends on X, we can take them all outside the integral signs, and cancel the determinants, to get

$$\overline{K}^{ij} = \frac{\partial x^i}{\partial \bar{x}^k}\frac{\partial x^j}{\partial \bar{x}^l}\frac{\int_I X^k X^l dX}{\int_I dX}$$
$$= \frac{\partial x^i}{\partial \bar{x}^k}\frac{\partial x^j}{\partial \bar{x}^l} K^{kl},$$

i.e. K^{ij} are the components of a twice-contravariant tensor. □

To see that this tensor is positive-definite, simply use the symmetry to choose a co-ordinate system $\tilde{x}$ in which $\tilde{K}^{ij}$ is diagonal, that is, one for which $\tilde{K}^{ij} = 0$ if $i \neq j$. Then

$$\tilde{K}^{ii} = \frac{\int_I \tilde{X}^i \tilde{X}^i d\tilde{X}}{\int_I d\tilde{X}} > 0,$$

so K^{ij} is positive-definite.

Since a symmetric, positive-definite, twice-covariant tensor is a metric tensor, we can see the K^{ij} as the (inverse) components of a Riemannian metric. Thus we have canonically associated to our Finsler manifold a Riemannian metric.

Definition 4. The metric $K^{ij} = (n+2)\frac{\int_I X^i X^j dX}{\int_I dX}$ on a Finsler space (M, F) is called the (inverse of the) *osculating Riemannian metric.*

The existence of such a metric (which has also arisen in a different context [BCS]) leads to very interesting questions. For example, we could simply define the Laplacian on the Finsler space (M, F) to be the Laplacian gotten from thinking of (M, F) as a Riemannian space with coefficients K^{ij}, because these coefficients determine their own Laplacian. Is this Laplacian the same as the one we defined in Section II? The fact that $(n+2)\frac{\int_I X^i X^j dX}{\int_I dX}$ are the leading coefficients in the differential operator Δ suggests the two are the same, but in fact we will see that they are different. Would the Riemannian Laplacian make a good Laplacian, then? Perhaps, but it seems not to say anything about the Finsler structure and to violate our motivation for the Finsler Laplacian.

Since a Riemannian space is also a Finsler space, the K^{ij} are defined canonically on any Riemannian manifold, and presumably they have some relation to the metric tensor. This relation is very simple (and very similar to a result of [Pinsky, Eq.(4-9)]).

Lemma 7. *Let M be a Riemannian manifold of dimension n with metric tensor g_{ij}. Then*

$$g^{ij} = (n+2)\frac{\int_I X^i X^j dX^1 dX^2 \dots dX^n}{\int_I dX^1 dX^2 \dots dX^n}.$$

Proof. The above statement is really a statement at a point p, though that argument has been suppressed. To simplify calculations, choose coordinates so that

$$g^{ij}(p) = g_{ij}(p) = \delta_{ij}(p), \tag{V-5}$$

i.e. so that I at p is the unit sphere D in $\mathbb{R}^n$, Then, by symmetry of D,

$$K^{ij} = 0 \quad \text{if} i \neq j, \tag{V-6}$$

and

$$K^{11} = K^{22} = \cdots = K^{nn}, \tag{V-7}$$

so just calculate

$$K^{nn} = (n+2)\frac{\int_D (X^n)^2 dX^1 dX^2 \dots dX^n}{\int_D dX^1 dX^2 \dots dX^n}$$

$$= (n+2)\frac{\int_{-1}^{1} (X^n)^2 \left(\int_{-\sqrt{1-(X^n)^2}}^{\sqrt{1-(X^n)^2}} \int \cdots \int dX^1 \dots dX^{n-2} dX^{n-1}\right) dX^n}{\int_{-1}^{1} \left(\int_{-\sqrt{1-(X^n)^2}}^{\sqrt{1-(X^n)^2}} \int \cdots \int dX^1 \dots dX^{n-2} dX^{n-1}\right) dX^n}$$

$$= (n+2)\frac{\int_{-1}^{1} (X^n)^2 \cdot \left((n-1)\text{-sphere of radius } \sqrt{1-(X^n)^2}\right) dX^n}{\int_{-1}^{1} \left((n-1)\text{-sphere of radius } \sqrt{1-(X^n)^2}\right) dX^n}$$

$$= (n+2)\frac{\int_{-1}^{1} (X^n)^2 \kappa_{n-1} \left(\sqrt{1-(X^n)^2}\right)^{n-1} dX^n}{\int_{-1}^{1} \kappa_{n-1} \left(\sqrt{1-(X^n)^2}\right)^{n-1} dX^n}$$

Let $X = \sin\theta$ and integrate:

$$= (n+2)\frac{2\int_0^{\frac{\pi}{2}} \sin^2\theta \cos^{n-1}\theta \cos\theta d\theta}{2\int_0^{\frac{\pi}{2}} \cos^n \theta d\theta}$$

$$= (n+2)\frac{\frac{1}{(n+2)} \cos^{n+1}\theta \sin\theta\Big|_0^{\frac{\pi}{2}} + \frac{1}{(n+2)}\int_0^{\frac{\pi}{2}} \cos^n\theta d\theta}{\int_0^{\frac{\pi}{2}} \cos^n\theta d\theta}$$

$$= 1. \tag{V-8}$$

By (5) to (8),

$$K^{ij} = \delta^{ij} = g^{ij}.$$

We know g^{ij} is a twice-contravariant tensor, and since Lemma 6 proved that K^{ij} is a twice-contravariant tensor, it must be that $g^{ij} = K^{ij}$ in any co-ordinate system, i.e.

$$g^{ij} = (n+2)\frac{\int_I X^i X^j dX}{\int_I dX}.$$

□

As remarked before, the terms $(n+2)\frac{\int_I X^i X^j dX}{\int_I dX}$ appear as the leading coefficients in Δ. If our space is Riemannian, they will also appear in the first-order coefficients, and this fact allows us to prove the following folk theorem.

Theorem 8. *Let (M, F) be a Riemannian space with its canonical Laplacian $L(f)$. Define*

$$\Delta f = 2(n+2)\lim_{\varepsilon\to 0}\frac{1}{\varepsilon^2}\left(\frac{\int_{\Sigma_\varepsilon} f dV}{\int_{\Sigma_\varepsilon} dV} - f\right).$$

Then

$$\Delta f = L(f). \tag{V-9}$$

Proof. In the case of a Riemannian space with metric $F(X) = \sqrt{g_{ij}X^i X^j}$, the associated Laplacian is written in co-ordinates as

$$L(f) = g^{ij}(f_{ij} - \Gamma_{ij}^k f_k), \tag{V-10}$$

(or equivalently

$$L(f) = \frac{1}{\sqrt{g}}\frac{\partial}{\partial x^i}\left(\sqrt{g}\, g^{ij} f_j\right),$$

where $g = \det[g_{ij}]$).

Now consider the co-ordinate expression IV-(1) for Δf:

$$\Delta f = (n+2)\Bigg(\frac{\int_I X^i X^j dX}{\int_I dX} f_{ij} + f_i\Big(\frac{2\omega_j}{\omega}\frac{\int_I X^i X^j dX}{\int_I dX} - \frac{2\int_I \Gamma_{jq}^q X^i X^j dX}{\int_I dX} - \frac{\int_I \Gamma_{jk}^i X^j X^k dX}{\int_I dX}\Big)\Bigg) \tag{V-11}$$

Since the Christoffel symbols Γ^i_{jk} don't depend on tangent vectors in a Riemannian space, we can take them outside the integral signs, which leaves a common factor $(n+2)\frac{\int_I X^i X^j dX}{\int_I dX}$:

$$= (n+2)\frac{\int_I X^i X^j dX}{\int_I dX}\left(f_{ij} + f_i\left(2\frac{\omega_j}{\omega} - 2\Gamma^q_{jq}\right) - f_k\Gamma^k_{ij}\right). \qquad \text{(V-12)}$$

In a Riemannian space, the volume form $dV = \omega(x)dx$ is

$$dV = \sqrt{g}dx, \qquad \text{(V-13)}$$

so

$$\frac{\omega_j}{\omega} = \frac{1}{\sqrt{g}}\frac{\partial\sqrt{g}}{\partial x^j}. \qquad \text{(V-14)}$$

There is, however, the Riemannian identity

$$\Gamma^q_{jq} = \frac{1}{\sqrt{g}}\frac{\partial\sqrt{g}}{\partial x^j}, \qquad \text{(V-15)}$$

so two of the terms in expression (12) cancel, leaving

$$\Delta f = (n+2)\frac{\int_I X^i X^j dX}{\int_I dX}(f_{ij} - \Gamma^k_{ij}f_k). \qquad \text{(V-16)}$$

Now use Lemma 7, which asserts the identity

$$g^{ij} = (n+2)\frac{\int_I X^i X^j dX}{\int_I dX}, \qquad \text{(V-17)}$$

to convert (16) to

$$\begin{aligned}\Delta f &= g^{ij}(f_{ij} - \Gamma^k_{ij}f_k)\\ &= L(f)\end{aligned}$$

□

6 Properties of the Laplacian

Although the Laplacian we have defined does not result from any Riemannian metric, it shares some important properties with Riemannian Laplacians. In particular, Δ is "self-adjoint" in the sense that it satisfies a version of Green's Theorem for Finsler spaces, and as a result of this, it can be written in a divergence form very similar to the divergence form of the Riemannian Laplacian.

Definition 5. A linear operator $\mathcal{D}$ is *self-adjoint* on a Finsler space (M, F) if

$$\int_M (u\mathcal{D}(v))\omega(y)dy = \int_M (v\mathcal{D}(u))\omega(y)dy,$$

for any compactly supported functions u and v.

Lemma 9. *Let M be a manifold with Finsler manifold F and Δ defined as above. Then Δ is self-adjoint, i.e.*

$$\int_M (u\Delta v - v\Delta u)\omega(y)dy = 0.$$

Proof. $\int_M (u\Delta v - v\Delta u)\omega(y)dy$

$$= \int_M \Bigg(u(y)\lim_{\varepsilon\to 0} \frac{\int_{\Sigma_\varepsilon(y)}(v(x) - v(y))\omega(x)dx}{\varepsilon^2 \int_{\Sigma_\varepsilon(y)}\omega(x)dx} - v(y)\lim_{\varepsilon\to 0} \frac{\int_{\Sigma_\varepsilon(y)}(u(x) - u(y))\omega(x)dx}{\varepsilon^2 \int_{\Sigma_\varepsilon(y)}\omega(x)dx} \Bigg)\omega(y)dy$$

(where $\Sigma_\varepsilon(y)$ (resp. $\Sigma_\varepsilon(x)$)

is the ball of radius ε around y (resp. x))

$$= \lim_{\varepsilon\to 0}\frac{1}{\varepsilon^2}\int_M \int_{\Sigma_\varepsilon(y)} \left(\frac{1}{\int_{\Sigma_\varepsilon(y)}\omega(z)dz} \right) \times \Big[u(y)v(x) - v(y)u(x) - u(y)v(y) + u(y)v(y)\Big]\omega(x)\omega(y)dxdy$$

$$= \lim_{\varepsilon\to 0}\frac{1}{\varepsilon^2}\Big(\int_M\int_{\Sigma_\varepsilon(y)}\left(\frac{1}{\int_{\Sigma_\varepsilon(y)}\omega(z)dz}\right)u(y)v(x)\omega(x)\omega(y)dxdy$$
$$-\int_M\int_{\Sigma_\varepsilon(y)}\left(\frac{1}{\int_{\Sigma_\varepsilon(y)}\omega(z)dz}\right)v(y)u(x)\omega(x)\omega(y)dxdy\Big)$$
$$= \lim_{\varepsilon\to 0}\frac{1}{\varepsilon^2}\Big(\int_M\int_{\Sigma_\varepsilon(y)}\left(\frac{1}{\int_{\Sigma_\varepsilon(y)}\omega(z)dz}\right)u(y)v(x)\omega(x)\omega(y)dxdy$$
$$-\int_M\int_{\Sigma_\varepsilon(x)}\left(\frac{1}{\int_{\Sigma_\varepsilon(y)}\omega(z)dz}\right)v(y)u(x)\omega(x)\omega(y)dydx\Big)$$

(because, along the "diagonal" of $M\times M$,

$$\int_M\int_{\Sigma_\varepsilon(x)}f(x,y)dydx = \int_M\int_{\Sigma_\varepsilon(y)}f(x,y)dxdy)$$
$$= \lim_{\varepsilon\to 0}\frac{1}{\varepsilon^2}\Big(\int_M\int_{\Sigma_\varepsilon(y)}\left(\frac{1}{\int_{\Sigma_\varepsilon(y)}\omega(z)dz}\right)u(y)v(x)\omega(x)\omega(y)dxdy$$
$$-\int_M\int_{\Sigma_\varepsilon(y)}\left(\frac{1}{\int_{\Sigma_\varepsilon(x)}\omega(z)dz}\right)u(y)v(x)\omega(x)\omega(y)dxdy\Big)$$

(renaming the parameters in the second term)

$$= \lim_{\varepsilon\to 0}\frac{1}{\varepsilon^2}\int_M\Big(\int_{\Sigma_\varepsilon(y)}u(y)v(x)\omega(x)\omega(y)\Big)$$
$$\times\left(\left(\frac{1}{\int_{\Sigma_\varepsilon(y)}\omega(z)dz}\right)-\left(\frac{1}{\int_{\Sigma_\varepsilon(x)}\omega(z)dz}\right)\right)dxdy$$
$$= \lim_{\varepsilon\to 0}\frac{1}{\varepsilon^2}\int_M\Big(\int_{\Sigma_\varepsilon(y)}u(y)v(x)\omega(x)\omega(y)\Big)$$
$$\times\left(\frac{1}{\kappa_n\varepsilon^n+c_{n+2}(y)\varepsilon^{n+2}+\dots}-\frac{1}{\kappa_n\varepsilon^n+c_{n+2}(x)\varepsilon^{n+2}+\dots}\right)dxdy$$

(using (III.4-6))

$$= \lim_{\varepsilon\to 0}\frac{1}{\varepsilon^2}\int_M\Big(\frac{1}{\varepsilon^n}\int_{\Sigma_\varepsilon(y)}u(y)v(x)\omega(x)\omega(y)\Big)$$
$$\times\left(\frac{\kappa_n(c_{n+2}(x)-c_{n+2}(y))\varepsilon^{n+2}+\dots}{\kappa_n{}^2\varepsilon^n}\right)dxdy$$
$$= \lim_{\varepsilon\to 0}\frac{\varepsilon^{n+2}}{\varepsilon^{n+2}}\,\frac{1}{\kappa_n}\int_M\Big(\frac{1}{\varepsilon^n}\int_{\Sigma_1(y)}u(y)v(x)\omega(x)\omega(y)\Big)(c_{n+2}(x)-c_{n+2}(y))dxdy$$
$$= 0.$$

The last line was obtained by letting $\varepsilon \to 0$. When this happens, $x \to y$. c_{n+2} is just the Puiseux curvature we defined in Section III, where we also proved it was smooth. Since c_{n+2} is smooth, $c_{n+2}(x) \to c_{n+2}(y)$. $\square$

Self-adjointness induces an equivalent condition on the coefficients of the differential operator $\mathcal{D}$. For a second-order, linear differential operator $\mathcal{D}$ with symmetric leading coefficients, i.e.

$$\begin{aligned} \mathcal{D}(f) &= A^{ij} f_{ij} + B^i f_i, \\ A^{ij} &= A^{ji}, \end{aligned} \tag{VI-1}$$

this is indeed the case. For the Riemannian Laplacian $L(f)$, we can write

$$L(f) = \frac{1}{\sqrt{g}} \frac{\partial}{\partial x^i} \left(f_j g^{ij} \sqrt{g} \right),$$

and give a name to this convenient form.

Definition 6. We will say an operator $\mathcal{D}$ is in *divergence form* if

$$\mathcal{D}(f) = \frac{1}{\omega} \frac{\partial}{\partial x^i} (f_j h^{ij} \omega),$$

where $\omega(x)dx$ is the canonical Finsler volume form dV, and where h^{ij} is some symmetric, contravariant 2-tensor on M.

It turns out that the two conditions of self-adjointness and divergence form are equivalent for the special class of operators we're interested in.

Lemma 10. *If $\mathcal{D}$ is a second-order linear differential operator of the form*

$$\mathcal{D}f = A^{ij} f_{ij} + B^i f_i$$

with symmetric leading coefficients, then

$$\mathcal{D} \quad \text{is self-adjoint} \quad \Longleftrightarrow \mathcal{D} \quad \text{is in divergence form.}$$

Proof. $\Longleftarrow$. Assume that $\mathcal{D}$ is in divergence form , i.e. that $\mathcal{D}(f) = \frac{1}{\omega} \frac{\partial}{\partial x^i} \left(f_j h^{ij} \omega \right)$, for some h^{ij}. Let u and v be any compactly supported smooth functions on the manifold M. Then

$$\begin{aligned}
\int_M u\mathcal{D}(v)\omega(x)dx &= \int_M u\frac{1}{\omega}\frac{\partial}{\partial x^i}\left(v_j h^{ij}\omega\right)\omega(x)dx \\
&= -\int_M u_i v_j h^{ij}\omega(x)dx \\
&\qquad \text{(integration by parts, integrating } \frac{\partial}{\partial x^i}\left(v_j h^{ij}\omega\right)dx) \\
&= \int_M v\frac{\partial}{\partial x^j}\left(u_i h^{ij}\omega\right)dx \\
&\qquad \text{(integration by parts, integrating } v_j dx) \\
&= \int_M v\frac{1}{\omega}\frac{\partial}{\partial x^j}\left(u_i h^{ij}\omega\right)\omega dx \\
&= \int_M v\mathcal{D}(u)\omega(x)dx.
\end{aligned}$$

This is just the definition of self-adjointness.

$\Longrightarrow$. Let $\mathcal{D}$ be written as in expression (1). Assume self-adjointness and choose two arbitrary compactly supported functions u and v. Then

$$\begin{aligned}
0 &= \int_M (u\mathcal{D}v - v\mathcal{D}u)\,\omega(x)dx \\
&= \int_M \left(uA^{ij}v_{ij} + uB^i v_i - vA^{ij}u_{ij} - vB^i u_i\right)\omega(x)dx && \text{(VI-2)} \\
&= \int_M \Big[-v_i\left(\frac{\partial A^{ij}}{\partial x^j}u\omega + A^{ij}u_j\omega + A^{ij}u\omega_j\right) && \text{(VI-3)} \\
&\quad + u_i\left(\frac{\partial A^{ij}}{\partial x^j}v\omega + A^{ij}v_j\omega + A^{ij}v\omega_j\right) && \text{(VI-4)} \\
&\quad + uB^i v_i\omega - vB^i u_i\Big]\omega(x)dx.
\end{aligned}$$

To achieve the three lines above, we used integration by parts on the first term in (2) (integrating $v_{ij}dx$) and on the third term in (2) (integrating $u_{ij}dx$). We can now use the symmetry of the coefficients A^{ij} to cancel the second term in (3) with the second term in (4).

$$\begin{aligned}
&= \int_M (u_i v - v_i u)\left(\frac{\partial A^{ij}}{\partial x^j} + A^{ij}\frac{\partial}{\partial x^j}(\ln\omega) - B^i\right)\omega(x)dx \\
0 &= \int_M v^2\frac{\partial}{\partial x^i}\left(\frac{u}{v}\right)\left(\frac{\partial A^{ij}}{\partial x^j} + A^{ij}\frac{\partial}{\partial x^j}(\ln\omega) - B^i\right)\omega(x)dx. \\
B^i &= \frac{\partial A^{ij}}{\partial x^j} + A^{ij}\frac{\partial}{\partial x^j}(\ln\omega) && (*)
\end{aligned}$$

because any function can be written as $v^2 \frac{\partial}{\partial x^i}\left(\frac{u}{v}\right)$. □

Lemma 11 (Sub-Lemma). *Recall that $\mathcal{D}f = A^{ij} f_{ij} + B^i f_i$. In this notation,*

$$\mathcal{D}f = \frac{1}{\omega}\frac{\partial}{\partial x^i}\left(f_j h^{ij}\omega\right) \Longleftrightarrow B^i = \frac{\partial A^{ij}}{\partial x^j} + A^{ij}\frac{\partial}{\partial x^j}(\ln\omega) \quad \text{and} \quad A^{ij} = h^{ij}.$$

Proof. $\Longrightarrow$,

$$\begin{aligned}
\mathcal{D}f &= A^{ij} f_{ij} + B^i f_i && (1)\\
&= \frac{1}{\omega}\frac{\partial}{\partial x^i}\left(f_j h^{ij}\omega\right)\\
&= f_{ij}h^{ij} + f_i\left(\frac{\partial h^{ij}}{\partial x^j} + h^{ij}\frac{\partial}{\partial x^i}(\ln\omega)\right)\\
A^{ij} &= h^{ij}, && \text{(VI-5)}\\
B^i &= \frac{\partial h^{ij}}{\partial x^j} + h^{ij}\frac{\partial}{\partial x^j}(\ln\omega)\\
&= \frac{\partial A^{ij}}{\partial x^j} + A^{ij}\frac{\partial}{\partial x^j}(\ln\omega) && \text{by (5)}
\end{aligned}$$

$\Longleftarrow$. Assume $B^i = \frac{\partial A^{ij}}{\partial x^j} + A^{ij}\frac{\partial}{\partial x^j}(\ln\omega)$.

$$\begin{aligned}
\text{Let} \qquad h^{ij} &= A^{ij}\\
\text{Then} \quad \frac{1}{\omega}\frac{\partial}{\partial x^i}\left(f_j h^{ij}\omega\right) &= \frac{1}{\omega}\frac{\partial}{\partial x^i}\left(f_j A^{ij}\omega\right)\\
&= A^{ij} f_{ij} + f_i\frac{\partial A^{ij}}{\partial x^j} + f_i A^{ij}\frac{\partial}{\partial x^j}(\ln\omega)\\
&= A^{ij} f_{ij} + B^i f_i\\
&= \mathcal{D}f.
\end{aligned}$$

(∗) now implies (by the Sub-lemma) that $\mathcal{D}f = \frac{1}{\omega}\frac{\partial}{\partial x^i}\left(f_j h^{ij}\omega\right)$ for some h^{ij}. In fact, we can see from the calculations that we must have $h^{ij} = A^{ij}$. □

Recall the co-ordinate expression for Δf:

$$\Delta f = (n+2)\left(\frac{\int_I X^i X^j dX}{\int_I dX} f_{ij} + f_i\left(\frac{2\omega_j}{\omega}\frac{\int_I X^i X^j dX}{\int_I dX} - \frac{2\int_I \Gamma^q_{jq} X^i X^j dX}{\int_I dX} - \frac{\int_I \Gamma^i_{jk} X^j X^k dX}{\int_I dX}\right)\right)$$

Lemma 6 shows that Δ is self-adjoint, and Lemma 7 gives conditions on the coefficients of Δ. In particular, the A^{ij} of Lemma 7 must just be

$$A^{ij} = K^{ij} = (n+2)\frac{\int_I X^i X^j dX}{\int_I dX}, \tag{VI-6}$$

i.e. the coefficients of the osculating Riemannian metric. The B^i, of course, are just

$$B^i = (n+2)\left(\frac{2\omega_j}{\omega}\frac{\int_I X^i X^j dX}{\int_I dX} - \frac{2\int_I \Gamma^q_{jq} X^i X^j dX}{\int_I dX} - \frac{\int_I \Gamma^i_{jk} X^j X^k dX}{\int_I dX}\right). \tag{VI-7}$$

The sub-lemma gives the condition

$$\begin{aligned} B^i &= \frac{\partial A^{ij}}{\partial x^j} + A^{ij}\frac{\partial}{\partial x^j}(\ln\omega) \\ &= \frac{\partial}{\partial x^j}\left((n+2)\frac{\int_I X^i X^j dX}{\int_I dX}\right) + (n+2)\frac{\int_I X^i X^j dX}{\int_I dX}\left(\frac{\partial}{\partial x^j}\ln\omega\right). \end{aligned}$$

Using $\omega = \frac{\kappa_n}{\int_I dX}$,

$$\begin{aligned} \frac{1}{n+2}B^i &= \frac{\frac{\partial}{\partial x^j}\int_I X^i X^j dX}{\int_I dX} \\ &\quad - \frac{\int_I X^i X^j dX \cdot \frac{\partial}{\partial x^j}\int_I dX}{\left(\int_I dX\right)^2} + \frac{\int_I X^i X^j dX}{\int_I dX}\frac{\partial}{\partial x^j}\left(\ln\left(\frac{\kappa_n}{\int_I dX}\right)\right) \\ &= \frac{\frac{\partial}{\partial x^j}\int_I X^i X^j dX}{\int_I dX} - \frac{\int_I X^i X^j dX \cdot \frac{\partial}{\partial x^j}\int_I dX}{\left(\int_I dX\right)^2} \\ &\quad - \frac{\int_I X^i X^j dX}{\int_I dX}\frac{1}{\int_I dX}\cdot\frac{\partial}{\partial x^j}\int_I dX \\ &= -2\frac{\partial}{\partial x^j}\left(\ln\left(\int_I dX\right)\right) \\ &\quad \times \frac{\int_I X^i X^j dX}{\int_I dX} + \frac{\frac{\partial}{\partial x^j}\int_I X^i X^j dX}{\int_I dX}. \end{aligned} \tag{VI-8}$$

Of course from (7), we already know

$$\frac{1}{n+2}B^i = \frac{2\omega_j}{\omega}\frac{\int_I X^i X^j dX}{\int_I dX} - \frac{2\int_I \Gamma^q_{jq} X^i X^j dX}{\int_I dX} - \frac{\int_I \Gamma^i_{jk} X^j X^k dX}{\int_I},$$

and if we again use $\omega = \frac{\kappa_n}{\int_I dX}$, then

$$\frac{1}{n+2}B^i = -2\frac{\partial}{\partial x^j}\left(\ln \int_I dX\right)$$
$$\times \frac{\int_I X^i X^j dX}{\int_I dX} - \frac{2\int_I \Gamma^q_{jq} X^i X^j dX}{\int_I dX} - \frac{\int_I \Gamma^i_{jk} X^j X^k dX}{\int_I dX}. \quad \text{(VI-9)}$$

Equating (8) and (9) and cancelling the common term, and multiplying through by $\int_I dX$ gives

$$-\frac{\partial}{\partial x^j}\int_I X^i X^j dX = \int_I \left(\Gamma^q_{jq}(X)X^i X^j + \frac{1}{2}\Gamma^i_{jk}(X)X^j X^k\right) dX. \quad \text{(VI-10)}$$

This, then, must be an identity in Finsler space. It would be very interesting to verify this identity directly (a fairly straightforward exercise in the Riemannian case).

If we go back to Lemma 7, we can see a way to write Δf in divergence form directly. The proof of Lemma 7 necessitates $h^{ij} = A^{ij} = K^{ij}$, the coefficients of the osculating Riemannian metric, and this is all that's needed to prove

Theorem 12. *Let M be a differentiable manifold with a Finsler metric F, and let $f : M \longrightarrow \mathbb{R}$ be a smooth function. Let $dV = \omega(x)dx$, where $\omega(x) = \frac{\kappa_n}{\int_I dX}$ is the coefficient of the canonical volume form, and $K^{ij} = (n+2)\frac{\int_I X^i X^j dX}{\int_I dX}$ is the osculating Riemannian metric. Then the Laplacian $\Delta f(p) = 2(n+2)\lim_{\epsilon\to 0}\frac{1}{\epsilon^2}\left(\frac{\int_{\Sigma_\epsilon} f dV}{\int_{\Sigma_\epsilon} dV} - f(p)\right)$ can also be written*

$$\Delta f(p) = \frac{1}{\omega(p)}\frac{\partial}{\partial x^i}(f_j K^{ij}\omega(p)). \quad \text{(VI-11)}$$

Thus Δ is a self-adjoint operator (that is, $\int_M (u\Delta v - v\Delta u)\omega(y)dy = 0$), in divergence form.

This provides an interesting interpretation of our Laplacian : we can see Δ as the Laplacian of the osculating Riemannian space with respect to a volume form which comes from the Finsler metric.

A natural question is whether the osculating Riemannian metric K^{ij} has a Riemannian volume form which is *equal* to the volume form of the Finsler space which K^{ij} osculates. An example shows that this is not generally the case.

Example: Let $M = \mathbb{R}^2$, and let F be the Minkowski metric whose every unit ball is actually a diamond. Since the terms K^{ij}, $\int_I dX$, and so on that we'll use depend solely on the indicatrix at a point, and not on any derivatives, we'll work just at the origin p and say that T_pM (which is canonically isomorphic to $\mathbb{R}^2$) has the diamond with vertices $(1,0)$, $(0,-1)$, $(-1,0)$, and $(0,1)$ as its indicatrix. Strictly speaking, this is not a Finsler unit ball, so we must bow all the edges very, very slightly, and round off the corners microscopically; since we will only be taking *integrals* over I, these adjustments will not significantly affect our final answers. Let us calculate the osculating Riemannian coefficients K^{ij}, $1 \le i, j \le 2$. By the symmetry of the indicatrix,

$$K^{12} = K^{21} = (n+2)\frac{\int_I X^1 X^2 dX}{\int_I dX} = 0, \qquad \text{(VI-12)}$$

and

$$K^{11} = K^{22}. \qquad \text{(VI-13)}$$

We'll calculate K^{11}:

$$\begin{aligned} K^{11} &= (n+2)\,\frac{\int_I X^1 X^1 dX}{\int_I dX} \\ &= \frac{4}{2}\cdot\left(4\int_0^1\int_0^{1-X^1}(X^1)^2 dX^2 dX^1\right) \\ &= 2\cdot 4\int_0^1 (X^1)^2 - (X^1)^3 dX^1 \\ &= 8\cdot\left(\frac{1}{3}(X^1)^3 - \frac{1}{4}(X^1)^4\right)\Big|_0^1 \qquad (2) \\ &= \frac{2}{3}\,. \qquad \text{(VI-14)} \end{aligned}$$

By (12), (13), and (14),

$$K^{ij} = \begin{bmatrix} \frac{2}{3} & 0 \\ 0 & \frac{2}{3} \end{bmatrix}.$$

K^{ij} are the conjugate components of a Riemannian metric. The coefficient of the volume form of that metric is

$$\begin{aligned} \kappa_n\sqrt{\det[K_{ij}]} &= \kappa_n\sqrt{\frac{1}{\det\left[K^{ij}\right]}} \\ &= \frac{\kappa_n}{\sqrt{\frac{4}{9}}} \\ &= \frac{3}{2}\,\kappa_n. \qquad \text{(VI-15)} \end{aligned}$$

The coefficient of the Finsler volume form, however, is given by

$$\frac{\kappa_n}{\int_I dX} = \frac{1}{2}\,\kappa_n. \tag{VI-16}$$

Since (15) and (16) are not equal, the Finsler space and the osculating Riemannian space have different volume forms, and therefore our Laplacian cannot in general be induced by a Riemannian metric, as was asserted earlier without proof.

An interesting question, which has not been investigated yet, is to characterize those Finsler metrics for which the Finsler volume form and the osculating Riemannian volume form coincide; one suspects that the only such metrics are Riemannian.

7 The Laplacian on Forms

So far, we have defined a Laplacian only for *functions* on a Finsler manifold, and not, as is done in Riemannian spaces, also for forms of any degree. In this section, we will extend our definition to arbitrary p-forms. Although the definitions we give are sensible, they are not as natural as in the case of functions. The reason is that it is not clear what properties of the Riemannian case we should seek to generalize. In the case of functions, we generalized the mean-value property of harmonic functions, but there is no similar property for p-forms. It seems natural to insist that the Laplacian comes from some kind of elliptic complex, so that we can get a Hodge Decomposition Theorem, but this condition is not very stringent, and many possible Laplacians satisfy it. The only other condition which generalizes easily is the Riemannian statement that a harmonic 1-form evaluated on a Killing field should be a constant [Yano, Chap.2, § 3]. The concept of Killing field generalizes almost automatically to Finsler manifolds, so we might expect the same statement to hold in a Finsler space. For a p-form α where p is arbitrary, we will arrive at the Laplacian indirectly by constructing a norm for p-forms. To define the norm for a p-form at a point, we will use an averaging process, which is in line with thinking of a Laplacian as a mean value. At a point x, we can imagine all possible unit p-vectors, that is, sets of p ordered vectors of unit length through the origin of T_xM. We can evaluate α (and square it) on any of these p-vectors. We'll define the norm of α at x as the "average" of its value over all p-vectors.

As in the Riemannian case, start in a given co-ordinate system with the p-form $\alpha = dx^{i_1} dx^{i_2} \dots dx^{i_p}$. We will evaluate this on the unit p-vectors. There are p unit vectors which can take any values in the unit ball, so we'll take as a region of integration p copies of the unit ball, and integrate over each in turn (scaling by the volumes, with a constant of $n+2$ which will be convenient later), with the integrand being the square of α evaluated on these p vectors (we take the square of α to make sure our norm is positive):

$$\begin{aligned}|\alpha| = &\int_I\int_I \cdots \int_I \Big(\alpha\,(X_1, X_2, \ldots, X_p)\Big)^2 \\ &\times \Big(\frac{n+2}{\int_I dX} dX_1\Big)\Big(\frac{n+2}{\int_I dX} dX_2\Big)\cdots\Big(\frac{n+2}{\int_I dX} dX_p\Big).\end{aligned} \tag{VII-1}$$

Now think of α as being written

$$\alpha = \alpha_{i_1 i_2 \ldots i_p}(x) dx^{i_1} dx^{i_2} \ldots dx^{i_p}, \tag{VII-2}$$

with the usual skew-symmetry in the components of α. Then

$$\begin{aligned}\Big(\alpha\,(X_1, X_2, \ldots, X_p)\Big)^2 &= \Big(\alpha_{i_1 i_2 \ldots i_p}(x) X_1^{i_1} X_2^{i_2} \ldots X_p^{i_p}\Big) \\ &\quad\times \Big(\alpha_{j_1 j_2 \ldots j_p}(x) X_1^{j_1} X_2^{j_2} \ldots X_p^{j_p}\Big) \\ &= \alpha_{i_1 i_2 \ldots i_p}(x)\alpha_{j_1 j_2 \ldots j_p}(x) X_1^{i_1} X_1^{j_1} X_2^{i_2} X_2^{j_2} \ldots X_p^{i_p} X_p^{j_p}.\end{aligned} \tag{VII-3}$$

If we substitute this into (1), then we can perform the integrations over $X_1, X_2, \ldots X_p$ *separately*, and, since the coefficients $\alpha_{i_1 i_2 \ldots i_p}(x)$ don't depend on tangent vectors, we can take them outside the integral to get

$$\begin{aligned}|\alpha|^2 &= \alpha_{i_1 i_2 \ldots i_p}\alpha_{j_1 j_2 \ldots j_p}(n+2)^p \\ &\quad\times \frac{\int_I X_1^{i_1} X_1^{j_1} dX_1}{\int_I dX_1}\frac{\int_I X_2^{i_2} X_2^{j_2} dX_2}{\int_I dX_2}\cdots\frac{\int_I X_p^{i_p} X_p^{j_p} dX_p}{\int_I dX_p} \\ &= \alpha_{i_1 i_2 \ldots i_p}\alpha_{j_1 j_2 \ldots j_p} K^{i_1 j_1} K^{i_2 j_2} \ldots K^{i_p j_p},\end{aligned} \tag{VII-4}$$

which bears a strong resemblance to the Riemannian definition $\alpha_{i_1 i_2 \ldots i_p}\alpha_{j_1 j_2 \ldots j_p} g^{i_1 j_1} g^{i_2 j_2} \ldots g^{i_p j_p}$. This gives the norm of α at a point x; we define the norm of α over the entire compact manifold M in the obvious way as

$$|\alpha|^2 = \int_M |\alpha|^2(x) dV \;\; = \int_M \alpha_{i_1 i_2 \ldots i_p}\alpha_{j_1 j_2 \ldots j_p} K^{i_1 j_1} K^{i_2 j_2} \ldots K^{i_p j_p}\omega(x) dx. \tag{VII-5}$$

We defined this norm in a particular co-ordinate system, so we should check that this definition is independent of co-ordinates. This follows from writing

$$\overline{\alpha}_{k_1k_2\ldots k_p} = \frac{\partial \overline{x}^{i_1}}{\partial x^{k_1}}\frac{\partial \overline{x}^{i_2}}{\partial x^{k_2}}\cdots\frac{\partial \overline{x}^{i_p}}{\partial x^{k_p}}\alpha_{i_1i_2\ldots i_p}, \tag{VII-6}$$

$$\overline{\alpha}_{l_1l_2\ldots l_p} = \frac{\partial \overline{x}^{j_1}}{\partial x^{l_1}}\frac{\partial \overline{x}^{j_2}}{\partial x^{l_2}}\cdots\frac{\partial \overline{x}^{j_p}}{\partial x^{l_p}}\alpha_{j_1j_2\ldots j_p}, \quad \text{VII-7)}$$

$$\overline{K}^{q_1l_1} = \frac{\partial x^{q_1}}{\partial \overline{x}^{i_1}}\frac{\partial x^{l_1}}{\partial \overline{x}^{j_1}}K^{i_1j_1}, \tag{VII-8}$$

in a barred co-ordinate system $\overline{x}^i = \overline{x}^1(x)$. Substituting (6), (7), and (8) into the integrand of (5) gives

$$\alpha_{i_1i_2\ldots i_p}\alpha_{j_1j_2\ldots j_p}K^{i_1j_1}K^{i_2j_2}\ldots K^{i_pj_p} = \overline{\alpha}_{k_1k_2\ldots k_p}\overline{\alpha}_{l_1l_2\ldots l_p}\overline{K}^{q_1l_1}\overline{K}^{q_2l_2}\ldots\overline{K}^{q_pl_p}, \tag{VII-9}$$

so that the integrand is an invariant function; clearly then, $|\alpha|$ both at a point and over the manifold, is an invariant. (As [BL, 3] remark in their paper, this invariance depends solely on the tensor (density) character of $K^{i_1j_1}K^{i_2j_2}\ldots K^{i_pj_p}\omega(x)$; they use a different tensor density to construct a norm, and indeed, there are many other possible densities, too.)

This norm contains all the information we need to construct a Laplacian on forms, with a method very similar to the Riemannian method. The key is to note that expression (8) lends itself readily to an *inner product* on p-forms. Specifically, if α and β are p-forms, with $\alpha = \alpha_{i_1i_2\ldots i_p}dx^{i_1}dx^{i_2}\ldots dx^{i_p}$, $\beta = \beta_{j_1j_2\ldots j_p}dx^{j_1}dx^{j_2}\ldots dx^{j_p}$,

$$< \alpha, \beta > |_x := \alpha_{i_1i_2\ldots i_p}\beta_{j_1j_2\ldots j_p}K^{i_1j_1}K^{i_2j_2}\ldots K^{i_pj_p},$$

$$< \alpha, \beta > := \int_M \alpha_{i_1i_2\ldots i_p}\beta_{j_1j_2\ldots j_p}K^{i_1j_1}K^{i_2j_2}\ldots K^{i_pj_p}\omega(x)dx. \tag{VII-10}$$

Now define a co-differential $\delta : \Lambda^p(M) \longrightarrow \Lambda^{p-1}(M)$, where $\Lambda^p(M)$ is the space of p-forms on M by requiring (for any p-form α and $(p-1)$-form γ)

$$< \alpha, d\gamma > = < \delta\alpha, \gamma > . \tag{VII-11}$$

As usual define $\delta f = 0$ for any *function* f. Now simply define, for any p-form α,

$$\Delta\alpha = d\delta\alpha + \delta d\alpha. \tag{VII-12}$$

This is our Laplacian on p-forms.

We should check that this new definition is compatible with the old one for the case of 0-forms, or functions.

Lemma 13. *Let $f : M \longrightarrow \mathbb{R}$ be a smooth function. Our two definitions of Laplacians agree, i.e.*

$$\delta df = \frac{1}{\omega(x)} \frac{\partial}{\partial x^j} \left(f_j K^{ij} \omega(x) \right).$$

Proof. We will calculate explicitly $\delta : \Lambda^1(M) \longrightarrow \Lambda^0(M)$. Let α be a 1-form. Condition (VII-11) requires

$$< \alpha, d\gamma > = < \delta\alpha, \gamma >,$$

where γ is a 0-form, i.e. γ is a function from M to $\mathbb{R}$. Use (VII-10), the definition of the inner product, to write

$$< \alpha, d\gamma > = \int_M \alpha_i (d\gamma)_j K^{ij} \omega(x) dx.$$

Since $d\gamma$ is just the exterior derivative of γ, $(d\gamma)_j = \gamma_j$. so

$$\begin{aligned} < \alpha, d\gamma > &= \int_M \alpha_i \gamma_j K^{ij} \omega(x) dx \\ &= \int_M \gamma \frac{\partial}{\partial x^i} (\alpha_i K^{ij} \omega(x)) dx \\ &\quad \text{(integration by parts, integrating } \gamma_j) \\ &= \int_M \gamma \frac{\partial}{\partial x^j} (\alpha_i K^{ij} \omega(x)) \omega(x) dx \\ &= < \frac{1}{\omega} \frac{\partial}{\partial x^j} (\alpha_i K^{ij} \omega(x)), \gamma > . \end{aligned}$$

Since (VII-11) determines δ uniquely, it must be that

$$\delta\alpha = \frac{1}{\omega} \frac{\partial}{\partial x^j} (\alpha_i K^{ij} \omega(x)).$$

Now let $\alpha = df$. Then $\alpha_i = f_i$, and so

$$\delta\alpha = \delta df = \frac{1}{\omega(x)} \frac{\partial}{\partial x^j} \left(f_j K^{ij} \omega(x) \right),$$

which was to be shown. □

We also note from (VII-11) that Δ as an operator on p-forms is self-adjoint. To see this, let α, β be two p-forms. Then

$$\begin{aligned} <\alpha,\Delta\beta> &=<\alpha,(d\delta+\delta d)\beta> \\ &=<\alpha,d\delta\beta>+<\alpha,\delta d\beta> \\ &=<\delta\alpha,\delta\beta>+<d\alpha,d\beta> \quad \text{(because } \delta \text{ and } d \text{ are self-adjoint)} \\ &=<d\delta\alpha,\beta>+<\delta d\alpha,\beta> \quad \text{(because } \delta \text{ and } d \text{ are self-adjoint)} \\ &=<(d\delta+\delta d)\alpha,\beta> \\ &=<\Delta\alpha,\beta>. \end{aligned}$$

In terms of integrals, we would write

$$\int_M \alpha_{i_1i_2\ldots i_p}(\Delta\beta)_{j_1j_2\ldots j_p}K^{i_1j_1}K^{i_2j_2}\ldots K^{i_pj_p}\omega(x)dx$$
$$= \int_M \beta_{i_1i_2\ldots i_p}(\Delta\alpha)_{j_1j_2\ldots j_p}K^{i_1j_1}K^{i_2j_2}\ldots K^{i_pj_p}\omega(x)dx,$$

which is a direct generalization of the condition given for functions.

A point on which the Finsler Laplacian disagrees with the Riemannian Laplacian is the case of n-forms. In the Riemannian case, the canonical volume form is always harmonic. In the Finsler case, we'll find a condition under which an n-form η is harmonic, and see that the volume form $dV = \omega(x)dx$ doesn't generally satisfy this condition. Assume.

$$\Delta\eta = 0.$$

Since $d\eta = 0$,

$$\Delta\eta = d\delta\eta = 0.$$

Then, if β is any n-form,

$$\begin{aligned} 0 &=<\beta,d\delta\eta> \\ &=<\delta\beta,\delta\eta> \\ &=<d\delta\beta,\eta> \\ &=\lambda(x)<dx,\eta(x)dx> \end{aligned}$$

for some function $\lambda(x): M \longrightarrow \mathbb{R}$, where $\lambda(x) = d(\delta\beta)$,

$$\begin{aligned} &= \int_M \lambda(x)\eta(x)\det[K^{ij}]\omega(x)dx \\ &= \int_M \eta(x)\det[K^{ij}]\omega(x)(\lambda(x)dx) \\ &= \int_M \eta(x)\det[K^{ij}]\omega(x)d(\delta\beta). \end{aligned}$$

If η is the canonical volume form $\omega(x)dx$, then

$$0 = \int_M \omega(x) \det[K^{ij}]\omega(x)d(\delta\beta). \tag{VII-13}$$

In the Riemannian case. $\det[K^{ij}] = \frac{1}{g}$, $\omega(x) = \sqrt{g}$, and $\omega(x)\det[K^{ij}]\omega(x) = 1$, so this expression is just the integral of a closed form, which must be 0. In the Finsler case, however, $\omega(x)\det[K^{ij}]\omega(x)$ is not in general 1, so there is no guarantee this integral is 0. It is *possible*, however, that even in a Finsler space, $\omega(x)^2 \det[K^{ij}] = 1$ (or some other constant). This again raises the question raised at the end of Section VI: For which Finsler metrics does the volume form $\omega(x)$ coincide with the volume form $\det[K_{ij}]$ of the osculating Riemannian metric K_{ij}?

References

[BCS] Bao, David, Chern, Shiing-Shen and Shen, Zhongmin (in preparation) An Introduction to Riemann-Finsler Geometry.

[BL] Bao, David and Lackey, Brad (1996) A Hodge Decomposition Theorem for Finsler Spaces, *C. R. Acad. Sc. Paris*, **223**, 51-56.

[Busemann] Busemann, Herbert (1947) Intrinsic Area, *Annals of Math.*, **48**, 234-267.

[GV] Gray, A., and Vanhecke, L. (1979) Riemannian Geometry as Determined by the Volumes of Small Geodesic Balls, *Acta Mathematica*, **142**, no.3-4, 157-198.

[Hotelling] Hotelling, Harold (1939) Tubes and Spheres in n-Spaces and a Class of Statistical Problems, *American Journal of Mathematics*, **61**,. 440-460.

[Pinsky] Pinsky, Mark A. (1976) Isotropic Transport Process on a Riemannian Manifold, *Trans. AMS*, **218**, 353-360.

[Rund] Rund, Hanno (1959) *The Differential Geometry of Finsler Spaces*, Springer-Verlag, Berlin.

[RWW] Ruse, H. S., Walker, A.G. and Willmore, T.J. (1961) *Harmonic Spaces*, Edizioni Cremonese, Rome.

[Yano] Yano, Kentaro (1970) *Integral Formulas in Riemannian Geometry*, Marcel Dekker, New York.

THE NON-LINEAR LAPLACIAN FOR FINSLER MANIFOLDS

ZHONGMIN SHEN

1 Introduction

For a Finsler manifold (M, F), there is a canonical energy function E defined on the Sobolev space. The variation of E gives rises to a non-linear Laplacian. Although this Laplacian is non-linear, it has a close relationship with curvatures and other geometric quantities. There are two curvatures involved. The first one is the Ricci curvature, which is a Riemannian quantity, and the second one is the mean tangent curvature defined in [S2]. The mean tangent curvature is a non-Riemannian quantity. In this report, we shall briefly describe the recent developments in the study of this non-linear Laplacian.

A Finsler manifold is a smooth manifold equipped with a function $F: TM \to [0, \infty)$, which has the following properties

(i) $F(tY) = |t|F(Y), \quad \forall t \in \mathbb{R},\ Y \in TM$;

(ii) F is C^∞ on $TM \setminus \{0\}$ and for any non-zero $Y \in T_xM$, the associated quadratic form g_Y is an inner product, where

$$g_Y(U, V) := \frac{1}{2}\frac{\partial^2}{\partial s \partial t}\Big[F^2(Y + sU + tV)\Big]\Big|_{s=t=0}, \qquad \forall U, V \in TM. \tag{1}$$

Note that the restriction F_x of F is a norm in T_xM. Let $S_xM \subset T_xM$ denote the unit tangent sphere at x.

The Finsler metric F induces a metric d in a natural way, and the canonical Hausdorff measure (denoted by Vol) takes the following form [B]

$$d\mathrm{Vol} = \sigma(x) dx^1 \wedge \cdots \wedge dx^n = \frac{\mathrm{Vol}[\mathbb{B}^n(1)]}{\mathrm{Vol}[\mathrm{B}^n_x(1)]} dx^1 \wedge \cdots \wedge dx^n, \tag{2}$$

where $\mathbb{B}^n(r)$ denotes the standard r-ball in the Euclidean space $\mathbb{R}^n$ and

$$\mathrm{B}^n_x(r) := \Big\{(y^i) \in \mathbb{R}^n : \ F\Big(y^i \frac{\partial}{\partial x^i}\Big|_x\Big) < r\Big\}.$$

We shall study the following canonical energy function

$$E(u) = \frac{\int_M F^*(du)^2 d\mathrm{Vol}}{\int_M u^2 d\mathrm{Vol}}, \qquad \forall 0 \neq u \in C^1(M),$$

where $F^* : T^*M \to [0, \infty)$ is the Finsler metric dual to F.

Let $\mathrm{W}^{1,2}(M)$ denote the Sobolev space consisting of measurable functions with $f, df \in \mathrm{L}^2(M)$. Let

1. $\mathcal{V} := \{u \in \mathrm{W}^{1,2}(M) : \int_M u d\mathrm{Vol} = 0\}$ if M is closed (compact without boundary);
2. $\mathcal{V} := \{u \in \mathrm{W}^{1,2}(M) : u_{|\partial M} = 0\}$ if M is compact with nonempty boundary $\partial M \neq \emptyset$ (the Dirichlet problem).

E can be extended to be a function on $\mathcal{V}$. Further, E is differentiable on $\mathcal{V}$.

For a C^1 function u, the gradient $Y = \mathrm{grad}\, u$ is determined by

$$du(V) = g_Y(Y, V), \qquad \forall V \in TM.$$

We have

$$D_u E(\varphi) = -2 \frac{\int_M \left(\Delta u + \lambda\, u\right) \varphi \, d\mathrm{Vol}}{\int_M u^2 \, d\mathrm{Vol}}, \qquad \forall \varphi \in \mathcal{V}, \tag{3}$$

where $E(u) = \lambda$ and Δu is given by

$$\int_M \varphi \Delta u \, d\mathrm{Vol} := -\int_M d\varphi(\mathrm{grad}\, u) d\mathrm{Vol}, \qquad \forall \varphi \in C_c^\infty(M) \tag{4}$$

Definition 1.1. *Critical values λ of E are called the eigenvalues of M and the corresponding critical points u are called the eigenfunctions of M.*

It is easy to see that

$$\lambda_1 = \inf_{u \in \mathcal{V} \setminus \{0\}} \frac{\int_M F^*(du)^2 d\mathrm{Vol}}{\int_M u^2 d\mathrm{Vol}} \tag{5}$$

is the smallest eigenvalue of M.

One can use the min-max method to define the (possibly all) eigenvalues of M, by choosing a particular dimension-like function on the subsets of $\mathcal{V}$. The dimension-like function in [Str] is called the *Krasnoselskii genus.*

It is not clear that the set of eigenvalues of M with Krasnoselskii genus contains all eigenvalues of M (i.e., the critical values of E).

In [G2], M. Gromov introduces several dimension-like functions. One of them is the *essential dimension.* He defines eigenvalues in a different way, using the spectral function $N(\lambda)$. According to [G2], $N(\lambda)$ can be bounded by other geometric quantities, provided that the dimension-like function is "subadditive". Note that the Krosanolsekii genus and the essential dimension are subadditive.

2 Regularity of Eigenfunctions

Let $u \in \mathcal{V}$ be an eigenfunction corresponding to λ. In virtue of (3), we can rewrite $D_u E = 0$ with $E(u) = \lambda$ as follows

$$\int_M \Big(\Delta u + \lambda u\Big)\varphi \, d\mathrm{Vol} = 0, \qquad \forall \varphi \in \mathcal{V}. \tag{6}$$

When M is compact without boundary, $\int_M \Delta u \, d\mathrm{Vol} = 0$ by (4). Thus (6) holds for $\varphi = constant$, then it holds for any $\varphi \in C_c^\infty(M)$. Now (6) can be simply written as

$$\Delta u + \lambda \, u = 0. \tag{7}$$

A natural question arises: Is every eigenfunction u smooth ?

Example [GS]: Let $u_o = \chi(|x|)$ be the eigenfunction corresponding to the first eigenvalue $\lambda_1(\mathbb{B}^n(r))$ of the standard r-ball $\mathbb{B}^n(r)$ in the Euclidean space $\mathbb{R}^n$, where $|\cdot|$ denotes the Euclidean norm. Let (R^n, F) be an arbitrary Minkowski space and $\mathrm{B}^n(r)$ denote the r-ball. Define

$$u(x) := \chi(F(x)).$$

We claim that $\lambda_1(\mathrm{B}^n(r)) = \lambda_1(\mathbb{B}^n(r))$ and u is an eigenfunction corresponding to $\lambda_1(\mathrm{B}^n(r))$.

First one can easily verify that

$$E(u) = \lambda_1(\mathbb{B}^n(r)).$$

Second, we have the Brunn inequality for functions with compact support on the norm space (R^n, F).

$$\left\{ \int_{\mathrm{R}^n} |f|^{\frac{n}{n-1}} d\mathrm{Vol} \right\}^{\frac{n-1}{n}} \leq C(n) \int_{\mathrm{R}^n} F^*(du) d\mathrm{Vol}, \qquad \forall u \in C_c^\infty(\mathrm{R}^n),$$

where $C(n) = \mathrm{Vol}(\mathbb{B}^n(1))^{\frac{n}{n-1}}/\mathrm{Vol}(\mathbb{S}^{n-1}(1))$. By the Brunn inequality and the special co-area formula for Minkowski spaces, one can easily extend the Faber-Krahn inequality to Minkowski spaces: For any compact domain $\Omega \subset (\mathrm{R}^n, F)$ and a ball $\mathbb{B}^n(r) \subset \mathbb{R}^n$ with $\mathrm{Vol}(\Omega) = \mathrm{Vol}(\mathbb{B}^n(r))$,

$$\lambda_1(\Omega) \geq \lambda_1(\mathbb{B}^n(r)). \tag{8}$$

Equality holds if and only if $\Omega = \mathrm{B}^n(r)$. Thus

$$\lambda_1(\mathrm{B}^n(r)) = \lambda_1(\mathbb{B}^n(r)),$$

and $u(x) = \chi(F(x))$ is indeed an eigenfunction corresponding to $\lambda_1(\mathrm{B}^n(r))$. Clearly, u is C^∞ at $x \neq 0$ and only $C^{1,1}$ at $x = 0$.

The above example shows that one can not expect a better regularity of eigenfunctions than $C^{1,1}$. At present, we can prove the following

Theorem 2.1. (Ge-Shen [GS]) *Let $u \in \mathcal{V}$ be a weak solution of (7). Then $u \in C^{1,\alpha}(M)$ for some $0 < \alpha < 1$, and $u \in C^\infty(M \setminus K)$ where $K = \{du = 0\}$. Further, if $\partial M \neq \emptyset$ and u is an eigenfunction corresponding to λ_1, then either $u > 0$ or $u < 0$ inside M.*

3 The Laplacian of Distance Functions

In order to study the relationship between eigenvalues and curvatures, we first investigate the Laplacian of distance functions.

Let $d_B(x) = d(B, x)$ be the distance function from a compact subset $B \subset M$. It is easy to show that $F(\mathrm{grad}\, d_B) = 1$ on the subset where d_B is smooth. A function ρ is called a *distance function* if $F(\mathrm{grad}\, \rho) = 1$.

Lemma 3.1. ([S3]) *Let $Y = \mathrm{grad}\, \rho$ be the gradient of a smooth distance function on an open subset $\mathcal{U}$. Then integral curves of Y in $\mathcal{U}$ are unit speed geodesic of F. Further, ρ is a distance function of g_Y and its gradient $\overline{\mathrm{grad}}\, \rho = \mathrm{grad}\, \rho$. Thus, the integral curves of Y are unit speed geodesics of g_Y, too. Here $\overline{\mathrm{grad}}\, \rho$ denotes the gradient of ρ with respect to g_Y.*

A unit vector field is called a *geodesic field* if its integral curves are all unit speed geodesics. By Lemma 3.1, the gradient of a distance function is a geodesic field. Conversely, one can show that every geodesic field is the gradient of a distance function.

Let $\mathcal{U}$ be an open subset and $i : N \to \mathcal{U}$ be an embedded hypersurface. At every point $x \in N$, there is a unique unit vector (up to a sign) $Y_x \in T_xM$ such that $\theta := g_{Y_x}(Y_x, \cdot)$ vanishing on T_xN. Y_x is called a *normal vector* to N. Let $\omega^1 = \omega, \omega^2, \cdots, \omega^n$ be an orthonormal co-frame for T_x^*M with respect to g_{Y_x} and put

$$d\mathrm{Vol} = \sigma_x \omega^1 \wedge \cdots \wedge \omega^n.$$

Then, the induced volume form dA on N is given by

$$dA = \sigma_x i^*(\omega^2 \wedge \cdots \wedge \omega^n). \tag{9}$$

Clearly, when F is Riemannian, dA is the volume form of the induced Riemann metric on N. In general, dA is not the volume form of the induced metric on N. However, it has the following important property. If $U_\varepsilon(N)$ is an ε-neighborhood of N, then

$$\mathrm{Vol}(N) = \lim_{\varepsilon \to 0^+} \frac{\mathrm{Vol}(U_\varepsilon(N))}{2\varepsilon}.$$

This shows that dA in (9) is a natural volume form for hypersurfaces. There are other reasons to support this view. For example, we still have the co-area formula for Finsler manifolds [S3]: For any $f \in C_c^0(M)$ and any $\chi \in C^1(M)$

$$\int_M f d\mathrm{Vol} = \int_{-\infty}^{\infty} \Big[\int_{\chi^{-1}(t)} \frac{f}{F(\mathrm{grad}\,\chi)} dA \Big] dt. \tag{10}$$

Suppose that ρ is a smooth distance function on $\mathcal{U}$ such that $N \subset N_0 = \rho^{-1}(0)$. Then $Y := \mathrm{grad}\,\rho$ restricted to N is a normal vector field. The integral curves of Y induce embeddings $\Phi_t : N \to N_t := \rho^{-1}(t)$ for small t. Let dA_t denote the induced volume form on N_t. Put

$$(\Phi_t)^* dA_t = \Theta(x,t) dA.$$

The *mean curvature* Π_x of N at x is then defined by

$$\Pi_x = \frac{d}{dt}\Big[\ln \Theta(x,t) \Big]\Big|_{t=0}. \tag{11}$$

Lemma 3.2. ([S3]) *Let* ρ *be a smooth distance function on* $\mathcal{U}$ *and* $N \subset N_0 = \rho^{-1}(0)$. *Then*

$$\Delta\rho(x) = \Pi_x, \quad \forall x \in N. \tag{12}$$

In [BP], Belletini and Paolini define the notion of mean curvature in a different way. They also establish the same identity as (12). Thus their definition coincides with (11).

We now define the mean tangent curvature (which is called the mean covariation in [S2]). Let $\gamma : (-\varepsilon, \varepsilon)$ be a constant speed geodesic with $\gamma(0) = x$ and $\frac{d\gamma}{dt}(0) = Y_x$. Let $\{(e_t)_i\}$ be an orthonormal basis along γ with respect to $g_{\dot\gamma(t)}$. Let $\{\omega_t^i\}$ be the dual co-frame along γ. Put

$$dVol|_{\gamma(t)} = \sigma(t)\omega_t^1 \wedge \cdots \wedge \omega_t^n.$$

The *mean tangent curvature* is defined by

$$\mathrm{H}(Y_x) := \frac{d}{dt}\Big(\ln \sigma(t)\Big)|_{t=0}. \tag{13}$$

Roughly speaking, the mean tangent curvature H tells us the average rate of changes of (T_xM, F_x).

Proposition 3.1. ([S2]) *For any Berwald metric,* $\mathrm{H} = 0$.

We do not know any Finsler metrics with $\mathrm{H} = 0$, other than Berwald metrics. We have

Lemma 3.3. ([S3]) *For any smooth distance function* ρ *on* $\mathcal{U}$ *with* $Y = \operatorname{grad}\rho$,

$$\Delta\rho - \bar\Delta\rho = \mathrm{H}(Y) \tag{14}$$

where $\bar\Delta$ *denotes the Laplacian of* g_Y.

4 Some Estimates on $\Delta\,\rho$

The identity (14) is important, by which one can estimate $\Delta\,\rho$ under various curvature bounds.

Let $Y_x \in T_xM$ be a unit vector. One can extend it to a unit geodesic field Y in a neighborhood of x. Let g_Y be the associated Riemann metric on $\mathcal{U}$. Y must be the geodesic field of g_Y, too. Let $\bar R$ denote the Riemann curvature tensor of g_Y. Define a linear operator $\mathrm{R}_{Y_x} : T_xM \to T_xM$ by

$$\mathrm{R}_{Y_x}(V) := \bar R(V, Y_x)Y_x, \qquad \forall V \in T_xM. \tag{15}$$

Lemma 4.1. ([S1]) R_{Y_x} *is well-defined, i.e., it is independent of the local extension* Y *of* Y_x.

The arguments for Lemma 4.1 show that R_{Y_x} is just the *flag curvature* of F in the direction of Y_x. We define the Ricci curvature $\mathrm{Ric}(Y_x)$ of F in the direction of Y_x by

$$\mathrm{Ric}(Y_x) = \overline{\mathrm{Ric}}(Y_x), \tag{16}$$

where $\overline{\text{Ric}}$ denotes the Ricci curvature of g_Y. There are non-Riemannian Finsler metrics on S^2 with $R = 1$, i.e., $R_Y(V) = V - g_Y(Y, V)Y$ (see [Br]). This is why we have to consider an extra quantity H. The author conjectures that Finsler metrics with $R = 1$ and $H = 0$ must be Riemannian.

Now we consider the Laplacian of distance functions. Let $\rho(z) := d(x, z)$ be the distance function from x. Let $\mathcal{C}_x$ denote the cutlocus of x and let $\Omega_x = M \setminus \mathcal{C}_x$. It is known that ρ is C^∞ in $\Omega_x \setminus \{x\}$. By Lemma 3.1, we know $Y = \text{grad}\,\rho$ is a unit geodesic field. Further, g_Y is a C^∞ Riemann metric with an isolated conical singular point x [S1]. Since g_Y has a nice singular point, we still have the following

Lemma 4.2. ([S3]) *Let $\rho(z) = d(x, z)$. At $z = \exp_x(\rho Y_x)$, where $Y_x \in S_x M$,*

$$\bar{\Delta}\rho(z) = \frac{n-1}{\rho} - \frac{1}{3}\overline{\text{Ric}}(Y_x)\,\rho + o(\rho). \tag{17}$$

Another important distance function is the distance function to the boundary N of a compact Finsler manifold M, that is, $\rho(z) := d(N, z)$. It is known that ρ is C^∞ on $\Omega_N := M \setminus \mathcal{C}_N$ for some closed subset $\mathcal{C}_N$ of zero Hausdorff measure. $Y := \text{grad}\,\rho$ is a unit geodesic field on Ω_N and it is normal to N. Let $\overline{\Pi}(:= \bar{\Delta}\rho|_N)$ denote the *Riemannian* mean curvature of N with respect to g_Y. Then it follows from (12)(14) that

$$\Pi_x = \overline{\Pi}_x + \text{H}(Y_x), \qquad \forall x \in N.$$

The following lemma is known in Riemann geometry.

Lemma 4.3. *Let N be the boundary of a compact Finsler manifold M, $\rho(z) = d(N, z)$. At $z = \exp_x(\rho Y_x)$, where Y_x is normal to N,*

$$\bar{\Delta}\rho(z) = \overline{\Pi}_x - \left[\|\bar{A}_x\|^2 + \overline{\text{Ric}}(Y_x)\right]\rho + o(\rho), \tag{18}$$

where $\bar{A}$ denotes the normal curvature of N with respect to g_Y.

In order to estimate the Laplacian of a distance function, one needs the following important inequality which follows from the Riccati equation for the Riemann metric g_Y. Namely,

$$Y(\bar{\Delta}\rho) + \frac{1}{n-1}(\bar{\Delta}\rho)^2 \le -\overline{\text{Ric}}(Y). \tag{19}$$

Let $\mathrm{s}_\lambda(t)$ denote the unique solution to

$$\mathrm{s}_\lambda''(t) + \lambda\,\mathrm{s}_\lambda(t) = 0, \qquad \mathrm{s}_\lambda(0) = 0,\ \mathrm{s}_\lambda'(0) = 1.$$

We get from (14), (17) and (19) the following

Theorem 4.1. ([S3]) *Let $\rho(z) = d(x, z)$ be the distance function from z. Suppose that the Ricci curvature satisfies* $\mathrm{Ric} \geq (n-1)\lambda$. *Then on $\Omega_x \setminus \{x\}$*

$$\Delta\rho \leq (n-1)\frac{\mathrm{s}'_\lambda(\rho)}{\mathrm{s}_\lambda(\rho)} + \mathrm{H}(\mathrm{grad}\,\rho). \tag{20}$$

In [S2], we extend the Bishop-Gromov volume comparison theorem to Finsler manifolds. In virtue of (11) and (12), one can also prove this volume comparison theorem by (10) and (20).

By the same argument as for Theorem 4.1, we get from (14), (18) and (19) the following

Theorem 4.2. ([S3]) *Let $\rho(z) = d(N, z)$ be the distance function from a compact hypersurface N. Suppose that the Ricci curvature satisfies* $\mathrm{Ric} \geq (n-1)\lambda$, *and the Riemannian mean curvature of N satisfies*

$$\overline{\Pi} \leq (n-1)\tau. \tag{21}$$

Then on Ω_N,

$$\Delta\rho \leq (n-1)\frac{\tau\,\mathrm{s}'_\lambda(\rho) - \lambda\mathrm{s}_\lambda(\rho)}{\tau\,\mathrm{s}_\lambda(\rho) + \mathrm{s}'_\lambda(\rho)} + \mathrm{H}(\mathrm{grad}\,\rho). \tag{22}$$

An easy application of (22) is a generalization of the Heintze-Karcher comparison theorem [HK].

Theorem 4.3. ([S3]) *Let (M, F) be a compact Finsler n-manifold with non-empty boundary N. Suppose that the Ricci curvature satisfies*

$$\mathrm{Ric} \geq (n-1)\lambda, \qquad |\mathrm{H}| \leq (n-1)\delta, \tag{23}$$

and the Riemannian mean curvature of N satisfies (21). Then

$$\mathrm{Vol}(M) \leq \mathrm{Vol}(N) \int_0^d e^{(n-1)\delta t}\Big(\tau\,\mathrm{s}_\lambda(t) + \mathrm{s}'_\lambda(t)\Big)^{n-1} dt, \tag{24}$$

where $d = \sup_{x\in M} d(x, N)$.

5 Some Estimates on λ_1

Based on the estimates on $\Delta\,\rho$, one can get lower and upper bounds on the first eigenvalue and other eigenvalues. We are particularly interested in the first eigenvalue, because it is the smallest eigenvalue.

S. Y. Cheng [Ch] proves that the first eigenvalue $\lambda_1(M)$ of a Riemann n-manifold M satisfying $\mathrm{Ric} \geq (n-1)\lambda$ and $\mathrm{Diam} = d$ is bounded from above by a constant depending only on n, λ and d. By estimates in §4, one can generalize his results to Finsler n-manifolds (M, F) satisfying

$$\mathrm{Ric} \geq (n-1)\lambda, \qquad |\mathrm{H}| \leq (n-1)\delta, \tag{25}$$

$$\mathrm{Diam} = d. \tag{26}$$

Let $\kappa = \kappa(n, \lambda, \delta, r) > 0$ be a number such that there is a decreasing function $\chi \in C^1[0,1]$, satisfying

$$\chi''(t) + (n-1)\Big[\frac{\mathbf{s}_\lambda'(t)}{\mathbf{s}_\lambda(t)} + \delta\Big]\chi'(t) + \kappa\, \chi(t) \geq 0$$

$$\chi(r) = 0, \qquad \chi'(0) = 0.$$

We have

Lemma 5.1. *Let $B(p, r)$ be an open r-ball in a complete Finsler n-manifold satisfying (25). Then*

$$\lambda_1(B(p,r)) := \inf_{u \in C_c^1(B(p,r))} \frac{\int_{B(p,r)} F^*(du)^2 \, d\mathrm{Vol}}{\int_{B(p,r)} u^2 \, d\mathrm{Vol}} \leq \kappa(n, \lambda, \delta, r). \tag{27}$$

We sketch the proof below. Let $\rho(x) := d(p, x)$ and

$$u(x) := \chi(\rho(x)). \tag{28}$$

Since $\chi' \leq 0$, we have from (20)

$$\Delta u = \chi''(\rho) + \chi'(\rho)\, \Delta\rho \geq \chi''(\rho) + (n-1)\Big[\frac{\mathbf{s}_\lambda'(\rho)}{\mathbf{s}_\lambda(\rho)} + \delta\Big]\chi'(\rho) \geq -\kappa\, u.$$

We may assume that $B(p, r)$ has a regular boundary, otherwise approximate u by functions in $C_c^\infty(B(p, r))$. We have

$$\begin{aligned}
\int_{B(p,r)} F^*(du)^2 d\mathrm{Vol} &= \int_{B(p,r)} du(\mathrm{grad}\, u) d\mathrm{Vol} \\
&= -\int_{B(p,r)} u \Delta u \, d\mathrm{Vol} \\
&\leq \kappa \int_{B(p,r)} u^2 \, d\mathrm{Vol}
\end{aligned}$$

This implies (27). □

Now we proceed to obtain an upper bound on the first eigenvalue of a closed Finsler n-manifold satisfying (25) and (26). Let $p_1, p_2 \in M$ with $d(p_1, p_2) = d$. Let $r = \frac{1}{2}d$ and $\kappa = \kappa(n, \lambda, \delta, \frac{1}{2}d)$ obtained in Lemma 5.1. Then $B(p_1, r) \cap B(p_1, r) = \emptyset$. Let $u_i(x) := \chi(d(p_i, x))$. We have

$$\int_{B(p_i,r)} F^*(du_i)^2 d\mathrm{Vol} \le \kappa \int_{B(p_i,r)} |u_i|^2 d\mathrm{Vol}. \tag{29}$$

There exist constants a_1, a_2 such that $\varphi := a_1 u_i + a_2 u_2 \neq 0$ and

$$\int_M \varphi d\mathrm{Vol} = 0. \tag{30}$$

Observe that

$$\begin{aligned}
\lambda_1(M) \int_M \varphi^2 d\mathrm{Vol} &\le \int_M F^*(d\varphi)^2 d\mathrm{Vol} \\
&= \sum_{i=1}^2 a_i^2 \int_{B(p_i,r)} F^*(du_i)^2 d\mathrm{Vol} \\
&\le \kappa \sum_{i=1}^2 a_i^2 \int_{B(p_i,r)} |u_i|^2 d\mathrm{Vol} \\
&= \kappa \int_M |\varphi|^2 d\mathrm{Vol}.
\end{aligned}$$

The equalities hold because that $\mathrm{supp}(u_1) \cap \mathrm{supp}(u_2) = \emptyset$. We conclude the following

Theorem 5.1. *Let (M, F) be a closed Finsler n-manifold (M, F) satisfying (25) and (26). Then the first eigenvalue of M satisfies*

$$\lambda_1(M) \le \kappa(n, \lambda, \delta, \frac{1}{2}d).$$

One can also obtain a lower bound on $\lambda_1(M)$ depending only on n, λ, δ, d. Li-Yau develop a technique to estimate the first eigenvalue of a Riemann manifold with (25) and (26) (in this case, $\mathrm{H} = 0$). The key idea is to apply the maximum principle to a special function constructed from the eigenfunction and its gradient (see [SY] and references there). However, their methods can not be carried out directly to estimating the first eigenvalue of a Finsler manifold, since there is no appropriate maximum principle for the non-linear Laplacian. Instead, one can use the isoperimetric method to achieve this goal. See [G1][BBG] for the arguments in the Riemannian case. It is believed that for any closed Finsler n-manifold (M, F) with (25) and (26), $\lambda_1(M) \ge c(n, \lambda, \delta, d) > 0$.

References

[B] Busemann, H. (1947) Intrinsic Area, *Ann. of Math.*, **48**, 234-267.

[BBG] Berard, P., Besson, G. et Gallot, S. (1985) Sur une inégalité isopérimétrique qui généralise celle de Paul Lévey-Gromov, *Invent. Math.*, **80**, 295-308.

[Br] Bryant, R. (1996) Finsler Structures on the 2-Sphere Satisfying $K = 1$, *Contemporary Mathematics*, **196**, 27-40.

[BP] Bellettini, G. and Paolini, M. (1996) Anisotropic Motion by Mean Curvature in the Context of Finsler Geometry, preprint.

[Ch] Cheng, S.Y. (1975) Eigenvalue Comparison Theorems and Its Geometric Applications, *Math. Z.*, **143**, 289-297.

[GS] Ge, Y. and Shen, Z. in preparation.

[G1] Gromov, M. (1980) Paul Levy's Isoperimetric Inequality, *IHES Preprint.*

[G2] Gromov, M. (1988) Dimension, Non-Linear Spectra and Width, *Lecture Notes in Mathematics*, **1317**, 132-185.

[HK] Heintze, E. and Karcher, H. (1978) A General Comparison Theorem with Applications to Volume Estimates for Submanifolds, *Ann. Sci. Ec. Norm. Super.*, **11**, 451-470.

[S1] Shen, Z. (1996) Finsler Manifolds of Constant Positive Curvature, *Contemporary Mathematics*, **196**, 83-92.

[S2] Shen, Z. (1997) Volume Comparison and Its Applications in Riemann-Finsler Geometry, *Advances in Mathematics*, **128**, 306-328.

[S3] Shen, Z. (June, 1997) Curvature, Distance and Volume in Finsler Geometry, *IHES preprint.*

[Str] Struwe, M. (1990) *Variational Methods*, Springer-Verlag.

[SY] Schoen, R. and Yau, S.T. (1994) Lectures on Differential Geometry, *Conference Proceedings and Lecture Notes in Geometry and Topology*, International Press Incorporated.

A BOCHNER VANISHING THEOREM FOR ELLIPTIC COMPLICES

BRAD LACKEY *

Throughout this paper, M will be a smooth manifold which we assume is compact, boundaryless, and oriented with volume form $\sqrt{G}\, dx$. Although we use the notation $\sqrt{G}\, dx$, this is not to suggest that this form arises from a metric structure. For our purposes, $\sqrt{G}$ is any positive tensor density.

When it comes time to discuss the connection theory of elliptic complices, we will also assume that M is equipped with an affine torsion-free connection which is compatible with the volume. It is important to note (and we will do so again later) that this connection is given *á priori*, but after all is done, it will play no role.

1 Elliptic Operators and Complices and the Hodge Decomposition Theorem

1.1 Preliminaries

All of the material in this section is standard. We present it here for the sake of completeness. See [Wa, We] for more complete treatments.

Let V, W be vector bundles over our manifold, M, and let $\Gamma(V), \Gamma(W)$ be the spaces of smooth sections of these bundles. Let $L : \Gamma(V) \to \Gamma(W)$ be a linear partial differential operator of order, say, ℓ. We wish to define what it means for L to be elliptic, and derive consequences thereof. For convenience, let us denote by $\Gamma^k(V)$ the C^k sections of V (thus $\Gamma^\infty(V) = \Gamma(V)$). Later we will discuss elliptic complices of operators.

1.2 Sobolev Spaces

Let $\{\rho_\alpha\}$ be a partition of unity subordinate to a covering of M by coordinate charts which locally trivialize V and T^*M. If $\psi \in \Gamma(V)$, then $\rho_\alpha\psi$

*Research partially supported by the Izaak Killam Trust. Parts of this material appear in the author's thesis.

is supported in a coordinate chart. Thus we can view $\rho_\alpha\psi$ as a vector valued function on $\mathbb{R}^n$ with support in the compact set $\overline{\mathrm{dom}(\rho_\alpha)}$. That is, we have $\rho_\alpha\psi \in C_c^\infty(\mathbb{R}^n, \mathbb{C}^d)$. So we may express $\rho_\alpha\psi$ in terms of its Fourier transform

$$\rho_\alpha\psi(x) = (2\pi)^{-\frac{n}{2}} \int_{\mathbb{R}^n} \widehat{\rho_\alpha\psi}(p)e^{ix\cdot p}\, dp\,. \tag{1}$$

This expression does not look particularly coordinate independent, and it is not. But this is not surprising: as the Fourier transform expresses an object in terms of its vibrational states, twisting and warping the coordinates do not treat the vibrational states kindly and hence not the transform either. Yet we achieved this expression by taking a piece of the manifold and using the diffeomorphism of it with a part of Euclidean space. So the local expression for the transform is heavily coordinate dependent, and to be fair, we should note this dependence. Since we are not at all interested in studying the transformational character of the Fourier integral, however, we will not.

One by-product of this last consideration is the identification of the p terms in (1). For fixed x, we have $T_x\mathbb{R}^n$ is canonically isomorphic to $\mathbb{R}^n$ itself, the term $x \cdot p$ is just a restatement of the natural pairing between vectors and covectors. Hence the p terms describe points in $T_x^*\mathbb{R}^n$. We still have in place the diffeomorphism between our neighborhood in M and an open set in $\mathbb{R}^n$, the pull-back of which is an isomorphism between $T_x^*\mathbb{R}^n$ and T_x^*M. Thus we may as well take the p's as coordinates of T_x^*M. For concreteness let us write out, for one time only, what the proper expression should be for (1) including all the dependence on coordinates. Say ϕ_α is the coordinate chart in question. Then (1) would, and perhaps should, read

$$\rho_\alpha\psi(x) = (2\pi)^{-\frac{n}{2}} \int_{T_x^*M} \widehat{\rho_\alpha\psi}(p, \phi_\alpha)e^{ix\cdot p}\, \phi_\alpha^* dp\,.$$

Admitting that our Fourier transform depends on the covering of M by coordinate charts, we fix one, and remove all the additional labels. We define $\hat{\psi}(p) = \sum_\alpha \widehat{\rho_\alpha\psi}(p)$, without concern for convergence as at each $x \in M$ will have only finitely many nonvanishing $\rho_\alpha\psi$ terms, hence the sum is finite, and we have the formula

$$\psi(x) = (2\pi)^{-\frac{n}{2}} \int_{T_x^*M} \hat{\psi}(p)e^{ix\cdot p}dp\,.$$

The object $\hat{\psi}$ is a (coordinate dependent) section of the bundle π^*V we get by pulling back V over T^*M. In each of our fixed coordinate systems, $\hat{\psi}$ is independent of x.

We define the sth Sobolev norm as

$$\|\psi\|_s^2 = \int_M \int_{T_x^*M} \|\hat{\psi}(p)\|_x^2 (1+\|p\|^2)^s dp\sqrt{G}\, dx$$

where the norm $\|\hat{\psi}(p)\|_x$ is that induced by pulling back the Hermitian inner product from V to π^*V. Similarly, $\|p\|$ is computed in a local coordinate system by pulling back the Euclidean norm from $T_x^*\mathbb{R}^n$. Thus we suppose and require no metric structure on T^*M itself. Although this Sobolev norm depends on the choice of partition of unity and coordinate cover, one can show that changing the partition and cover yields an equivalent norm.

Definition 1. The sth Sobolev space, $W^s(V)$ is the completion of $\Gamma(V)$ in the norm $\|\cdot\|_s$.

The two standard tools in the theory of Sobolev spaces are the Sobolev embedding theorem, and the Rellich compactness theorem. We state these without proof; proofs can be found in [Wa] or [We] for manifolds, or in [R] for $\mathbb{R}^n$.

Lemma 2 (Sobolev). *For $s \geq [\frac{n}{2}] + k + 1$, we have $W^s(V) \hookrightarrow \Gamma^k(V)$.*

Lemma 3 (Rellich). *For $t < s$, $W^s(V) \hookrightarrow W^t(V)$ completely continuously.*

A less known estimate is the "Peter-Paul" inequality.

Proposition 4. *Let $r < s < t$ and $\epsilon > 0$ be given. Then there exists a constant $k > 0$ such that $\|\psi\|_s^2 \leq \epsilon\|\psi\|_t^2 + k\|\psi\|_r^2$ for all $\psi \in W^t(V)$.*

Proof. ([Wa]) For any $\beta \geq 0$, we have $1 \leq \beta^{t-s} + (\frac{1}{\beta})^{s-r}$. Set $\beta = \epsilon^{\frac{1}{t-s}}(1 + \|p\|^2)$. Thus

$$\begin{aligned}(1+\|p\|^2)^s &\leq (1+\|p\|^2)^s[\epsilon(1+\|p\|^2)^{t-s} + \epsilon^{\frac{s-r}{s-t}}(1+\|p\|^2)^{r-s}] \\ &= \epsilon(1+\|p\|^2)^t + \epsilon^{\frac{s-r}{s-t}}(1+\|p\|^2)^r .\end{aligned}$$

Set $k = \epsilon^{\frac{s-r}{s-t}}$; multiplying both sides by $\|\hat{\psi}(p)\|_x^2$ and integrating gives the desired inequality. □

We will write $W^\infty(V) = \bigcap_{t\in\mathbb{R}} W^t(V)$ and $W^{-\infty}(V) = \bigcup_{t\in\mathbb{R}} W^t(V)$.

1.3 Elliptic Operators

Our differential operator, in local coordinates, will have the form

$$L = \sum_{|I|\le \ell} A^I(x)\partial_I \ ,$$

where we have adopted a multi-index notation: if $I = [i_1, \ldots, i_p]$ then $|I| = p$ and $\partial_I = \frac{\partial^p}{\partial x^{i_1}\cdots\partial x^{i_p}}$. In this equation, each A^I is a bundle morphism between V and W, that is, it is a matrix at each point $x \in M$.

If we apply this operator to (1), then we get

$$L\rho_\alpha\psi(x) = (2\pi)^{-\frac{n}{2}} \int_{\mathbb{R}^n} \sum_{|I|\le\ell} i^{|I|} A^I(x) p_I \widehat{\rho_\alpha\psi}(p) e^{ix\cdot p}\, dp \ .$$

Note that inside the integral, all derivatives have disappeared. What remains is a linear map, but one which depends both on x and p. That is, what has appeared is a bundle map between the pull-back spaces π^*V and π^*W over the cotangent space T^*M.

Using the Sobolev norms, we get the following inequality.

$$\begin{aligned}
\|L\psi\|_s^2 &= \int_M \int_{T_x^*M} \| \sum_{|I|\le\ell} i^{|I|} A^I p_I \hat{\psi}\|_x^2 (1+\|p\|^2)^s dp\sqrt{G}\, dx \\
&\le \int_M \int_{T_x^*M} \sum_{|I|\le\ell} \|A^I\|_x^2 \|p\|^{2|I|} \|\hat{\psi}\|_x^2 (1+\|p\|^2)^s dp\sqrt{G}\, dx \\
&\le \sum_{j=0}^{\ell} C_j \int_M \int_{T_x^*M} \|p\|^{2j} \|\hat{\psi}\|_x^2 (1+\|p\|^2)^s dp\sqrt{G}\, dx \\
&\le \sum_{j=0}^{\ell} C_j \int_M \int_{T_x^*M} (1+\|p\|^2)^j \|\hat{\psi}\|_x^2 (1+\|p\|^2)^s dp\sqrt{G}\, dx \\
&= \sum_{j=0}^{\ell} C_j \|\psi\|_{s+j}^2 \\
&\le \text{constant } \|\psi\|_{s+\ell}^2 \ .
\end{aligned}$$

In the last step, we used Rellich's lemma. Thus, $L : W^{s+\ell}(V) \to W^s(W)$ is bounded.

This estimate, we used Rellich's result to 'dispose' of dependence on the lower order terms. Note that the dependence on p of the bundle map indicated above is polynomial as well; thus, we have a special interest in its highest order term, which we call the symbol of L (many authors study all the terms in the bundle map, so they would use the term 'leading order symbol'). We will denote the symbol as $\sigma L(p) = i^\ell \sum_{|I|=\ell} A^I p_I$. The symbol dictates most of the properties of a differential operator.

We see that the symbol is a matrix valued polynomial in p which is homogeneous of degree ℓ. Let us reformulate this in a more geometric way. Let us write $T^*M \setminus \{0\}$ for the cotangent space of M with the origin of each fibre removed (often called the 'slit cotangent bundle'). Given the natural projection, $\pi : T^*M \setminus \{0\} \to M$, we can construct the pull-back bundles π^*V, π^*W. Then the symbol is just a bundle map $\sigma L : \pi^*V \to \pi^*W$.

Definition 5. A differential operator, $L : \Gamma(V) \to \Gamma(W)$, is elliptic if its symbol map $\sigma L : \pi^*V \to \pi^*W$ is a vector bundle isomorphism.

Note that the zero section is removed, so ellipticity can be restated as $\sigma L(p)$ is invertible for every nonzero $p \in T^*M$. Clearly, an operator can only be elliptic if the vector bundles on which it lives are of the same rank. With the additional assumption of ellipticity, we may prove a reverse inequality to the one above.

Proposition 6. *Let $L : \Gamma(V) \to \Gamma(W)$ be an elliptic operator, then*

$$\|\psi\|_{s+\ell} \leq \text{constant} \cdot (\|L\psi\|_s + \|\psi\|_s) \ . \tag{2}$$

Proof. ([Wa]) Say L is of order ℓ, and let us denote by $\tilde{L}$ the differential operator which consists only of the ℓ-order terms of L. Thus $\sigma\tilde{L} = \sigma L$. So as $\sigma\tilde{L}$ is an isomorphism as well and as M is compact, there exists a constant $c > 0$ so that

$$\|\sigma\tilde{L}(x,p)\xi\|_x^2 \geq c\|\xi\|_x^2$$

for all $x \in M$, $\|p\| = 1$, and $\xi \in \pi^*V$. But $\sigma\tilde{L}$ is homogeneous of degree ℓ in p so that

$$\|\sigma\tilde{L}\xi\|_x^2 \geq c\|p\|^{2\ell}\|\xi\|_x^2$$

Now,

$$\begin{aligned}
(\|\tilde{L}\psi\|_s^2 + \|\psi\|_s^2) &= \int_M \int_{T_x^*M} \left[\|\sigma\tilde{L}\hat{\psi}\|_x^2 + \|\hat{\psi}\|_x^2\right] (1+\|p\|^2)^s dp\sqrt{G}\, dx \\
&\geq \int_M \int_{T_x^*M} \left[c\|p\|^{2\ell}\|\hat{\psi}\|_x^2 + \|\hat{\psi}\|_x^2\right] (1+\|p\|^2)^s dp\sqrt{G}\, dx \\
&\geq K \int_M \int_{T_x^*M} \|\hat{\psi}\|_x^2 (1+\|p\|^2)^{s+\ell} dp\sqrt{G}\, dx \\
&= K\, \|\psi\|_{s+\ell}^2 \,,
\end{aligned}$$

where $K > 0$ is constant. Thus,

$$\|\psi\|_{s+\ell}^2 \leq \text{constant}\,(\|\tilde{L}\psi\|_s^2 + \|\psi\|_s^2) \leq \text{constant}\,(\|\tilde{L}\psi\|_s + \|\psi\|_s)^2 \;.$$

But we have, $\|\tilde{L}\psi\|_s \leq \|L\psi\|_s + \|(\tilde{L} - L)\psi\|_s \leq \|L\psi\|_s + r\|\psi\|_{s+\ell-1}$ for some constant $r > 0$ since $\tilde{L} - L$ is a differential operator of order $\ell - 1$. By the Peter-Paul inequality, we get for each small ϵ

$$\|\psi\|_{s+\ell-1}^2 \leq \epsilon^2\|\psi\|_{s+\ell}^2 + k^2\|\psi\|_s^2 \leq (\epsilon\|\psi\|_{s+\ell} + k\|\psi\|_s)^2$$

for some constant k. Inserting all this into the above,

$$\|\psi\|_{s+\ell} \leq \text{constant}\,(\|L\psi\|_s + \epsilon r\|\psi\|_{s+\ell} + kr\|\psi\|_s) \;.$$

We choose $\epsilon < \frac{1}{r\cdot\text{constant}}$, subtract the $\|\psi\|_{s+\ell}$ term to the left side of this inequality, and reset our constant to get the desired result. □

Corollary 7. *Let L be elliptic and $L\psi = \xi$ with $\xi \in W^s(W)$. If $\psi \in W^{-\infty}(V)$ then $\psi \in W^{s+\ell}(V)$.*

Proof. Let $\psi \in W^{-\infty}(V)$, that is $\psi \in W^t(V)$ for some $t \in \mathbb{R}$. If $t \geq s$, then

$$\|\psi\|_{s+\ell} \leq \text{constant}\,(\|L\psi\|_s + \|\psi\|_s) = \text{constant}\,(\|\xi\|_s + \|\psi\|_s) \;.$$

Hence $\psi \in W^{s+\ell}(V)$ as desired.

If $t < s$ then we have

$$\|\psi\|_{t+\ell} \leq \text{constant}\,(\|\xi\|_t + \|\psi\|_t) \;.$$

So $\psi \in W^{t+\ell}(V)$. Iterating, we eventually find that $t + n\ell > s$ for some integer n, so we return to the case above. □

Corollary 8. *Let L be elliptic and $L\psi = 0$. Then $\psi \in \Gamma(V)$.*

Proof. We have $\psi \in W^\infty(V)$ from above, and $W^\infty(V) \hookrightarrow \Gamma^k(V)$ for all k by the Sobolev embedding theorem. Thus $\psi \in \Gamma(V)$. □

Corollary 9. *The kernel of an elliptic operator, L, is finite dimensional.*

Proof. For $\psi \in \ker(L)$ and $s = 0$, inequality (2) reads

$$\|\psi\|_\ell \leq \text{constant}\,\|\psi\|_0 \;.$$

So, any orthonormal basis of $\ker(L)$ is bounded in $W^\ell(V)$. Yet by Rellich's lemma, $W^\ell(V) \hookrightarrow W^0(V)$ completely continuously. Therefore, any orthonormal basis has a Cauchy subsequence in $W^0(V)$, which can only occur if it is finite. □

1.4 Elliptic Complices

In the previous section we studied elliptic operators, yet our primary interest is in elliptic complices. The reader should beware that we will strictly deal with the case of covector bundles to make the closest contact with our primary example, the deRham complex (that is, the exterior algebra with the exterior derivative). We will also use freely the terminology of forms, although our bundles are completely general.

Definition 10. Let $\{V^p\}_{p\in\mathbf{Z}}$ be a collection of covector bundles over a compact and boundaryless, orientable manifold. A complex over this collection is a selection of linear differential operators,

$$\cdots \to \Gamma(V^{p-1}) \xrightarrow{d} \Gamma(V^p) \xrightarrow{d} \Gamma(V^{p+1}) \to \cdots ,$$

such that $d^2 = 0$. A complex is said to be elliptic if the associated sequence of pull-back covector bundles over $T^*M \setminus \{0\}$ induced by the symbol map,

$$\cdots \to \pi^*V^{p-1} \xrightarrow{\sigma d} \pi^*V^p \xrightarrow{\sigma d} \pi^*V^{p+1} \to \cdots$$

is an exact sequence.

Many authors require that only finitely many of the covector bundles be nonzero. This is assumed so they may "roll-up" the complex and still have finite dimensional vector bundles. We will not need to do this, so we need not assume only finitely many of the bundles are nonzero.

Now suppose that V^p is given an metric, G_p. We continue the tradition by calling the object (V^p, G_p) a Riemannian covector bundle. For such, we have the usual $\mathcal{L}_2$ inner product on $\Gamma(V^p)$, defined by

$$\langle\theta, \psi\rangle = \int_M G_p(\theta, \psi)\sqrt{G}\, dx .$$

Example 11 (Riemannian Geometry). Suppose we are given a Riemannian metric on M, say $g = g_{jk}(x)dx^j \otimes dx^k$. On p-forms, one then has the naturally induced metric

$$G_p(\theta, \psi) = \frac{1}{p!} g^{j_1 k_1} \cdots g^{j_p k_p} \theta_{j_1 \cdots j_p} \psi_{k_1 \cdots k_p} ,$$

where g^{jk} is the inverse of g_{jk}. Traditionally, one takes the prefered volume to be $\sqrt{G} = \sqrt{\det(g_{jk})}$.

Example 12 (Finsler Geometry). Suppose we are given a Finsler structure $F(x, y)$. By the standard theory, [BCS, BL, BS], the projective sphere bundle of M, SM, is equipped with a Riemannian metric and a volume form such that the volume of each fibre of the bundle is one. The metric is given by

$$g_{jk}(x,y)dx^j \otimes dx^k + h_{jk}(x,y)\delta y^j \otimes \delta y^k \,,$$

where $g_{jk}(x,y) = \frac{1}{2}\frac{\partial^2 F^2}{\partial y^j \partial y^k}$ is the fundamental tensor, $h_{jk}(x,y) = F\frac{\partial^2 F}{\partial y^j \partial y^k}$ is the angular metric tensor, and $\delta y^j = \frac{1}{F}(dy^j + N^j{}_k dx^k)$ is the vertical coframe induced by a nonlinear connection N.

This Riemannian metric defines a volume form on SM. Yet, the volume of the each S_xM, will in general vary from point to point, [BS]. If we write $Vol(x)$ for the volume of S_xM, then we may construct the normalized volume form of SM, which we write as

$$\zeta_x \wedge dx^1 \wedge \cdots \wedge dx^n \,,$$

where $\zeta_x = \frac{(-1)^{n-1}\sqrt{g}}{Vol(x)} \sum_{j=1}^n (-1)^j y^j \, dy^1 \wedge \cdots \wedge \widehat{dy^j} \wedge \cdots \wedge dy^n$.

Now, as the natural projection, $\pi : SM \to M$, is a submersion, the pull-back $\pi^* : \Omega^*(M) \to \Omega^*(SM)$ is an injection. Yet, SM is a Riemannian manifold with a given volume form, so $\Omega^*(SM)$ has a natural $\mathcal{L}^2$ structure. By identifying $\Omega^*(M)$ with it image under π^*, $\Omega^*(M)$ is a subspace of $\Omega^*(SM)$ which we then equip with the induced inner product. Writing this out, we have

$$\langle \theta, \psi \rangle = \frac{1}{p!}\int_{SM} \theta_{j_1 \cdots j_p} \psi_{k_1 \cdots k_p} \, g^{j_1 k_1} \cdots g^{j_p k_p} \sqrt{g}\, \zeta_x \wedge dx^1 \wedge \cdots \wedge dx^n \,.$$

Let $\{\rho_\alpha\}$ be a partition of unity subordinate to a trivialization of SM, and let $f, h \in \Omega^0(M)$. Then

$$\begin{aligned}
&\int_{SM} fh\,\sqrt{g}\,\zeta_x \wedge dx^1 \wedge \cdots \wedge dx^n \\
&\quad = \sum_\alpha \int_{\mathrm{dom}(\rho_\alpha)} \int_{S_xM} \rho_\alpha fh\,\sqrt{g}\,\zeta_x \wedge dx^1 \wedge \cdots \wedge dx^n \\
&\quad = \sum_\alpha \int_{\mathrm{dom}(\rho_\alpha)} fh\,\left(\int_{S_xM} \rho_\alpha \sqrt{g}\,\zeta_x\right) dx^1 \wedge \cdots \wedge dx^n \\
&\quad = \int_M fh\,\sqrt{G}\,dx^1 \wedge \cdots \wedge dx^n \,,
\end{aligned}$$

where we have introduced the tensor density on M,

$$\sqrt{G}(x) = \int_{S_xM} \sqrt{g}\,\zeta_x \,.$$

As in Riemannian geometry, we are to use this as the canonical volume form for our manifold.

This argument carries over to p-forms identically. Namely,

$$\begin{aligned}\int_{SM} \theta_{j_1\ldots j_p}\psi_{k_1\ldots k_p} g^{j_1k_1}\cdots g^{j_pk_p}\sqrt{g}\,\zeta_x \wedge dx^1\wedge\cdots\wedge dx^n \\ = \int_M \theta_{j_1\ldots j_p}\psi_{k_1\ldots k_p} G^{j_1k_1\cdots j_pk_p}\sqrt{G}\,dx\end{aligned}$$

where we now introduce the tensor on M,

$$G^{j_1k_1\cdots j_pk_p}(x) = \frac{1}{\sqrt{G(x)}}\int_{S_xM} g^{j_1k_1}\cdots g^{j_pk_p}\sqrt{g}\,\zeta_x\,. \tag{3}$$

We adopt a multi-index notation, say $I = [i_1,\ldots,i_p]$. Then the induced inner product on p-forms is,

$$\langle\theta,\psi\rangle = \frac{1}{p!}\int_M \theta_J\psi_K G^{JK}\sqrt{G}\,dx\,.$$

Recall that we normalized the volume of each fibre of SM. Thus if our Finsler space is Riemannian, $G^{j_1k_1\cdots j_pk_p} = g^{j_1k_1}\cdots g^{j_pk_p}$ recovering the Riemannian construction. Yet in the properly Finslerian case, $G^{j_1k_1\cdots j_pk_p} \neq G^{j_1k_1}\cdots G^{j_pk_p}$.

1.5 The Hodge Decomposition Theorem

The Hodge decomposition theorem is a very general theorem about elliptic complices. Yet when applied to the case of the deRham complex, the ethereal complex cohomologies are identified as the deRham cohomologies, hence carry topological information about the manifold.

Specifically, let

$$\cdots\to\Gamma(V^{p-1})\xrightarrow{d}\Gamma(V^p)\xrightarrow{d}\Gamma(V^{p+1})\to\cdots$$

be our elliptic complex of interest. Let us write

$$\begin{aligned}Z^p &= \ker(d:\Gamma(V^p)\to\Gamma(V^{p+1}))\\ B^p &= \operatorname{im}(d:\Gamma(V^{p-1})\to\Gamma(V^p))\end{aligned}$$

for the closed p-forms and exact p-forms respectively (recall that form just means a section of the appropriate covector bundle). Then, $d^2 = 0$ implies that $B^p \subseteq Z^p$. We define the cohomologies of this complex by $H^p = Z^p/B^p$ where the quotient is as vector spaces.

Theorem 13. *Let $(V^p, G_p)_{p\in\mathbf{Z}}$ be a collection of Riemannian covector bundles over a compact and boundaryless, oriented manifold with volume form $\sqrt{G}\,dx$. Let*

$$\cdots \to \Gamma(V^{p-1}) \xrightarrow{d} \Gamma(V^p) \xrightarrow{d} \Gamma(V^{p+1}) \to \cdots$$

be an elliptic complex of linear differential operators, and define the associated Laplacians $\Delta = d^*d + dd^*$ where d^* is the $\mathcal{L}_2$ adjoint of d. Then the following hold.

(a) H^p is finite dimensional;

(b) $H^p \cong \ker(\Delta : \Gamma(V^p) \to \Gamma(V^p))$;

(c) $\Gamma(V^p) = \operatorname{im} d \oplus \operatorname{im} d^* \oplus \ker \Delta$ where the summands are orthogonal.

Proof. Let us write $\mathcal{H}^p = \ker(\Delta : \Gamma(V^p) \to \Gamma(V^p))$ (usually called the harmonic p-forms). We claim that $\mathcal{H}^p$ is a finite dimensional space of smooth forms. This will follow immediately by corollaries 8 and 9 once we establish that Δ is elliptic. Fix a nonzero $\xi \in T^*M$. For ease of reading, let us denote

$$\begin{aligned} A &= \sigma(d : \Gamma(V^{p-1}) \to \Gamma(V^p))(\xi) \\ B &= \sigma(d : \Gamma(V^p) \to \Gamma(V^{p+1}))(\xi) \end{aligned}$$

for the two symbols. One can easily see that $\sigma\Delta(\xi) = AA^* + B^*B$. As we are working with an elliptic complex, we have that $\operatorname{im} A = \ker B$. Now, over each $x \in M$

$$V^p_x = \operatorname{im} A_x \oplus (\operatorname{im} A_x)^\perp = \operatorname{im} A_x \oplus (\ker B_x)^\perp = \operatorname{im} A_x \oplus \operatorname{im} B^*_x\,.$$

On $\operatorname{im} A$, $\ker A^* = (\operatorname{im} A)^\perp$ and $\ker B = \operatorname{im} A$ imply that AA^* is injective and B^*B vanishes. Identically, on $\operatorname{im} B^*$, we have AA^* vanishes and B^*B is injective. Therefore $\sigma\Delta(\xi)$ is injective at each point, and hence an isomorphism as desired.

As we now know that $\mathcal{H}^p$ is finite dimensional, (a) follows immediately from (b). To prove (b), let us take $\theta \in Z^p$ and consider the element $\theta + B^p \in H^p$ as an affine subset of $\Gamma(V^p)$. If we take closures in $W^0(V^p)$, then $\overline{\theta + B^p}$ is a closed and convex subset of the Hilbert space $\overline{Z^p}$. Therefore, this set contains a unique element of least norm, θ_0 (see for instance lemma 3.1.6 of [P]). We claim that θ_0 is in fact harmonic. To this end, let $\psi \in \Gamma(V^{p-1})$. Then by construction, the function $f(\epsilon) = \|\theta_0 + \epsilon d\psi\|^2$ has a unique minimum at $\epsilon = 0$. Yet,

$$f(\epsilon) = \|\theta_0\|^2 + 2\epsilon\langle\theta_0, d\psi\rangle + \|d\psi\|^2\,.$$

To have a minimum at $\epsilon = 0$, we must have $0 = \langle \theta_0, d\psi \rangle = \langle d^* \theta_0, \psi \rangle$. Yet, $\psi \in \Gamma(V^{p-1})$ is arbitrary, and $\Gamma(V^{p-1})$ is dense in $W^0(V^{p-1})$. Thus, $d^*\theta_0 = 0$. We already know that $d\theta_0 = 0$, therefore $\theta_0 \in \mathcal{H}^p$. We note that the map $\theta + B^p \mapsto \theta_0$ is well defined, linear, and injective by construction. To see surjectivity, just note that if θ_0 is harmonic, then $\theta_0 + B^p \mapsto \theta_0$. Therefore, (b) is established.

Finally for (c), note that Δ is self-adjoint, hence $\ker \Delta = (\operatorname{im} \Delta)^{\perp}$. Thus, $(\ker \Delta)^{\perp} = \overline{\operatorname{im} \Delta}$. Yet, we have seen that $\ker \Delta$ is finite dimensional, so $\operatorname{im} \Delta$ has finite codimension and is therefore closed. Hence, $W^0(V^p) = \mathcal{H}^p \oplus \operatorname{im} \Delta$. If we take a $\theta \in \Gamma(V^p)$ then this decomposition give an $\eta \in \mathcal{H}^p$ and $\xi \in W^0(V^p)$ such that

$$\theta = \eta + \Delta\xi \, .$$

Yet, as both θ, η are smooth, so is $\Delta\xi$ and therefore so is ξ by corollary 8. Thus $\theta = d(d^*\xi) + d^*(d\xi) + \eta$ where $d^*\xi$ and $d\xi$ are smooth. Finally, noting that $\operatorname{im} d = (\ker d^*)^{\perp}$, we see that these three terms are orthogonal and live in the appropriate summands. □

2 A Weitzenböck Formula for an Elliptic Complex

2.1 Preliminaries

Let M be our compact and boundaryless, oriented manifold with volume form $\sqrt{G}\, dx$. As usual, the volume is completely general. Suppose also that we are given a collection of Riemannian covector bundles $(V^p, G_p)_{p \in \mathbf{Z}}$, and an elliptic complex,

$$\cdots \to \Gamma(V^{p-1}) \xrightarrow{d} \Gamma(V^p) \xrightarrow{d} \Gamma(V^{p+1}) \to \cdots$$

where the connecting maps are now first order partial differential operators. As before, we will consistently abuse terminology and call the elements of $\Gamma(V^p)$, p-forms.

We will consistently denote a bundle morphism by matrix multiplication on the left. Hence our index locations will be skew from the traditional placements. Unnecessary in the analytical case, we must take extreme care as to the geometric nature of objects in this section. We will adopt an index convention to support this: we use lower case Latin indices for objects intrinsic to the base manifold, and upper case Latin indices for objects in our bundles, either covectors or forms. We will not distinguish between indices in the different grades, the context should make their use clear. Of course, we adopt this convention to also mimic the multi-index formulation of our previous examples.

Along this note, let us consider a connecting map $d : \Gamma(V^p) \to \Gamma(V^{p+1})$. For $\theta \in \Gamma(V^p)$, this map will have the local structure of

$$d\theta = B^j \partial_j \theta + S\theta ,$$

or with indices inserted as

$$d\theta = (B^j)_J{}^K \partial_j \theta_K + S_J{}^K \theta_K .$$

As $\partial_j \theta \in \Gamma(V^p \otimes T^* M)$, we see that the symbol term $B : \Gamma(V^p \otimes T^* M) \to \Gamma(V^{p+1})$. Since functions will pull through this symbol, we have $B^j(x) \in L(V^p, V^{p+1})_x \otimes T_x M$. That is, $B^j(x)$ is a vector-valued matrix on M. We see that such an object carries both the geometric character of the bundle and of the manifold.

A slightly more complicated object to consider is the symbol of the Laplacian. We have already noted that for a nonzero $\xi \in T^* M$, $\sigma\Delta(\xi)$ is homogeneous quadratic in the components of ξ. Moreover, the elliptic complex condition guarantees that $\sigma\Delta(\xi)$ is invertible (in fact, it is easy to see that it is positive definite). A natural construction is to take two covectors $\xi, \psi \in T^* M$ and polarize the symbol:

$$\sigma\Delta(\xi, \psi) = \frac{1}{4} \left(\sigma\Delta(\xi + \psi) - \sigma\Delta(\xi - \psi) \right) .$$

This object is a symmetric bilinear form on $T^* M$ with values in $\mathrm{End}(V^p)$. One might wonder what sort of positivity such an object may possess. We pause for a moment to consider this question in the abstract.

2.2 Complete Positivity

Let (V, G) be a finite dimensional real vector space with an inner product G, and let W be another finite dimensional real vector space. These are to serve the roles of the fibre of V^p and $T^* M$ above. Now let $\Sigma \in W^* \otimes W^* \otimes L(V)$ be a matrix-valued tensor. We will assume that for all $\xi, \psi \in W$, we have $\Sigma(\xi, \psi) = \Sigma(\psi, \xi)$ and that $\Sigma(\xi, \xi)$ is a positive definite G-symmetric matrix whenever $\xi \neq 0$. This tensor is like the symbol of the Laplacian.

Definition 14. A symbol, Σ, as above is *completely positive* if for any n, and collection of n linearly independent vectors, $\{\xi_\alpha\} \in W$, then $\Sigma(\xi_\alpha, \xi_\beta)$ is positive definite as an element of $M_n(L(V)) \cong L(V^n)$. Here, V^n has the usual norm, $\|(v_\alpha)\|^2 = \sum_\alpha G(v_\alpha, v_\alpha)$.

Note that for $n = 1$, this is just the usual definition of positivity which the Laplacian satisfies. We call this extended notion 'completely positive' in analogy with the situation in operator space theory. As a principle submatrix of a positive definite matrix is positive definite, it clearly suffices just to consider the case $n = \dim(W)$ in which case $\{\xi_\alpha\}$ forms a basis of W. One can also easily verify that if Σ is completely positive for a particular basis of W then it is for all bases of W. Therefore, complete positivity is a property of Σ alone.

Let us define an bilinear pairing on $W \otimes V$ by

$$\langle \xi \otimes v, \psi \otimes w \rangle_\Sigma = G(v, \Sigma(\xi, \psi)w)$$

for elementary tensors, and extending bilinearly. Clearly, from the bilinearity of both Σ and G, this pairing is well defined.

Lemma 15. *$\langle \cdot, \cdot \rangle_\Sigma$ is an inner product if and only if Σ is completely positive.*

Proof. Bilinearity is built into the definition. Symmetry is also clear:

$$\begin{aligned} \langle \xi \otimes v, \psi \otimes w \rangle_\Sigma &= G(v, \Sigma(\xi, \psi)w) = G(\Sigma(\xi, \psi)w, v) \\ &= G(w, \Sigma(\psi, \xi)v) = \langle \psi \otimes w, \xi \otimes v \rangle_\Sigma \,. \end{aligned}$$

To prove positive definiteness, let us take $\sum_\alpha \xi_\alpha \otimes v_\alpha \in W \otimes V$. Without loss of generality, we may assume that the ξ_α are linearly independent (otherwise we reduce the expression to this case). Then

$$\langle \sum_\alpha \xi_\alpha \otimes v_\alpha, \sum_\beta \xi_\beta \otimes v_\beta \rangle_\Sigma = \sum_{\alpha\beta} G(v_\alpha, \Sigma(\xi_\alpha, \xi_\beta)v_\beta) \geq 0$$

as $\Sigma(\xi_\alpha, \xi_\beta)$ is positive definite, and $\sum_{\alpha\beta} G(v_\alpha, \Sigma(\xi_\alpha, \xi_\beta)v_\beta) = 0$ if and only if $v_\alpha \equiv 0$.

Conversely, if Σ is not completely positive, then for some choice of ξ_α, v_α (not all zero), $\sum_{\alpha\beta} G(v_\alpha, \Sigma(\xi_\alpha, \xi_\beta)v_\beta) \leq 0$. Then

$$\langle \sum_\alpha \xi_\alpha \otimes v_\alpha, \sum_\beta \xi_\beta \otimes v_\beta \rangle_\Sigma \leq 0$$

as well. Therefore, this bilinear form is not positive definite, so not an inner product. □

Proposition 16. *If Σ is completely positive, then, for $\xi \otimes B, \psi \otimes C \in W \otimes L(V)$, the formula*

$$\langle \xi \otimes B, \psi \otimes C \rangle = \mathrm{Tr}_G(B^t \Sigma(\xi, \psi)C)$$

extends bilinearly to an inner product on $W \otimes L(V)$.

Proof. Again, bilinearity and symmetry are clear, we need only show positivity. To this end, let $\sum_\alpha \xi_\alpha \otimes B_\alpha \in W \otimes L(V)$ be such that the ξ_α are linearly independent. Now note that $\sum_{\alpha\beta} B^t_\alpha \Sigma(\xi_\alpha, \xi_\beta) B_\beta$ is a symmetric element of $L(V)$. Let v be an eigenvector (of unit length) of this matrix with corresponding eigenvalue λ. Then

$$\begin{aligned}
\lambda &= G(v, \sum_{\alpha\beta} B^t_\alpha \Sigma(\xi_\alpha, \xi_\beta) B_\beta v) \\
&= \sum_{\alpha\beta} G(B_\alpha v, \Sigma(\xi_\alpha, \xi_\beta) B_\beta v) \\
&= \| \sum_\alpha \xi_\alpha \otimes B_\alpha v \|^2_\Sigma \geq 0 .
\end{aligned}$$

As the eigenvectors are complete, the spectrum of $\sum_{\alpha\beta} B^t_\alpha \Sigma(\xi_\alpha, \xi_\beta) B_\beta$ is nonnegative. Thus $\mathrm{Tr}_G(\sum_{\alpha\beta} B^t_\alpha \Sigma(\xi_\alpha, \xi_\beta) B_\beta) \geq 0$. If equality holds then all the eigenvalues of this matrix must vanish; that is $\sum_{\alpha\beta} B^t_\alpha \Sigma(\xi_\alpha, \xi_\beta) B_\beta = 0$. But from above, we see that for any $v \in V$, $\| \sum_\alpha \xi_\alpha \otimes B_\alpha v \|^2_\Sigma = 0$. Thus $\sum_\alpha \xi_\alpha \otimes B_\alpha v = 0$ in $W \otimes V$; but the ξ_α are linearly independent, so $B_\alpha v = 0$ for all α. Therefore $B_\alpha \equiv 0$ as required. □

2.3 A divergence lemma

We take another very brief aside to discuss a useful tool we will need. As our manifold M is compact, boundaryless and orientable, the normal divergence theorem applies:

$$\int_M \partial_j(v^j)\, dx = 0 ,$$

for any global vector field v.

Now, suppose we choose any connection on M such that the connection coefficients satisfy

$$\Gamma^j{}_{kj} = \partial_k \log \sqrt{G} .$$

Then for any global vector field v we have,

$$\begin{aligned}
0 &= \int_M \partial_j(v^j \sqrt{G})\, dx \\
&= \int_M \left[\partial_j(v^j)\sqrt{G} + \partial_j(\log\sqrt{G}) v^j \sqrt{G} \right] dx \\
&= \int_M \left(\partial_j(v^j) + \Gamma^j{}_{kj} v^k \right) \sqrt{G}\, dx \\
&= \int_M \nabla_j(v^j)\sqrt{G}\, dx .
\end{aligned}$$

We restate this formula as a lemma.

Lemma 17. *Let M be a compact, boundaryless and oriented manifold with volume form $\sqrt{G}\,dx$. With respect to any connection which satisfies ${\Gamma^j}_{kj} = \partial_k \log \sqrt{G}$ the following formula holds for all global vector fields:*

$$\int_M \nabla_j(v^j)\sqrt{G}\,dx = 0 \ .$$

This is the 'volume compatibility' condition that was alluded to earlier. We will suppose throughout that our manifold is equipped with such a connection. As we will discover, this choice of connection will play no role in the final result.

2.4 Torsion on an Elliptic Complex

Recall our elliptic complex,

$$\cdots \to \Gamma(V^{p-1}) \xrightarrow{d} \Gamma(V^p) \xrightarrow{d} \Gamma(V^{p+1}) \to \cdots \ ,$$

To avoid unnecessary notational difficulties, we adopt the following notation:

$$\begin{aligned} d = (A^j\partial_j + S^-) &\ : \ \Gamma(V^{p-1}) \to \Gamma(V^p) \ , \\ d = (B^j\partial_j + S^+) &\ : \ \Gamma(V^p) \to \Gamma(V^{p+1}) \ . \end{aligned}$$

As noted before, each of the A and B and the two S terms are matrices; and, without any further assumptions on the two symbol terms, the S matrices will not be covariant as we shall shortly see.

Suppose we fix a connection on V^p. Say, in a coordinate neighborhood of M the connection coefficient matrices are $\omega_j^{(p)}$. Then, as we have chosen covector bundles, tradition has the covariant derivative given by

$$\nabla_j\theta = \partial_j\theta - \omega_j^{(p)}\theta \ ,$$

where it understood that a matrix multiplication is occurring. We then covariantize d as follows:

$$\begin{aligned} d\theta &= B^j(\nabla_j + \omega_j^{(p)})\theta + S^+\theta \\ &= B^j\nabla_j\theta + (B^j\omega^{(p)} + S^+)\theta \\ &= B^j\nabla_j\theta + T^+_{(p)}\theta \ . \end{aligned}$$

Here we have defined $T^+_{(p)} = B^j\omega_j^{(p)} + S^+$. Notice that as $T^+_{(p)}$ is covariant by construction, S^+ will be covariant if and only if $B^j\omega_j^{(p)}$ is as well. Of course, the latter is not covariant without any special assumptions on B^j. Thus we are forced to consider these lower order terms just to properly covariantize the equation.

Identically, we define $T^+_{(p-1)} = A^j\omega_j^{(p-1)} + S^-$, in which case $d\eta = A^j\nabla_j + T^+_{(p-1)}\eta$. Yet this term is related to the connection on V^{p-1}. To tie this with the covariant derivative on $\Gamma(V^p)$, we demand the covariant version of the lower d is given by

$$d\eta = \nabla_j(A^j\eta) + T^-_{(p)}\eta \ .$$

From this we easily compute T^-.

$$\begin{aligned} d\eta &= \nabla_j(A^j\eta) + T^-_{(p)}\eta \\ &= (\nabla_j A^j)\eta + A^j\nabla_j\eta + T^-_{(p)}\eta \\ &= (\nabla_j A^j)\eta + d\eta - T^+_{(p-1)}\eta + T^-_{(p)}\eta \ , \end{aligned}$$

or

$$(\nabla_j A^j)\eta = T^+_{(p-1)}\eta - T^-_{(p)}\eta \ .$$

This last equation holds for all $\eta \in \Gamma(V^{p-1})$, hence

$$(\nabla_j A^j) = T^+_{(p-1)} - T^-_{(p)} \ .$$

Now spelling out the covariant derivative, recalling that A is a matrix-valued vector field, and inserting the T^+ term, we get

$$\partial_j A^j - \omega_j^{(p)} A^j + A^j\omega_j^{(p-1)} + A^j\partial_j \log\sqrt{G} = A^j\omega_j^{(p-1)} + S^- - T^-_{(p)} \ .$$

Thus we have proven the following.

Proposition 18. *For all $\theta \in \Gamma(V^p)$ and $\eta \in \Gamma(V^{p-1})$, the following formulae hold:*

1. $d\theta = B^j\nabla_j\theta + T^+_{(p)}\theta$,
2. $d\eta = \nabla_j(A^j\eta) + T^-_{(p)}\eta$,
3. $\nabla_j A^j = T^+_{(p-1)} - T^-_{(p)}$,
4. $\nabla_j B^j = T^+_{(p)} - T^-_{(p+1)}$,

where

- $T^+_{(p)} = B^j \omega_j^{(p)} + S^+$,
- $T^-_{(p)} = -\partial_j A^j + \omega_j^{(p)} A^j - A^j \partial_j \log \sqrt{G} + S^-$.

We see that covariantizing d introduces two types of lower order terms, which we call torsions. At each grade of the complex these torsions are computed solely by the symbol terms, the zero order terms, and the connection on that grade. The covariant derivative of the symbol terms then intermix these torsions. Again note that although the connection Γ was needed to compute these covariant derivatives, it does not enter into the definition of the torsions.

2.5 The Adjoint Operator

Now we wish to compute a covariant formula for $d^* : V^{p+1} \to V^p$. Yet, there are two forms for $d : V^p \to V^{p+1}$ as seen in the previous section. We treat each form individually. Consider,

$$\begin{aligned}\langle \eta, d\theta \rangle &= \int_M G^{IJ} \eta_I \left((B^j)_J{}^K \nabla_j \theta_K + (T^+_{(p)})_J{}^K \theta_K \right) \sqrt{G}\, dx \\ &= -\int_M \nabla_j \left(G^{IJ} \eta_I (B^j)_J{}^K \right) \theta_K \sqrt{G}\, dx + \int_M G^{IJ} \eta_I (T^+_{(p)})_J{}^K \theta_K \sqrt{G}\, dx\,.\end{aligned}$$

Careful consideration of the above formulae denotes the need to introduce the transpose map,

$$(A^t)^I{}_L = G_{KL}\, A_J^K\, G^{IJ}\,.$$

Although the transpose as defined is just the adjoint of a bundle morphism, we will use reserve the term adjoint and its notation for differential operators. Using this notation, we rewrite the above expressions as

$$\begin{aligned}\langle \eta, d\theta \rangle &= -\int_M \nabla_j \left(G^{JK} ((B^j)^t)^I{}_J \eta_I \right) \theta_K \sqrt{G}\, dx \\ &\quad + \int_M G^{JK} ((T^+_{(p)})^t)^I{}_K \eta_J \theta_K \sqrt{G}\, dx \\ &= \int_M G^{JK} \left[-\nabla_j \left(((B^j)^t)^I{}_J \eta_I \right) + ((T^+_{(p)})^t)^I{}_J \eta_I \right] \theta_K \sqrt{G}\, dx\,,\end{aligned}$$

where we have used that $\nabla_j G^{JK} = 0$; that is, our connection we will assume to be metric compatible on V^p. Now, since we have $\langle \eta, d\theta \rangle = \langle d^*\eta, \theta \rangle$, we read off from the above expression,

$$d^*\eta = -\nabla_j \left((B^j)^t \eta \right) + (T^+_{(p)})^t \eta\,,$$

where as usual, this equation has matrix multiplication implied.

For the other case, consider

$$\begin{aligned}\langle\eta, d\theta\rangle &= \int_M G^{IJ}\eta_I\left[\nabla_j\left((B^j)_J^K\theta_K\right) + (T^-_{(p+1)})_J^K\theta_K\right]\sqrt{G}\,dx \\ &= -\int_M G^{IJ}(B^j)_J^K(\nabla_j\eta_I)\theta_K\sqrt{G}\,dx + \int_M G^{IJ}\eta_I(T^-_{(p+1)})_J^K\theta_K\sqrt{G}\,dx\,.\end{aligned}$$

As above, we have used that $\nabla_j G^{IJ} = 0$. Hence, we will assume that all the connections on our vector bundles are metric compatible. Again we use the transpose map and rewrite our expressions as

$$\langle\eta, d\theta\rangle \quad = \quad \int_M G^{JK}\left[-((B^j)^t)^I{}_J\nabla_j\eta_I + ((T^-_{(p+1)})^t)^I{}_J\eta_I\right]\theta_K\sqrt{G}\,dx\,.$$

We summarize our computations. Although we could weaken the conditions to get a stronger result, the following will suffice for our purposes.

Proposition 19. *For any pair of metric connections on V^p and V^{p+1}, the following formulae hold for $\eta \in \Gamma(V^{p+1})$:*

1. $d^*\eta = -\nabla_j\left((B^j)^t\eta\right) + (T^+_{(p)})^t\eta$,

2. $d^*\eta = -(B^j)^t\nabla_j\eta + (T^-_{(p+1)})^t\eta$.

We see that $\nabla_j^* = -\nabla_j$ as long as we assume our connection is metric. We also have the following obvious lemma which will be useful later.

Lemma 20. *For any pair of metric connections on V^p, V^{p+1},*

$$\nabla_j(B^k)^t = (\nabla_j B^k)^t\,.$$

2.6 A Preliminary Formula

Our present goal is to compute a covariant form for the Laplace operator of our elliptic complex, $\Delta = d^*d + dd^*$. This calculation is a bit cumbersome, so we compute the two terms separately. First,

$$\begin{aligned}dd^*\theta &= d\left(-(A^k)^t\nabla_k\theta + (T^-_{(p)})^t\theta\right) \qquad (4)\\ &= -\nabla_j\left(A^j(A^k)^t\nabla_k\theta\right) + \nabla_j\left(A^j(T^-_{(p)})^t\theta\right) \\ &\quad -T^-_{(p)}(A^j)^t\nabla_j\theta + T^-_{(p)}(T^-_{(p)})^t\theta\,.\end{aligned}$$

Second,

$$\begin{aligned} d^* d\theta &= d^* \left(B^k \nabla_k \theta + T^+_{(p)}\theta\right) \qquad (5) \\ &= -\nabla_j \left((B^j)^t B^k \nabla_k \theta\right) - \nabla_j \left((B^j)^t T^+_{(p)}\theta\right) \\ &\quad + (T^+_{(p)})^t B^j \nabla_j \theta + (T^+_{(p)})^t T^+_{(p)}\theta \, . \end{aligned}$$

Looking at the first two terms in each expression, we will find the following notation useful,

$$(B^j)^t B^k + A^j (A^k)^t = \sigma\Delta^{jk} + \alpha\Delta^{jk} \, ,$$

where

$$\begin{aligned} \sigma\Delta^{jk} &= \frac{1}{2}\left((B^j)^t B^k + (B^k)^t B^j + A^j (A^k)^t + A^k (A^j)^t\right) \, , \\ \alpha\Delta^{jk} &= \frac{1}{2}\left((B^j)^t B^k - (B^k)^t B^j + A^j (A^k)^t - A^k (A^j)^t\right) \, . \end{aligned}$$

The notation for $\sigma\Delta$ serves a dual role: it is the symmetric part of $B^t B + AA^t$, and it is the symbol of the Laplacian, considered as a matrix valued symmetric bilinear form.

Example 21. Returning to Finsler geometry (for which Riemannian geometry is a special case), for each $\Lambda^p(T^*M)$, we have the metric $G^{j_1 k_1 \cdots j_p k_p}$ given by formula (3). The exterior derivative has the following local formula: for $\theta = \theta_{j_1 \cdots j_p} \frac{1}{p!} dx^{j_1} \wedge \cdots dx^{j_p}$,

$$d\theta = \frac{1}{p!} \epsilon^{j j_1 \cdots j_p}_{k_1 \cdots k_{p+1}} \partial_j \theta_{j_1 \cdots j_p} \frac{1}{(p+1)!} dx^{k_1} \wedge \cdots \wedge dx^{k_{p+1}} \, ,$$

where the Kronecker-ϵ is symbol is defined as

$$\epsilon^{j_1 \cdots j_p}_{k_1 \cdots k_1} = \begin{cases} +1 & (j_1, \cdots, j_p) \text{ an even permutation of } (k_1, \cdots, k_p) \\ -1 & (j_1, \cdots, j_p) \text{ an odd permutation of } (k_1, \cdots, k_p) \\ 0 & \text{otherwise} \end{cases} \, .$$

Using multi-indices, we get quaint expressions for the metric, G^{JK}, and

$$d\theta = \frac{1}{p!} \epsilon^{(jJ)}_K \partial_j \theta_J \frac{1}{(p+1)!} dx^K \, .$$

Here we have used the notation (jJ) to represent concatenation of a p-multi-index and a single index into a $p+1$-multi-index. We identify $(B^j)^J_K =$

$\frac{1}{p!}\epsilon_K^{(jJ)}$. One can also compute the symbol of the Laplacian, see [BL],

$$
\begin{aligned}
(\sigma\Delta^{jk})_K^J &= \frac{1}{2}\Bigg(G^{C(jB)}G_{BK}\,\frac{1}{p!}\epsilon_C^{(kJ)} + \frac{1}{(p-1)!}\epsilon_K^{(jE)}G^{J(kD)}G_{DE} \\
&\qquad + G^{C(kB)}G_{BK}\,\frac{1}{p!}\epsilon_C^{(jJ)} + \frac{1}{(p-1)!}\epsilon_K^{(kE)}G^{J(jD)}G_{DE}\Bigg)\,.
\end{aligned}
$$

If the Finsler space is Riemannian, then this expression reduces to $\frac{1}{p!}g^{jk}\epsilon_K^J$. But note that $\frac{1}{p!}$ times the Kronecker-ϵ is just the identity on p-forms. Clearly, the symbol is completely positive in the Riemannian case, but to what extent this is true in the Finsler case is unknown.

Anyhow, we consider only the sum of the highest order terms in (4) and (5):

$$
\begin{aligned}
-\nabla_j\left([(B^j)^tB^k + A^j(A^k)^t]\theta\right) &= -\nabla_j(\sigma\Delta^{jk}\nabla_k) - \nabla_j(\alpha\Delta^{jk}\nabla_k\theta) \\
&= \nabla_j^*(\sigma\Delta^{jk}\nabla_k) - \nabla_j(\alpha\Delta^{jk})\nabla_k\theta + \frac{1}{2}\alpha\Delta^{jk}\Omega_{jk}\theta\,.
\end{aligned}
$$

In this last step we have used $\alpha\Delta^{jk}\nabla_j\nabla_k = \frac{1}{2}\alpha\Delta^{jk}(\nabla_j\nabla_k - \nabla_k\nabla_j)$ and again that our vector bundles are transforming covariantly, so that the curvature has action $(\nabla_j\nabla_k - \nabla_k\nabla_j)\theta = -\Omega_{jk}\theta$, together with the torsion freeness of Γ so that the contributions from correcting the j and k indices cancel.

Putting everything together,

$$
\begin{aligned}
\Delta\theta &= \nabla_j^*(\sigma\Delta^{jk}\nabla_k) \\
&\quad + \left[\nabla_k(\alpha\Delta^{jk}) + A^j(T^-_{(p)})^t - T^-_{(p)}(A^j)^t + (T^+_{(p)})^tB^j - (B^j)^tT^+_{(p)}\right]\nabla_j\theta \\
&\quad + \Bigg[\frac{1}{2}\alpha\Delta^{jk}\Omega_{jk} + \nabla_j\left(A^j(T^-_{(p)})^t\right) - \nabla_j\left((B^j)^tT^+_{(p)}\right) \\
&\qquad + T^-_{(p)}(T^-_{(p)})^t + (T^+_{(p)})^tT^+_{(p)}\Bigg]\,\theta\,.
\end{aligned}
$$

Thus we have shown:

Proposition 22. *Suppose the connections on V^{p-1}, V^p, and V^{p+1} are metric compatible, and suppose that in addition, the connection on V^p satisfies*

$$
\nabla_k(\alpha\Delta^{jk}) + A^j(T^-_{(p)})^t - T^-_{(p)}(A^j)^t + (T^+_{(p)})^tB^j - (B^j)^tT^+_{(p)} = 0\,. \qquad (6)
$$

Then

$$\begin{aligned}\Delta\theta &= \nabla_j^*(\sigma\Delta^{jk}\nabla_k\theta) + \left[\frac{1}{2}\alpha\Delta^{jk}\Omega_{jk} + \nabla_j\left(A^j(T^-_{(p)})^t\right)\right.\\ &\quad \left. -\nabla_j\left((B^j)^t T^+_{(p)}\right) + T^-_{(p)}(T^-_{(p)})^t + (T^+_{(p)})^t T^+_{(p)}\right]\theta\,.\end{aligned}$$

2.7 Existence and Uniqueness of a Connection

The preceding formula requires the vanishing of a complicated expression. At this point it is not clear what implications this may have on the connection.

Before we continue, let us pause for a moment and again consider metric compatibility. In index-free notation, we have the metric compatibility condition:

$$0 = (\nabla_k G)G^{-1} = (\partial_k G)G^{-1} - \omega_k^{(p)} - (\omega_k^{(p)})^t\,.$$

Let us abbreviate $P_k = (\partial_k G)G^{-1}$ so that the above equation reads

$$P_k = \omega_k^{(p)} + (\omega_k^{(p)})^t\,.$$

With this notation, we also have,

$$\begin{aligned}\partial_k(L^t) &= \partial_k(GLG^{-1}) = GL\partial_k(G^{-1}) + (\partial_k L)^t + (\partial_k G)LG^{-1}\\ &= P_k L^t + (\partial_k L)^t - L^t P_k\,.\end{aligned}$$

Of course this last equation could also be derived by metric compatibility directly. Notice as well that $(P_k)^t = P_k$ for all k.

Back to our computations. We need a more descriptive notation than above, thus we will move to the notation of our primary examples. Let us denote the metric structure as (V^p, G_p) for each grade of our complex. We will also write $P_k^{(p)} = (\partial_k G_p)G_p^{-1}$.

To begin let us consider the following:

$$\begin{aligned}&\nabla_k\left((B^j)^t B^k\right) - 2(B^j)^t T^+_{(p)}\\ &= \partial_k\left((B^j)^t B^k\right) - \omega_k^{(p)}(B^j)^t B^k + (B^j)^t B^k\omega_k^{(p)}\\ &\quad +(B^l)^t B^k\Gamma^j{}_{lk} + (B^j)^t B^l\Gamma^k{}_{lk} - 2(B^j)^t\left(B^k\omega_k^{(p)} + S^+\right)\\ &= P_k^{(p)}(B^j)^t B^k + (\partial_k B^j)^t B^k - (B^j)^t P_k^{(p+1)} B^k + (B^j)^t\partial_k B^k\\ &\quad -\omega_k^{(p)}(B^j)^t B^k - (B^j)^t B^k\omega_k^{(p)} + (B^l)^t B^k\Gamma^j{}_{lk}\\ &\quad +(B^j)^t B^k\partial_k\log\sqrt{G} - 2(B^j)^t S^+\,.\end{aligned}$$

If we take the transpose of the above equation, and use metric compatibility, we get

$$
\begin{aligned}
&\nabla_k \left((B^k)^t B^j\right) - 2(T^+_{(p)})^t B^j \\
&\quad = (B^k)^t B^j P^{(p)}_k + (B^k)^t \partial_k B^j - (B^k)^t P^{(p+1)}_k B^j + (\partial_k B^k)^t B^j \\
&\qquad -(B^k)^t B^j (\omega^{(p)}_k)^t - (\omega^{(p)}_k)^t (B^k)^t B^j + (B^k)^t B^l \Gamma^j{}_{lk} \\
&\qquad +(B^k)^t B^j \partial_k \log\sqrt{G} - 2(S^+)^t B^j \\
&\quad = -P^{(p)}_k (B^k)^t B^j + (B^k)^t \partial_k B^j - (B^k)^t P^{(p+1)}_k B^j + (\partial_k B^k)^t B^j \\
&\qquad +(B^k)^t B^j \omega^{(p)}_k + \omega^{(p)}_k (B^k)^t B^j + (B^k)^t B^l \Gamma^j{}_{lk} \\
&\qquad +(B^k)^t B^j \partial_k \log\sqrt{G} - 2(S^+)^t B^j .
\end{aligned}
$$

Subtracting the second equation from the first yields:

$$
\begin{aligned}
&\nabla_k \left((B^j)^t B^k - (B^k)^t B^j\right) + 2(T^+_{(p)})^t B^j - 2(B^j)^t T^+_{(p)} - 2(B^j)^t S^+ + 2(S^+)^t B^j \\
&\quad = P^{(p)}_k \left((B^j)^t B^k + (B^k)^t B^j\right) + \left((B^j)^t B^k - (B^k)^t B^j\right) \partial_k \log\sqrt{G} \\
&\qquad +(B^j)^t \partial_k B^k - (B^k)^t \partial_k B^j + (\partial_k B^j)^t B^k - (\partial_k B^k)^t B^j \\
&\qquad -(B^j)^t P^{(p+1)}_k B^k + (B^k)^t P^{(p+1)}_k B^j - \omega^{(p)}_k \left((B^j)^t B^k + (B^k)^t B^j\right) \\
&\qquad - \left((B^j)^t B^k + (B^k)^t B^j\right) \omega^{(p)}_k - 2(B^j)^t S^+ + 2(S^+)^t B^j .
\end{aligned}
$$

Now we perform the identical calculation for the A terms.

$$
\begin{aligned}
&\nabla_k \left(A^j (A^k)^t\right) + 2A^j (T^-_{(p)})^t \\
&\quad = \partial_k \left(A^j (A^k)^t\right) - \omega^{(p)}_k A^j (A^k)^t + A^j (A^k)^t \omega^{(p)}_k + A^l (A^k)^t \Gamma^j{}_{lk} \\
&\qquad +A^j (A^k)^t \partial_k \log\sqrt{G} - 2A^j \left(\partial_k A^k - \omega^{(p)}_k A^k + A^k \partial_k \log\sqrt{G} - S^-\right)^t \\
&\quad = (\partial_k A^j)(A^k)^t + A^j P^{(p-1)}_k (A^k)^t - A^j (\partial_k A^k)^t + A^j (A^k)^t P^{(p)}_k \\
&\qquad -\omega^{(p)}_k A^j (A^k)^t - A^j (A^k)^t \omega^{(p)}_k - A^j (A^k)^t \partial_k \log\sqrt{G} \\
&\qquad +A^l (A^k)^t \Gamma^j{}_{lk} + 2A^j (S^-)^t .
\end{aligned}
$$

Again applying the transpose

$$\begin{aligned}
&\nabla_k \left(A^k(A^j)^t\right) + 2T^-_{(p)}(A^j)^t \\
&\quad = A^k(\partial_k A^j)^t + A^k P_k^{(p-1)}(A^j)^t - (\partial_k A^k)(A^j)^t + P_k^{(p)} A^k(A^j)^t \\
&\qquad -(\omega_k^{(p)})^t A^k(A^j)^t - A^k(A^j)^t(\omega_k^{(p)})^t - A^k(A^j)^t \partial_k \log\sqrt{G} \\
&\qquad +A^k(A^l)^t \Gamma^j{}_{lk} + 2S^-(A^j)^t \\
&\quad = A^k(\partial_k A^j)^t + A^k P_k^{(p-1)}(A^j)^t - (\partial_k A^k)(A^j)^t - A^k(A^j)^t P_k^{(p)} \\
&\qquad +\omega_k^{(p)} A^k(A^j)^t + A^k(A^j)^t \omega_k^{(p)} - A^k(A^j)^t \partial_k \log\sqrt{G} \\
&\qquad +A^k(A^l)^t \Gamma^j{}_{lk} + 2S^-(A^j)^t \;.
\end{aligned}$$

Subtracting the second from the first gives

$$\begin{aligned}
&\nabla_k \left(A^j(A^k)^t - A^k(A^j)^t\right) + 2A^j(T^-_{(p)})^t - 2T^-_{(p)}(A^j)^t \\
&\quad = \left(A^j(A^k)^t + A^k(A^j)^t\right) P_k^{(p)} - \left(A^j(A^k)^t - A^k(A^j)^t\right) \partial_k \log\sqrt{G} \\
&\qquad +(\partial_k A^j)(A^k)^t - A^k(\partial_k A^j)^t - A^j(\partial_k A^k)^t + (\partial_k A^k)(A^j)^t \\
&\qquad +A^j P_k^{(p-1)}(A^k)^t - A^k P_k^{(p-1)}(A^j)^t - \omega_k^{(p)} \left(A^j(A^k)^t + A^k(A^j)^t\right) \\
&\qquad - \left(A^j(A^k)^t + A^k(A^j)^t\right) \omega_k^{(p)} + 2A^j(S^-)^t - 2S^-(A^j)^t \;.
\end{aligned}$$

Finally putting everything together, we find that the left side of equation (6) is

$$\begin{aligned}
&P_k^{(p)} \left((B^j)^t B^k + (B^k)^t B^j\right) + \left(A^j(A^k)^t + A^k(A^j)^t\right) P_k^{(p)} \\
&\quad + \left((B^j)^t B^k - (B^k)^t B^j - A^j(A^k)^t + A^k(A^j)^t\right) \partial_k \log\sqrt{G} \\
&\quad +(B^j)^t \partial_k B^k - (B^k)^t \partial_k B^j + (\partial_k B^j)^t B^k - (\partial_k B^k)^t B^j \\
&\quad +(\partial_k A^j)(A^k)^t - A^k(\partial_k A^j)^t - A^j(\partial_k A^k)^t + (\partial_k A^k)(A^j)^t \\
&\quad -(B^j)^t P_k^{(p+1)} B^k + (B^k)^t P_k^{(p+1)} B^j + A^j P_k^{(p-1)}(A^k)^t - A^k P_k^{(p-1)}(A^j)^t \\
&\quad -2\omega_k^{(p)} \sigma\Delta^{jk} - 2\sigma\Delta^{jk}\omega_k^{(p)} \\
&\quad -2(B^j)^t S^+ + 2(S^+)^t B^j + 2A^j(S^-)^t - 2S^-(A^j)^t \;.
\end{aligned}$$

But even this is not the final form we desire. We write,

$$\omega_k^{(p)} \sigma\Delta^{jk} = -\left(\sigma\Delta^{jk}\omega_k^{(p)}\right)^t + P_k^{(p)} \sigma\Delta^{jk}$$

in which case we discover that the term above vanishes if and only if

$$\begin{aligned}\sigma\Delta^{jk}\omega_k^{(p)} - \left(\sigma\Delta^{jk}\omega_k^{(p)}\right)^t &= (S^+)^t B^j - (B^j)^t S^+ + A^j(S^-)^t - S^-(A^j)^t \\ &+\frac{1}{2}\Big[\left(A^j(A^k)^t + A^k(A^j)^t\right)P_k^{(p)} - P_k^{(p)}\left(A^j(A^k)^t + A^k(A^j)^t\right) \\ &+ \left((B^j)^t B^k - (B^k)^t B^j - A^j(A^k)^t + A^k(A^j)^t\right)\partial_k \log\sqrt{G} \\ &+(B^j)^t\partial_k B^k - (B^k)^t\partial_k B^j + (\partial_k B^j)^t B^k - (\partial_k B^k)^t B^j \\ &+(\partial_k A^j)(A^k)^t - A^k(\partial_k A^j)^t - A^j(\partial_k A^k)^t + (\partial_k A^k)(A^j)^t \\ &-(B^j)^t P_k^{(p+1)} B^k + (B^k)^t P_k^{(p+1)} B^j \\ &+A^j P_k^{(p-1)}(A^k)^t - A^k P_k^{(p-1)}(A^j)^t\Big] \ .\end{aligned}$$

Note that this is an equation between n antisymmetric matrices. Thus if r is the rank of V^p as a vector bundle, then the above statement is $n\cdot\frac{r(r-1)}{2}$ linear equations in the components of $\omega^{(p)}$. Yet as metric compatibility specified the symmetric part of $\omega^{(p)}$, only the skew part remains. Hence there are precisely $n \cdot \frac{r(r-1)}{2}$ variables. Another point of interest is that the dependence on the manifold connection Γ has disappeared; we have required it to be torsion free with the volume compatibility condition, but it plays no other role in this theory. Oddly enough as well, it is the symbol of the Laplacian which is determining the solvability of this equation; the antisymmetric terms, $\alpha\Delta$, do not appear. For convenience, let us define the right side of the previous expression as K^j, in which case the preceding formula has the form

$$\sigma\Delta^{jk}\omega_k^{(p)} - \left(\sigma\Delta^{jk}\omega_k^{(p)}\right)^t = K^j \ .$$

Lemma 23. *Suppose the symbol of the Laplacian is completely positive. Then there exists a unique solution to*

$$\sigma\Delta^{jk}\omega_k^{(p)} - \left(\sigma\Delta^{jk}\omega_k^{(p)}\right)^t = K^j \tag{7}$$

among the collection of antisymmetric matrix valued forms $\omega_k^{(p)}dx^k$.

Proof. We have already seen that this is a well-posed linear system in the components of $\omega^{(p)}$. Thus we need only show that the left side of (7) is injective for it to be invertible. That is, we must show that the only solution to

$$\sigma\Delta^{jk}\omega_k^{(p)} - \left(\sigma\Delta^{jk}\omega_k^{(p)}\right)^t = 0 \tag{8}$$

is $\omega^{(p)} = 0$. So suppose (8), and let us premultiply with $\omega_j^{(p)}$ and trace with respect to G_p. We get

$$\begin{aligned} 0 &= \operatorname{Tr}_G \left(\omega_j^{(p)} \sigma \Delta^{jk} \omega_k^{(p)} + \left(\sigma \Delta^{jk} \omega_k^{(p)} \omega_j^{(p)} \right)^t \right) \\ &= -2 \operatorname{Tr}_G \left((\omega_j^{(p)})^t \sigma \Delta^{jk} \omega_k^{(p)} \right) . \end{aligned}$$

But from proposition 16, this trace is a norm, hence $\omega^{(p)} = 0$ as desired. □

Now we can rephrase the intermediate formula from before.

Theorem 24. *Let Γ be a torsion-free connection on M which satisfies the volume compatibility condition, $\Gamma^j{}_{kj} = \partial_k \log \sqrt{G}$. Suppose the connections on V^{p-1} and V^{p+1} are metric compatible and the symbol of the Laplacian on V^p is completely positive. Then there exists a unique connection on V^p (independent of Γ), which satisfies*

$$\nabla_k(\alpha\Delta^{jk}) + A^j(T^-_{(p)})^t - T^-_{(p)}(A^j)^t + (T^+_{(p)})^t B^j - (B^j)^t T^+_{(p)} = 0 .$$

For this connection, we have

$$\begin{aligned} \Delta\theta &= \nabla_j^*(\sigma\Delta^{jk}\nabla_k\theta) + \left[\frac{1}{2}\alpha\Delta^{jk}\Omega_{jk} + \nabla_j\left(A^j(T^-_{(p)})^t\right) - \nabla_j\left((B^j)^t T^+_{(p)}\right) \right. \\ &\qquad \left. + T^-_{(p)}(T^-_{(p)})^t + (T^+_{(p)})^t T^+_{(p)} \right] \theta . \end{aligned}$$

2.8 A Weitzenböck Formula

To achieve our final formula, we apply a Bianchi identity to the complicated zeroth order term in the formula above. Now, our connection is metric compatible and satisfies the complicated vanishing criterion of the previous theorem. Covariant differentiation of the metric compatibility equation only yield the usual skew symmetry of the curvature form, so we focus on the second of the defining equations.

We covariantly differentiate and trace over the free index in (6) yielding

$$\begin{aligned} 0 &= \nabla_j\nabla_k(\alpha\Delta^{jk}) + \nabla_j(A^j(T^-_{(p)})^t) - \nabla_j(T^-_{(p)}(A^j)^t) \\ &\qquad + \nabla_j((T^+_{(p)})^t B^j) - \nabla_j((B^j)^t T^+_{(p)}) . \end{aligned}$$

Yet

$$\begin{aligned} \nabla_j\nabla_k(\alpha\Delta^{jk}) &= \frac{1}{2}(\nabla_j\nabla_k - \nabla_k\nabla_j)\alpha\Delta^{jk} \\ &= \frac{1}{2}\left(\alpha\Delta^{jk}\Omega_{jk} - \Omega_{jk}\alpha\Delta^{jk}\right) , \end{aligned}$$

where we have again used the torsion-freeness of the manifold connection Γ. So the final form of our Bianchi identity is

$$\begin{aligned} 0 &= \frac{1}{2}\left(\alpha\Delta^{jk}\Omega_{jk} - \Omega_{jk}\alpha\Delta^{jk}\right) + \nabla_j(A^j(T^-_{(p)})^t) - \nabla_j(T^-_{(p)}(A^j)^t) \quad (9) \\ &\quad + \nabla_j((T^+_{(p)})^t B^j) - \nabla_j((B^j)^t T^+_{(p)}) \, . \end{aligned}$$

Note that this term is precisely the antisymmetric part of the zeroth order term in the Weitzenböck formula. This leads us to the following definition.

Definition 25. The curvature endomorphism of the connection $\omega^{(p)}$ is the symmetric matrix $\mathbf{R}^{(p)} \in \mathrm{End}(V^p, V^p)$ given by

$$\begin{aligned} \mathbf{R}^{(p)} &= \frac{1}{4}\alpha\Delta^{jk}\Omega_{jk} + \frac{1}{4}\Omega_{jk}\alpha\Delta^{jk} + \frac{1}{2}\nabla_j\left(A^j(T^-_{(p)})^t + T^-_{(p)}(A^j)^t\right) \\ &\quad - \frac{1}{2}\nabla_j\left((B^j)^t T^+_{(p)} + (T^+_{(p)})^t B^j\right) + T^-_{(p)}(T^-_{(p)})^t + (T^+_{(p)})^t T^+_{(p)} \, . \end{aligned}$$

Corollary 26. *Let M be a compact oriented manifold with volume form $\sqrt{G}\ dx$, and Γ be any torsion-free connection on M such that ${\Gamma^j}_{kj} = \partial_k \log\sqrt{G}$. Take any elliptic complex of first order differential operators on Riemannian covector bundles, $((V^p, G_p), d)$, whose Laplacian $\Delta = d^*d + dd^*$ has completely positive symbol. Then there exists a unique metric compatible connection on each V^p, $\omega^{(p)}$, such that for $\theta \in \Gamma(V^p)$*

$$\Delta\theta = \nabla_j^*(\sigma\Delta^{jk}\nabla_k\theta) + \mathbf{R}^{(p)}\theta \, ,$$

where $\mathbf{R}^{(p)}$ is the curvature endomorphism of $\omega^{(p)}$.

3 A Bochner Vanishing Theorem

Considering the previous results, in this section, we will take $(V^p, G_p)_{p\in\mathbf{Z}}$ as a collection of Riemannian covector bundles, connected by a first order differential operator, d, which turns this collection into a elliptic complex whose associated Laplacians have completely positive symbol.

Recall that the space of harmonic forms $\mathcal{H}^*$ is isomorphic to the cohomology of any elliptic complex, thus measuring the lack of exactness. Also, we have seen that on our elliptic complex, there is a unique special connection on each covector bundle for which a Weitzenböck formula holds. Naturally, we will always choose this to be our prefered connection.

We begin with the fundamental inequality.

Lemma 27. *Let $\theta \in \mathcal{H}^p$. Then $\langle\theta, \mathbf{R}^{(p)}\theta\rangle \le 0$ with equality holding if and only if $\nabla\theta = 0$.*

Proof. From the Weitzenböck formula, we have,

$$\begin{aligned}\langle \theta, \Delta\theta \rangle &= \langle \theta, \nabla_j^*(\sigma\Delta^{jk}\nabla_k\theta) + \mathbf{R}^{(p)}\theta \rangle \\ &= \langle \nabla_j\theta, \sigma\Delta^{jk}\nabla_k\theta \rangle + \langle \theta, \mathbf{R}^{(p)}\theta \rangle .\end{aligned}$$

Hence if $\theta \in \mathcal{H}^p$ then

$$\langle \theta, \mathbf{R}^{(p)}\theta \rangle = -\langle \nabla_j\theta, \sigma\Delta^{jk}\nabla_k\theta \rangle .$$

The symbol of the Laplacian is completely positive. Hence $\langle \theta, \mathbf{R}^{(p)}\theta \rangle \leq 0$ as desired. Now $\langle \theta, \mathbf{R}^{(p)}\theta \rangle = 0$ if and only if $\langle \nabla_j\theta, \sigma\Delta^{jk}\nabla_k\theta \rangle = 0$ if and only if $\nabla\theta = 0$, again as the symbol term is completely positive. □

With the lemma proven, we are able to prove the main result.

Theorem 28. *Given the situation as in corollary 26, if $\mathbf{R}^{(p)}$ is positive semidefinite then all harmonic forms are covariantly constant. If $\mathbf{R}^{(p)}$ is in fact positive definite, then $\mathcal{H}^p = 0$.*

Proof. If $\mathbf{R}^{(p)}$ is positive semidefinite, then from the fundamental inequality, $\langle \theta, \mathbf{R}^{(p)}\theta \rangle = 0$ for all harmonic forms. Hence again by the lemma, all harmonic forms are covariantly constant. If $\mathbf{R}^{(p)}$ is positive definite, then $\langle \theta, \mathbf{R}^{(p)}\theta \rangle = 0$ implies $\theta = 0$. Hence $\mathcal{H}^p = 0$. □

References

[BCS] Bao, D., Chern, S.S. and Shen, Z. (to appear) *An Introduction to Riemann-Finsler Geometry*, Springer.

[BL] Bao, D. and Lackey, B. (1996) A Hodge Decomposition Theorem for Finsler Spaces, *Comp. Ren. Acad. Sci. Paris*, **323**, 51-56.

[BS] Bao, D. and Shen, Z. (1994) On the Volume of Unit Tangent Spheres in a Finsler Manifold, *Results in Math.*, **26**, 1-17.

[EGH] Eguchi, T., Gilkey, T.B. and Hanson, A.J. (1980) Gravitation, Gauge Theories and Differential Geometry, *Phys. Rep.* **66** No. 6, 213-393.

[P] Pedersen, G.K. (1989) *Analysis NOW*, Springer.

[R] Rauch, J. (1991) *Partial Differential Equations*, Springer.

[Wa] Warner, F.W. (1983) *Foundations of Differentiable Manifolds and Lie Groups*, Springer.

[We] Wells, Jr., R.O. (1980) *Differential Analysis of Complex Manifolds*, Springer.

A LICHNEROWICZ VANISHING THEOREM FOR FINSLER SPACES

BRAD LACKEY *

1 Introduction

Hirzebruch's construction of the signature complex, [H] (or [N], for an easy to read exposition), allowed for a proof of the Lichnerowicz Vanishing Theorem on a Riemannian manifold which may not be spin. The construction uses the Hodge star to define self-dual and anti-self-dual forms, for which the Hodge-deRham operator serves as a first order elliptic operator with the signature as its index. Then a traditional Weitzenböck formula for the Levi-Civita connection allows one to show a relationship between the scalar curvature and the (vanishing of) the signature of the manifold in question. We wish to generalize this construction to the case of Finsler spaces.

Let (M, F) be a compact and boundaryless, orientable Finsler manifold. Then as seen in [BL1, BL2], the Finsler structure naturally induces a metric structure on each grade of forms. Specifically, on the slit tangent bundle $TM \setminus \{0\}$, there is the canonical Riemannian structure given by the Sasaki lift:

$$g_{jk}(x,y)dx^j \otimes dx^k + g_{jk}(x,y)\delta y^j \otimes \delta y^k \ .$$

Here, $\delta y^j = \frac{1}{F}dy^j + N^j_{\ k}dx^k$ is the vertical coframe induced by Cartan's nonlinear connection, and $g_{jk}(x,y) = \frac{1}{2}\frac{\partial^2 F^2}{\partial y^j \partial y^k}$ is the fundamental tensor. This metric gives us the volume form on the slit tangent bundle, $TM \setminus \{0\}$:

$$dVol = g(x,y)dx^1 \wedge \cdots \wedge dx^n \wedge \delta y^1 \wedge \cdots \wedge \delta y^n$$

where $g = \det(g_{jk})$. Note that this volume form is actually independent of the nonlinear connection, thus the use of Cartan's nonlinear connection was not necessary.

---

*Research partially supported by the Izaak Killam Trust. Parts of this material appears in the author's thesis. This is the second part of the paper "A Bochner Vanishing Theorem for Elliptic Complices". The reader is advised to consult that work for proofs of several of the results contained herein.

If we contract the volume form above with the radial direction, we get the Riemannian volume form of the projective sphere bundle,

$$dVol_{SM} = (-1)^{n-1} dVol(y^j \frac{\partial}{\partial y^j}, \cdot) .$$

The $(-1)^{n-1}$ occurs because we have taken our horizontal forms before our vertical forms as the prescribed orientation. The volume of each fibre of the projective sphere bundle, $S_x M$, may vary from point to point, [BS]; let us denote $Vol(x)$ for the volume of the fibre above x. We define a form ζ by

$$\frac{1}{Vol(x)} dVol_{SM} = \zeta_x \wedge dx^1 \wedge \cdots \wedge dx^n .$$

Note that for each $x \in M$, ζ_x is the volume form of the projective sphere above x, normalized to give a probability measure on each fibre.

Now, we can define a volume form on M by $\sqrt{G}\, dx$, where the tensor density

$$\sqrt{G}(x) = \int_{S_x} \sqrt{g}\, \zeta_x .$$

The metric tensor for p-forms we define as

$$G^{j_1 k_1 \cdots j_p k_p}(x) = \frac{1}{\sqrt{G}} \int_{S_x M} g^{j_1 k_1} \cdots g^{j_p k_p} \sqrt{g}\, \zeta_x .$$

This construction is not arbitrary; in fact, if we identify a form on M with its pull-back on SM, then these metrics are precisely those induced by the normalized Riemannian structure of SM:

$$\begin{aligned} \langle \theta, \psi \rangle &= \int_{SM} \frac{(-1)^{n-1}}{Vol(x)} \theta \wedge *\eta \\ &= \frac{1}{p!} \int_{SM} \frac{1}{Vol(x)} \theta_{j_1 \cdots j_p} \psi_{k_1 \cdots k_p} g^{j_1 k_1} \cdots g^{j_p k_p}\, dVol_{SM} \\ &= \frac{1}{p!} \int_M \theta_{j_1 \cdots j_p} \psi_{k_1 \cdots k_p} G^{j_1 k_1 \cdots j_p k_p} \sqrt{G}\, dx . \end{aligned}$$

The most important feature of this theory is that in general $G^{j_1 k_1 \cdots j_p k_p} \neq G^{j_1 k_1} \cdots G^{j_p k_p}$. This aspect of Finsler geometry forces us to consider the exterior algebra of M in a different light: the metrics on each grade of the algebra are unrelated. In fact, as we will never again refer to the explicit structure of the metric tensors, all of what follows will hold if one arbitrarily chooses metrics for each grade of forms. (Note, however, that there is a positivity condition to be satisfied by our selection of metrics, whether arbitrary or Finslerian.)

When it comes time to consider covariant derivatives, we must suppose that M is equipped with a connection Γ which is torsion free and volume compatible, $\nabla\sqrt{G}\,dx = 0$ (equivalently, $\Gamma^k{}_{jk} = \frac{\partial}{\partial x^j}\log\sqrt{G}$). Although we need this connection to define covariant derivatives of many of our objects, this connection will in the end, play no role in the theory (see [BL2, L] for details).

2 The Hodge Star

Let $\epsilon_{j_1\cdots j_n}$ be the totally antisymmetric tensor. Note that $\epsilon^{j_1\cdots j_n}$ is exactly the same object, but transforms contravariantly. Let us define the star map $* : \Lambda^p \to \Lambda^{n-p}$ by

$$*(\frac{1}{p!}dx^{j_1}\wedge\cdots\wedge dx^{j_p}) = \frac{\sqrt{G}}{p!}G^{j_1k_1\cdots j_pk_p}\epsilon_{k_1\cdots k_n}\frac{1}{(n-p)!}dx^{k_{p+1}}\wedge\cdots\wedge dx^{k_n}$$

and extend linearly.

The most useful property this map possesses is its relationship to the inner product. For ψ and θ in Λ^p we compute,

$$\begin{aligned}
\psi\wedge *\theta &= \frac{\sqrt{G}}{p!}\psi_{l_1\cdots l_p}\theta_{j_1\cdots j_p}G^{j_1k_1\cdots j_pk_p}\epsilon_{k_1\cdots k_n}\frac{1}{(n-p)!}\frac{1}{p!}dx^{l_1}\wedge\cdots\wedge dx^{l_p}\\
&\qquad\wedge dx^{k_{p+1}}\wedge\cdots\wedge dx^{k_n}\\
&= \frac{1}{(p!)^2}\frac{1}{(n-p)!}\psi_{l_1\cdots l_p}\theta_{j_1\cdots j_p}G^{j_1k_1\cdots j_pk_p}\epsilon_{k_1\cdots k_n}\epsilon^{l_1\cdots l_pk_{p+1}\cdots k_n}\sqrt{G}\,dx\\
&= \frac{1}{(p!)^2}\psi_{l_1\cdots l_p}\theta_{j_1\cdots j_p}G^{j_1k_1\cdots j_pk_p}\epsilon^{l_1\cdots l_p}_{k_1\cdots k_p}\sqrt{G}\,dx\\
&= \frac{1}{p!}\psi_{k_1\cdots k_p}\theta_{j_1\cdots j_p}G^{j_1k_1\cdots j_pk_p}\sqrt{G}\,dx\\
&= G_p(\psi,\theta)\sqrt{G}\,dx\ .
\end{aligned}$$

Here we have used the block symmetry of G and the following technical lemma.

Lemma 2.1. $\epsilon_{k_1\cdots k_pk_{p+1}\cdots k_n}\epsilon^{l_1\cdots l_pk_{p+1}\cdots k_n} = (n-p)!\,\epsilon^{l_1\cdots l_p}_{k_1\cdots k_p}$.

Proof. For the moment fix one collection of distinct indices $\{k_{p+1}\cdots k_n\}$. Then $\epsilon^{l_1\cdots l_pk_{p+1}\cdots k_n}$ vanishes unless $\{l_1\cdots l_p\}$ forms a collection of distinct indices complimentary to $\{k_{p+1}\cdots k_n\}$. The same holds for $\{k_1\cdots k_p\}$. Thus the left side of the equation is zero unless $\{l_1\cdots l_p\}=\{k_1\cdots k_p\}$. Both terms on the left will be positive or both will be negative if $\{l_1\cdots l_p\}$ is an even permutation of $\{k_1\cdots k_p\}$, in which case the left side is $+1$. If $\{l_1\cdots l_p\}$ is an odd permutation of $\{k_1\cdots k_p\}$, then one term will be negative and the other positive, whence the left side will be -1. Yet this is precisely the definition of the right side. There are $(n-p)!$ terms in the summation over $\{k_{p+1}\cdots k_n\}$. □

This map serves the same role as the Hodge star in Riemannian geometry, so we adopt the same name and notation. Clearly, star is an isomorphism between Λ^p and Λ^{n-p}. Yet as it appears, it need not have any other useful properties.

Definition 2.2. G_p and G_{n-p} are said to be compatible if star is an isometry. If all such pairs of metrics are compatible, the complex is said to be algebraically normal.

If we consider the Hodge star as a map from $\Lambda^p\oplus\Lambda^{n-p}$ into itself, and write G for the induced metric on this space, then

$$\begin{aligned}
G(*\psi,\theta) &= G_p(*\psi_{n-p},\theta_p)+G_{n-p}(*\psi_p,\theta_{n-p})\\
&= *\psi_{n-p}\wedge *\theta_p + *\psi_p\wedge *\theta_{n-p}\\
&= (-1)^{p(n-p)}[*\theta_p\wedge *\psi_{n-p} + *\theta_{n-p}\wedge *\psi_p]\\
&= (-1)^{p(n-p)}[G_{n-p}(*\theta_p,\psi_{n-p})+G_p(*\theta_{n-p},\psi_p)]\\
&= (-1)^{p(n-p)}G(*\theta,\psi)=(-1)^{p(n-p)}G(\psi,*\theta)\ ,
\end{aligned}$$

where $\psi,\theta\in\Lambda^p\oplus\Lambda^{n-p}$ with $\psi=\psi_p+\psi_{n-p}$, $\theta=\theta_p+\theta_{n-p}$, and $\psi_j,\theta_j\in\Lambda^j$. Note that the Hodge star is symmetric if either p or $n-p$ is even, yet antisymmetric if both are odd. But then,

$$G_{n-p}(*\psi,*\theta)=(-1)^{p(n-p)}G_p(\psi,**\theta)$$

for all $\theta,\psi\in\Lambda^p$. Thus G_{n-p} and G_p are compatible if and only if $**=(-1)^{p(n-p)}$. In any case, by taking $\psi=\theta$ in the above equation,

$$0\le\|*\theta\|^2=(-1)^{p(n-p)}G_p(\theta,**\theta)\ ,$$

with equality holding if and only if $\theta=0$. Therefore $**$ is either positive definite or negative definite depending on the parities of p and $n-p$. There are two distinct cases to consider; the easier is when n is odd, which we will do now. We will examine the even case shortly.

Suppose n is odd. Then either p or $n-p$ is even, so $**$ is always positive definite, and unity in the case of compatible metrics. From above, we see that $*$ is symmetric on $\Lambda^p \oplus \Lambda^{n-p}$. Thus we may decompose $\Lambda^p \oplus \Lambda^{n-p} = \Lambda_+^{p,n-p} \oplus \Lambda_-^{p,n-p}$ into the eigenspaces of the positive and negative eigenvalues of $*$. That is,

$$\Lambda_+^{p,n-p} = \text{span}\{\xi \in \Lambda^p \oplus \Lambda^{n-p} : *\xi = \lambda\xi,\ \lambda > 0\} ;$$
$$\Lambda_-^{p,n-p} = \text{span}\{\xi \in \Lambda^p \oplus \Lambda^{n-p} : *\xi = \lambda\xi,\ \lambda < 0\} .$$

These spaces will be orthogonal as $*$ is symmetric. In particular, if $\theta = \theta_p + \theta_{n-p} \in \Lambda_+^{p,n-p}$ be an eigenvector corresponding to eigenvalue λ. Then

$$\lambda\theta_p + \lambda\theta_{n-p} = *\theta = *\theta_p + *\theta_{n-p} .$$

So $*\theta_p = \lambda\theta_{n-p}$ and $*\theta_{n-p} = \lambda\theta_p$. If we consider $\tilde{\theta} = \theta_p - \theta_{n-p}$, then

$$\begin{aligned} *\tilde{\theta} &= *\theta_p - *\theta_{n-p} = \lambda\theta_{n-p} - \lambda\theta_p \\ &= (-\lambda)\tilde{\theta} . \end{aligned}$$

So for every positive eigenvalue-eigenvector pair (λ, θ) we have a corresponding negative eigenvalue-eigenvector pair $(-\lambda, \tilde{\theta})$. Clearly the converse holds. Hence the spectrum of $*$ is symmetric about zero, and $\Lambda_+^{p,n-p} \cong \Lambda_-^{p,n-p}$.

So we see that the Hodge star on an odd dimensional real vector space is very well behaved. Yet, odd dimensional manifolds do not have a notion of a signature, so most of this is useless for our purposes.

Now let us consider the even dimensional case: $\Lambda^*(\mathbb{R}^n)$ for $n = 2m$. This case has the distinguished feature that the Hodge star maps Λ^m back into itself. This fact is the basis for the notion of the signature. For now let us consider the case of $p \neq m$. From above, we have

$$G(*\psi, \theta) = (-1)^{p(2m-p)} G(\psi, *\theta) = (-1)^p G(\psi, *\theta) ,$$

as $p^2 = p(\text{mod } 2)$. Also,

$$0 \leq \| * \theta\|^2 = (-1)^p G_p(\theta, * * \theta) ,$$

again with equality if and only if $\theta = 0$. For the even grades, $*$ is symmetric and $**$ is positive, so we could repeat the arguments above. Yet for the odd grades, we need something new. Rather than separate into cases again, we proceed in the classical way.

Let us complexify our algebra. On $\Lambda^p \otimes \mathbb{C}$ we define

$$G_p^*(\psi, \theta) = \overline{\psi} \wedge *\theta .$$

This sesquilinear form serves as a Hermitian inner product. Define $\# = i^{p(p-1)+m}*$. Now, consider for $\psi, \theta \in \Lambda^p \otimes \mathbb{C}$,

$$\begin{aligned} G(\#\psi,\theta) &= \overline{\#\psi_{n-p}} \wedge *\theta_p + \overline{\#\psi_p} \wedge *\theta_{n-p} \\ &= (-i)^{(2m-p)(2m-p-1)+m}\overline{*\psi_{n-p}} \wedge *\theta_p + (-i)^{p(p-1)+m}\overline{*\psi_p} \wedge *\theta_{n-p}] \\ &= (-1)^p[(-i)^{(2m-p)(2m-p-1)+m} * \theta_p \wedge \overline{*\psi_{n-p}} + \\ &\qquad (-i)^{p(p-1)+m} * \theta_{n-p} \wedge \overline{*\psi_p}] \\ &= \overline{(-1)^p\left[i^{(2m-p)(2m-p-1)+p(p-1)+2m}(\overline{\#\theta_p} \wedge *\psi_{n-p} + \overline{\#\theta_{n-p}} \wedge *\psi_p)\right]} \\ &= \overline{G(\#\theta,\psi)} \\ &= G(\psi,\#\theta) \ . \end{aligned}$$

Here we have used that $i^{(2m-p)(2m-p-1)+p(p-1)+2m} = (-1)^{p^2} = (-1)^p$. Thus $\#$ is always Hermitian. Now the same arguments as before work for the operator $\#$. Namely,

$$0 \leq \|\#\theta\|^2 = G_p^*(\theta, \#\#\theta) \ ,$$

with equality if and only if $\theta = 0$. Thus $(\#)^2$ is positive definite, and an isometry with respect to G^* if and only if $(\#)^2 = 1$. Yet $\#\# = (-1)^p * *$. Hence we have proven the following.

Proposition 2.3. *G_p and G_{n-p} are compatible if and only if $\#$ is an isometry on the complexification. In this case, $(\#)^2 = 1$.*

In any case, we may orthogonally diagonalize $\#$ as before. We form the positive and negative eigenspaces,

$$\begin{aligned} \Lambda_+^{p,n-p} &= \mathrm{span}\{\xi \in (\Lambda^p \oplus \Lambda^{n-p}) \otimes \mathbb{C} : \#\xi = \lambda\xi,\ \lambda > 0\} \ ; \\ \Lambda_-^{p,n-p} &= \mathrm{span}\{\xi \in (\Lambda^p \oplus \Lambda^{n-p}) \otimes \mathbb{C} : \#\xi = \lambda\xi,\ \lambda < 0\} \ . \end{aligned}$$

For which we have $(\Lambda^p \oplus \Lambda^{n-p}) \otimes \mathbb{C} = \Lambda_+^{p,n-p} \oplus \Lambda_-^{p,n-p}$ with these summands being orthogonal. Let us write $\pi_\pm : (\Lambda^p \oplus \Lambda^{n-p}) \otimes \mathbb{C} \to \Lambda_\pm^{p,n-p}$ for the projections.

Identically, we find that for each positive eigenvalue-eigenvector pair (λ, ξ) we have that $(-\lambda, \tilde{\xi})$ is a negative eigenvalue-eigenvector pair. Therefore, the spectrum of $\#$ is symmetric about 0, and $\Lambda_+^{p,n-p} \cong \Lambda_-^{p,n-p}$.

Example 2.4. On Λ^0 and Λ^n, $n = 2m$, we see that $*1 = \sqrt{G}\, dx$ and $*\sqrt{G}\, dx = \frac{G}{G^n}$ where $G^n = G_{1j_1\cdots nj_n}\epsilon^{j_1\cdots j_n}$. Thus these metrics are compatible if and only if $G^n = G$; for Finsler surfaces ($n = 2$), this condition is strong enough to imply that the space is Riemannian, [BL2]. Nonetheless, in the basis $\{1, \sqrt{G}\,dx\}$ of $\Lambda^0 \oplus \Lambda^n$ (which is orthogonal but not orthonormal), we have

$$\# = \begin{pmatrix} 0 & i^m \\ (-i)^m \frac{G}{G^n} & 0 \end{pmatrix} .$$

The spectrum of $\#$ is $\{\pm(\frac{G}{G^n})^{\frac{1}{4}}\}$ with eigenvectors $\xi_\pm = (\frac{G}{G^n})^{-\frac{1}{4}} \pm (-i)^m \sqrt{G}\,dx$ (which are not unit length either).

If G_p and G_{n-p} are not compatible, we may normalize them in the following way. Recall that $\#\# = (-1)^p **$. This tells us that $(\#)^2$ is positive definite, but also that it is real, and hence symmetric. So we may take $(\#)^2$ as a mapping from Λ^p back into itself and construct the operator $\sqrt{\#\#} : \Lambda^p \to \Lambda^p$ which is also positive definite and symmetric. Let us define $\tilde{*} : \Lambda^p \to \Lambda^{n-p}$ by $\tilde{*} = *(\sqrt{\#\#})^{-1}$, and

$$\tilde{G}_p(\theta, \psi) = \theta \wedge \tilde{*}\psi .$$

The same procedure will yield the normalized $\tilde{G}_{n-p}$ as well. With this new star operator, we can construct $\tilde{\#}$ as before, which will have eigenspaces $\tilde{\Lambda}^{p,n-p}_\pm$. Considering the previous proposition, and the fact that $\#$ and $\tilde{\#}$ commute on $(\Lambda^p \oplus \Lambda^{n-p}) \otimes \mathbb{C}$ we have the following result.

Proposition 2.5. *$\tilde{G}_p$ and $\tilde{G}_{n-p}$ are compatible. Moreover $\tilde{\Lambda}^{p,n-p}_\pm = \Lambda^{p,n-p}_\pm$. If G_p and G_{n-p} are compatible to start, then $\tilde{G}_p = G_p$ and $\tilde{G}_{n-p} = G_{n-p}$.*

Finally we consider the case of $p = m$. First note that Λ^m is an even dimensional space regardless of the parity of m. Here $\# : \Lambda^m \otimes \mathbb{C} \to \Lambda^m \otimes \mathbb{C}$ is a Hermitian operator, so we may form its positive and negative eigenspaces as usual, $\Lambda^m \otimes \mathbb{C} = \Lambda^m_+ \oplus \Lambda^m_-$. As above, if $\#$ is not an isometry, then we can normalize it. The resulting operator $\tilde{\#}$ is then an isometry with respect to the normalized metric $\tilde{G}_m$, and the spaces $\Lambda^m_\pm$ are unchanged. Thus to analyze $\Lambda^m_\pm$, we may work with the normalized metric.

Proposition 2.6. $\Lambda^m_+ \cong \Lambda^m_-$.

Proof. Let $\{e^j\}$ be a $\tilde{G}$-orthonormal basis such that $\{e^1, \cdots, e^r\} \subset \Lambda^m_+$ and $\{e^{r+1}, \cdots, e^s\} \subset \Lambda^m_-$ where $s = \dim(\Lambda^m)$. Then the matrix representation for $\tilde{\#}$ in this basis is $A^{jk} = \tilde{G}^*_m(e^j, \tilde{\#}e^k)$. Noting that the eigenvalues of $\tilde{\#} = \pm 1$, the matrix A is really $A = \begin{pmatrix} I_r & 0 \\ 0 & -I_{s-r} \end{pmatrix}$. Clearly, the trace has $\mathrm{tr}(A) = 2r - s$.

Yet, $A^{jk} = \tilde{G}^*_m(e^j, \tilde{\#}e^k) = \overline{e^j} \wedge \tilde{*}\tilde{\#}e^k = (-i)^{m^2}\overline{e^j} \wedge e^k$. So let us compute the trace of A again, yet let us choose for the basis $\{e^j\}$, a natural basis $\{b^{j_1} \wedge \cdots \wedge b^{j_m}\}$. Now we have that,

$$(b^{j_1} \wedge \cdots \wedge b^{j_m}) \wedge (b^{j_1} \wedge \cdots \wedge b^{j_m}) = \pm(b^{j_1} \wedge b^{j_1} \wedge \cdots \wedge b^{j_m} \wedge b^{j_m}) = 0 \;,$$

as $\xi \wedge \xi = 0$ for any one form. Thus we see that $\overline{e^j} \wedge e^j = 0$ for all $j = 1, \cdots, s$. Thus $\mathrm{tr}(A) = \sum_j (-i)^{m^2} e^j \wedge e^j = 0$. Therefore, $2r = s$ and Λ^m_+ and Λ^m_- are of the same dimension. □

3 The Signature Complex

Suppose M is 4ℓ-dimensional. Then the signature of M, denoted $\tau(M)$, is the signature of the symmetric bilinear form on $H^{2\ell}(M)$ given by

$$([\omega], [\psi]) = \int_M \omega \wedge \psi \;.$$

Our short term goal is to show that $\tau(M)$ can also be realized as the index of an elliptic operator. So for this section, let us assume that M is 4ℓ-dimensional.

In the previous section we constructed a Hodge star on $\Lambda^*(\mathbb{R}^n)$ when each grade is equipped with a metric. We may extend the Hodge star to the exterior algebra of our Finsler space, $\Omega^*(M)$, by defining the action pointwise. As we have seen, if our manifold is even dimensional (which is the case of interest), we need to allow our forms to have complex values. Let us write $\Omega^*_{\mathbb{C}}(M)$ for the complex valued forms on M.

Perhaps the nicest feature of the Hodge star is the convenient expression for the codifferential. For the following, let $\theta \in \Omega^p_{\mathbb{C}}(M)$ and $\psi \in \Omega^{p+1}_{\mathbb{C}}(M)$. Then,

$$\begin{aligned}\langle \theta, d^*\psi \rangle &= \langle d\theta, \psi \rangle = \int_M \overline{d\theta} \wedge *\psi \\ &= \int_M d(\overline{\theta} \wedge *\psi) - (-1)^p \overline{\theta} \wedge d * \psi \\ &= (-1)^{p+1} \int_M \overline{\theta} \wedge * *^{-1} d * \psi \;.\end{aligned}$$

Therefore, $d^* = (-1)^{p+1} *^{-1} d * \psi$. For an even dimensional manifold, $n = 2m$, we express this in terms of $\#$ by

$$d^*\psi = (-1)^{p+1}(i^{p(p-1)+m}\#^{-1})d((-i)^{(p+1)p+m}\#)\psi = (-1)^{m+1}\#^{-1}d\#\psi \;.$$

If our space is 4ℓ-dimensional and algebraically normal, then $d^*\psi = -\#d\#\psi$.

To begin, let us assume our exterior algebra is algebraically normal. Since our space is 4ℓ-dimensional, our modified Hodge star $\# = i^{p(p-1)+2\ell}*$ is real despite its appearance. Therefore, when we diagonalize, we can take our eigenforms to be real valued, and define $\Lambda_{\pm}^{p,n-p}$ as we did before, but using the span over $\mathbb{R}$. We decompose the exterior algebra as $\Omega^p \oplus \Omega^{n-p} = \Omega_+^{p,n-p} \oplus \Omega_-^{p,n-p}$ where $\Omega_{\pm}^{p,n-p} = \pi_{\pm}(\Omega^p \oplus \Omega^{n-p})$ and $\pi_{\pm}$ is the pointwise projections onto $\Lambda_{\pm}^{p,n-p}$ extended to forms. We construct $\Omega_{\pm}^{2\ell}$ identically.

Let us consider the following spaces:

$$\Omega_+ = \Omega_+^{2\ell} \oplus \bigoplus_{0\le p<2\ell} \Omega_+^{p,n-p} \ ,$$

$$\Omega_- = \Omega_-^{2\ell} \oplus \bigoplus_{0\le p<2\ell} \Omega_-^{p,n-p} \ .$$

We have already seen that the fibres at each point of these two spaces are of the same dimension, so the situation is perfect for an elliptic operator between them. In particular, we take the Hodge-deRham operator: $d + d^*$.

Suppose $\omega \in \Omega_+$. Then

$$(d + d^*)\omega = d\omega - \#d\#\omega = (1 - \#)d\omega$$

is in Ω_-. Identically, $d+d^* : \Omega_- \to \Omega_+$. As the leading order symbol of the Hodge-deRham operator has $(\sigma(d + d^*))^2 = \sigma((d + d^*)^2) = \sigma\Delta$, which is injective for nonzero covectors, $\sigma(d + d^*)$ is injective for nonzero covectors as well. The dimensions of the two spaces coincide, therefore we see that this first order operator is elliptic.

Lemma 3.1. *As an elliptic operator* $d + d^* : \Omega_+ \to \Omega_-$, $\mathrm{index}_{\mathbb{R}}(d + d^*) = \tau(M)$.

Proof. Essentially, we have already proven this. For any $p \neq 2\ell$, the Hodge-deRham operator acting on $\Omega^p \oplus \Omega^{n-p}$ has kernel $\ker(d+d^*) = \mathcal{H}^p \oplus \mathcal{H}^{n-p}$. Let $\omega \in \pi_+(\mathcal{H}^p \oplus \mathcal{H}^{n-p})$ be written as $\omega = \omega_p + \omega_{n-p}$ as usual. Then,

$$\omega_p + \omega_{n-p} = \#\omega_p + \#\omega_{n-p} \ ,$$

so $\#\omega_p = \omega_{n-p}$ and $\#\omega_{n-p} = \omega_p$. Thus, $\tilde{\omega} = \omega_p - \omega_{n-p} \in \pi_-(\mathcal{H}^p \oplus \mathcal{H}^{n-p})$. If we write $b_{\pm}^{p,n-p} = \dim\ker(d+d^*)$ on $\Omega_{\pm}^{p,n-p}$, we see that $b_+^{p,n-p} = b_-^{p,n-p}$.

Now let $\psi \in \pi_+(\mathcal{H}^{2\ell})$ be nonzero. Then $\#\psi = *\psi = \psi$, so

$$0 < \langle \psi, \psi \rangle = \int_M \psi \wedge *\psi = \int_M \psi \wedge \psi \ .$$

Identically, if $\theta \in \pi_-(\mathcal{H}^{2\ell})$,then

$$0 < \langle \theta, \theta \rangle = \int_M \theta \wedge *\theta = - \int_M \theta \wedge \theta \, .$$

As $H^{2\ell}(M) \cong \mathcal{H}^{2\ell}$, we may take the harmonic representatives in each cohomology class. From the formulae above, we see that a self-dual harmonic form belongs to the positive eigenspace of the signature form, and the anti-self-dual forms are in the negative eigenspace. That is, $\tau(M) = b_+^{2\ell} - b_-^{2\ell}$ where we have continued the notation $b_\pm^{2\ell} = \dim \ker(d + d^*)$ on $\Omega_\pm^{2\ell}$.

Finally, on Ω_+ we have

$$\dim \ker(d + d^*) = b_+^{2\ell} + \sum_{0 \le p < 2\ell} b_+^{p,n-p}$$

and on Ω_- we have

$$\dim \ker(d + d^*) = b_-^{2\ell} + \sum_{0 \le p < 2\ell} b_-^{p,n-p} \, .$$

Whence,

$$\begin{aligned} \operatorname{index}_{\mathbf{R}}(d + d^*) &= \left(b_+^{2\ell} + \sum_{0 \le p < 2\ell} b_+^{p,n-p} \right) - \left(b_-^{2\ell} + \sum_{0 \le p < 2\ell} b_-^{p,n-p} \right) \\ &= b_+^{2\ell} - b_-^{2\ell} + \sum_{0 \le p < 2\ell} (b_+^{p,n-p} - b_-^{p,n-p}) = \tau(M) \, . \end{aligned}$$

□

We define the Hodge-deRham operator on a non-algebraically normal exterior algebra to be the usual Hodge-deRham operator of its algebraic normalization. To avoid the natural confusion here, we introduce the notation $\not{d}$ for the Hodge-deRham operator on an arbitrary exterior algebra; in the algebraically normal case, $\not{d} = d + d^*$. By proposition 2.3, the spaces $\Lambda_\pm^*$ are invariant under the normalization process, so we have $\not{d} : \Omega_+ \to \Omega_-$ as above, with index $\tau(M)$. Note, however, that the forms used above are harmonic for the normalized metric, not for the original metric. Gathering all this together, we have proven the following.

Theorem 3.2. *Let (M, F) be a Finsler manifold. Then the Hodge star induces spaces of self-dual and anti-self-dual forms in the exterior algebra, $\Omega_\pm$, for which the Hodge-deRham operator acts $\not{d} : \Omega_\pm \to \Omega_\mp$. Moreover, if $\dim(M) = 4\ell$,*

$$\operatorname{index}_{\mathbf{R}} \not{d} = \tau(M) \, .$$

One may be concerned with the fact that we have used an alternative metric to define our elliptic operator. Yet, we approach this with the idea that it is the spaces of self-dual and anti-self-dual forms that are fundamental to the theory. Once a Finsler structure is given, then these spaces are fixed, regardless of the normalization procedure. Naturally, unless the exterior algebra is algebraically normal, the kernel of the Hodge-deRham operator will not consist of harmonic forms. We consider harmonic forms, however, associated to the deRham complex. As we are considering an different complex altogether, we do not consider this facet to be problematical.

To avoid the natural confusion here, we will call a form in the kernel of the Hodge-deRham operator a signature form, and denote the collection of such as $\mathcal{S}^{\pm}$, according to self or anti-self duality. Thus, signature forms decompose into harmonic forms in the algebraically normal case, but in general do not.

4 A Lichnerowicz Vanishing Theorem

The Hodge-deRham operator is a first order differential operator serving as a connecting map of the elliptic complex

$$0 \rightarrow \Omega_{+} \xrightarrow{\not{d}} \Omega_{-} \rightarrow 0 \, .$$

As we have seen, $\not{d}^2 : \Omega_{\pm} \rightarrow \Omega_{\pm}$ is just the Laplace-Beltrami operator for the algebraically normalized deRham complex, restricted to the space of self-dual (respectively anti-self-dual) forms. As we will never consider the Laplacian of our given complex, we will use the notation $\not{d}^2 = \Delta$.

The leading order symbol of Hodge-deRham operator, $\sigma\not{d}$, is purely complex. We introduce the gamma matrices by writing

$$\sigma\not{d}(\xi) = i\gamma^j \xi_j \, .$$

Note that each of these matrices are real, and $\gamma^j : \Omega_{\pm} \rightarrow \Omega_{\mp}$. In fact, in a coordinate system, we can write our elliptic operator as $\not{d} = \gamma^j \frac{\partial}{\partial x^j} + S$, where S is some lower order term. If we polarize the statement $\sigma(\not{d})^2 = \sigma(\Delta)$ we get

$$\begin{aligned} \frac{1}{2}(\gamma^j\gamma^k + \gamma^k\gamma^j) &= -\sigma\Delta^{jk} \\ \frac{1}{2}(\gamma^j\gamma^k - \gamma^k\gamma^j) &= -\alpha\Delta^{jk} \, . \end{aligned}$$

In the first of these equations, the use justifies the notation: $\sigma\Delta$ is the matrix-valued tensor generated by the leading order symbol of $\not{d}^2$. Yet, in the second equation, one should merely take $\alpha\Delta$ as notation. If we had a Riemannian space, then not only is the exterior algebra normal, but $\sigma\Delta^{jk} = g^{jk} \cdot 1$. Thus the first of these equations is just the Clifford relation. In that case, the second equation tells us that $\alpha\Delta$ is the matrix which implements Clifford multiplication by two forms. We will see soon that even in the properly Finslerian case the character of $\alpha\Delta$ is still an analogue of Clifford multiplication.

As $\not{d}$ is self-adjoint with respect to the normalized metrics, we have

$$i\gamma^j \xi_j = \sigma\not{d}(\xi) = \sigma(\not{d}^*)(\xi) = \overline{\sigma\not{d}(\xi)} = -i(\gamma^j)^t \xi_j\,,$$

where $(\cdot)^t$ is the transpose of a matrix with respect to the normalized metric $\tilde{G}$. Thus $(\gamma^j)^t = -\gamma^j$ which will much simplify our expressions.

Warning: *Our use of the transpose map is unconventional. We compute the transpose with respect to the normalized metric $\tilde{G}$. Moreover, when we consider the symmetric and antisymmetric parts of matrix, it will always be with respect to $\tilde{G}$.*

Following [L], we define the torsion of a connection on $\Omega_\pm$ by the lowest order term in a covariant formula for $\not{d}$. In particular, we have four torsions to consider: $T^\pm_\pm$. These are defined by the following relations.

- For $\not{d} : \Omega_+ \to \Omega_-$,
  - $\not{d}\theta = \gamma^j \nabla^+_j \theta + T^+_+ \theta,$
  - $\not{d}\theta = \nabla^-_j (\gamma^j \theta) + T^-_+ \theta.$
- For $\not{d} : \Omega_- \to \Omega_+$,
  - $\not{d}\theta = \gamma^j \nabla^-_j \theta + T^+_- \theta,$
  - $\not{d}\theta = \nabla^+_j (\gamma^j \theta) + T^-_- \theta.$

Equivalently, one may compute two of the torsions directly:

$$\begin{aligned} T^+_+ &= \gamma^j \omega^+_j + S \\ T^-_+ &= -\frac{\partial}{\partial x^j}\gamma^j + \omega^-_j \gamma^j - \gamma^j \frac{\partial}{\partial x^j} \log\sqrt{G} + S \end{aligned}$$

where $\omega^\pm_k$ is the matrix of the self-dual (resp. anti-self-dual) connection forms. The other two we will find indirectly. Note that we have the curious relation

$$\nabla_j \gamma^j = T^+_+ - T^-_+\,.$$

Now, we will restrict our attention to $\tilde{G}$-metric compatible connections on $\Omega_\pm$. We may use [L, Prop'n 2.6] to compute the adjoint of $\not{d}\theta = \gamma^j\nabla_j^+\theta + T_+^+\theta$. The result is, for $\eta \in \Omega_-$,

$$\not{d}^*\eta = -\nabla_j^+((\gamma^j)^t\eta) + (T_+^+)^t\eta = \nabla_j^+(\gamma^j\eta) + (T_+^+)^t\eta\,.$$

Yet comparing this to the fourth of the equations above, we discover that $(T_+^+)^t = T_-^-$. Performing an identical computation for the third equation, and comparing the result to the second, we find $(T_+^-)^t = T_-^+$.

As an aside, we point out that 'torsion-free' is not a reasonable to condition to require. Two torsion matrices $T_+^\pm$ are both $2^{n-1}\times 2^{n-1}$; their vanishing yields 2^n linear equations in the connection forms. Yet, as we have already assume $\tilde{G}$-compatibility, there are $n^2(n-1)$ free variables left in the connection matrices. Thus, in general, requiring the torsions to vanish is not a well-posed linear system. Instead, we will look for a different condition to be satisfied.

We are prepared to formulate a Weitzenböck formula. For $\not{d}^2 : \Omega_+ \to \Omega_+$ direct computation yields

$$\begin{aligned}
\not{d}^2\theta &= \nabla_j^+\left[\gamma^j(\gamma^k\nabla_k^+\theta + T_+^+\theta)\right] + T_-^-\left(\gamma^k\nabla_k^+\theta + T_+^+\theta\right) \\
&= \nabla_j^+(\gamma^j\gamma^k\nabla_k^+\theta) + \nabla_j^+(\gamma^j T_+^+\theta) + T_-^-\gamma^k\nabla_k^+\theta + T_-^-T_+^+\theta \\
&= -\nabla_j^+(\sigma\Delta^{jk}\nabla_k^+\theta) + \left[\nabla_k^+(\alpha\Delta^{jk}) + \gamma^j T_+^+ + T_-^-\gamma^j\right]\nabla_j^+\theta \\
&\quad + \left[\frac{1}{2}\alpha\Delta^{jk}\Omega_{jk}^+ + \nabla_j^+(\gamma^j T_+^+) + T_-^-T_+^+\right]\theta\,.
\end{aligned}$$

We have used that $\alpha\Delta^{jk}\nabla_j^+\nabla_k^+ = -\frac{1}{2}\alpha\Delta^{jk}\Omega_{jk}^+$ where Ω^+ is the curvature of ∇^+; the minus sign is taken as we are working on covector bundles. Note that such a term in effect 'multiplies' a self-dual form by the curvature of our connection; this solidifies the use of $\alpha\Delta$ as the analogue of Clifford multiplication by two forms.

Note that the coefficient of the first order term is exactly as in [L], as expected. The idea is to choose the connection on Ω_+ so that this first order term vanishes. To understand the basic lemma toward this end, we must introduce the idea of a completely positive symbol.

Definition 4.1. Consider the matrix valued bilinear form generated by polarizing the leading order symbol of the Laplacian:

$$\sigma\Delta(\xi,\psi) = \frac{1}{4}\left(\sigma\Delta(\xi+\psi) - \sigma\Delta(\xi-\psi)\right)\,.$$

Then the symbol of the Laplacian is said to be completely positive (resp. completely negative), if for any basis, $\{\xi_j\}$, of T^*M the $n\times n$ matrix of matrices $(\sigma\Delta(\xi_j,\xi_k))_{j,k=1}^n$ is positive definite (resp. negative definite).

In Riemannian geometry, we have $\sigma\Delta(\xi,\psi) = g(\xi,\psi)\cdot 1$. In this case it is clear that the symbol is completely positive. Yet, in the properly Finslerian case, it is not clear when this condition is satisfied, see [BL2] for some progress on Finsler surfaces.

Lemma 4.2 ([L, Lemma 2.10]). *Consider the equation*

$$\sigma\Delta^{jk}\omega_k - \left(\sigma\Delta^{jk}\omega_k\right)^t = K^j \tag{1}$$

where $\sigma\Delta$ is the leading order symbol of the Laplacian, K is a fixed collection of skew-symmetric matrices, and $\omega_k\, dx^k$ is a skew-symmetric matrix-valued one form to be determined by this equation. Then, there exists a unique solution to (1) if the symbol of the Laplacian is completely positive.

One may rewrite the equation $\nabla_k^+(\alpha\Delta^{jk}) + \gamma^j T_+^+ + T_-^-\gamma^j = 0$ as $\sigma\Delta^{jk}\omega_k - \left(\sigma\Delta^{jk}\omega_k\right)^t = K^j$. Here, ω_k is the skew-symmetric part of the connection form (the symmetric part is determined by metric compatibility), and K^j is a collection of skew-symmetric matrices completely determined by γ^j, $\tilde{G}$, and $\sqrt{G}$ (see [L] for a formula). It is important to note that the arbitrarily assigned volume compatible connection Γ does not appear in the K terms. In light of the lemma above, we will assume the symbol of the Laplacian is completely positive, and thus there exists a unique connection (independent of Γ) for which the first order term in the Weitzenböck formula vanishes.

Taking the equation $\nabla_k^+(\alpha\Delta^{jk}) + \gamma^j T_+^+ + T_-^-\gamma^j = 0$, we covariantly differentiate and trace on the j index, yielding the Bianchi identity,

$$-\frac{1}{2}\Omega_{jk}^+\alpha\Delta^{jk} + \frac{1}{2}\alpha\Delta^{jk}\Omega_{jk}^+ + \nabla_j^+(\gamma^j T_+^+) + \nabla_j^+(T_-^-\gamma^j) = 0\ .$$

The left side of this equation is twice the antisymmetric part of the zeroth order term in our formula, which then vanishes. We define the self-dual curvature as twice the symmetric part of the zeroth order term.

Performing the calculation for $\not{d}^2 : \Omega_- \to \Omega_-$ is identical. In fact, the computation only involves interchanging some $+$ and $-$ signs, hence is left to the reader. In conclusion we have the following result.

Theorem 4.3. *Let (M,F) be a Finsler manifold with whose algebraically normalized deRham complex has completely positive Laplacian. Then there exist unique self-dual and anti-self-dual connections such that the following formulae hold for $\not{d} : \Omega_\pm \to \Omega_\pm$.*

$$\not{d}^2\theta = -\nabla_j^\pm(\sigma\Delta^{jk}\nabla_k^\pm\theta) + \mathbf{R}^\pm\theta\ ,$$

where

$$\mathbf{R}^{\pm} = \frac{1}{4}\alpha\Delta^{jk}\Omega^{\pm}_{jk} + \frac{1}{4}\Omega^{\pm}_{jk}\alpha\Delta^{jk} + \frac{1}{2}\nabla^{\pm}_{j}(\gamma^{j}T^{+}_{+} - T^{-}_{-}\gamma^{j}) + T^{\mp}_{-}T^{\pm}_{+}$$

are the self-dual and anti-self-dual curvature endomorphisms.

We follow the same line of logic as in [L]. Our fundamental estimate is the following.

Lemma 4.4. *Let $\theta \in \mathcal{S}^{\pm}$ be a self-dual (respectively anti-self-dual) signature form. Then*

$$\langle \theta, \mathbf{R}^{\pm}\theta \rangle \leq 0 \ ,$$

with equality holding if and only if $\nabla^{\pm}\theta = 0$.

Proof. We show the self-dual case, the anti-self-dual is identical. From the Weitzenböck formula, for $\theta \in \mathcal{S}^{+}$,

$$0 = \langle \theta, \not{d}^{2}\theta \rangle = \langle \nabla^{+}_{j}\theta, \sigma\Delta^{jk}\nabla^{+}_{k}\theta \rangle + \langle \theta, \mathbf{R}^{+}\theta \rangle \ .$$

As the symbol of the Laplacian is completely positive, the first term on the right is nonnegative. We conclude that $\langle \theta, \mathbf{R}^{+}\theta \rangle \leq 0$ with equality if and only if $\langle \nabla^{+}_{j}\theta, \sigma\Delta^{jk}\nabla^{+}_{k}\theta \rangle = 0$ as well. Yet this implies that $\nabla^{+}\theta = 0$. □

Note that our curvature endomorphisms are *not* symmetric with respect to the given metric G. Yet, they are symmetric with respect to $\tilde{G}$, hence are diagonalizable over $\mathbb{R}$. Therefore, positivity of the curvature endomorphisms is not a vacuous idea. Finally we achieve the main result.

Theorem 4.5. *Let (M, F) be a compact and boundaryless, oriented manifold Finsler 4ℓ-manifold whose algebraically normalized deRham complex has completely positive Laplacian. Then if the self-dual curvature endomorphism is positive definite, the signature of M is nonpositive. If the anti-self-dual curvature endomorphism is positive definite, then the signature of M is nonnegative. Hence if both curvatures are positive, the signature of M vanishes.*

Proof. From the previous section, $\tau(M) = \operatorname{index} \not{d} = \dim \mathcal{S}^{+} - \dim \mathcal{S}^{-}$. From our estimate, if $\mathbf{R}^{+}$ is positive definite, then $\mathcal{S}^{+} = 0$. Whence, $\tau(M) = -\dim \mathcal{S}^{-}$ is nonpositive. The rest of the theorem is clear. □

If our Finsler space is Riemannian, then many things simplify in this construction. First, Riemannian spaces are always algebraically normal, so there is no need for the normalization process. Moreover, signature forms will decompose into harmonic forms. The symbol of the Laplacian is just $g \otimes 1$ which is clearly completely positive. As well, the Christoffel (Levi-Civita) connection satisfies the conditions of theorem 4.3, so by uniqueness, this connection is to serve the role as both the self-dual and anti-self dual connections in the theory. Thus, both the self-dual and anti-self-dual curvature endomorphisms coincide (they are both equal to $\frac{s}{4} \cdot 1$ where s is the scalar curvature of Riemannian metric). Therefore, in the final theorem, positivity of one of the curvatures implies positivity of both, hence a vanishing signature. Thus, Riemannian geometry is rigid in the sense that the sign of the signature cannot be detected by curvature, only it vanishing.

To be fair, we should point out that the complete positivity condition is, although easy to state, very complicated to verify in general. It is still unknown whether there are Finsler spaces for which the symbol of the Laplacian is not completely positive; see [BL2] for some progress on this issue for Finsler surfaces.

Also, although theorem 4.3, guarantees a unique connection when the symbol of the Laplacian is completely positive, it is not clear how to solve (1) to find a formula for this connection. Note that we discovered that these connections were just the Christoffel connection in Riemannian geometry from the form of the Wietzenböck formula. I would seem a daunting task to verify (1) directly.

Finally, the author would like to thank Prof. P.L. Antonelli, Prof. D. Bao, and Prof. S.S. Chern for their continued interest in this theory, and thank the Izaak W. Killam trust for their financial support.

References

[BL1] Bao, D. and Lackey, B. (1996) A Hodge Decomposition Theorem for Finsler Spaces, *C. R. Acad. Sci. Paris*, **323**, 51-56.

[BL2] Bao, D. and Lackey, B. (to appear) A Geometric Inequality and a Weitzenböck Formula for Finsler Surfaces, *(this volume)*.

[BS] Bao, D. and Shen, Z. (1994) On the Volume of Unit Tangent Spheres in a Finsler Manifold, *Results in Math.* **26**, 1-17.

[H] Hirzebruch, F. (1966) *Topological Methods in Algebraic Geometry*, 3rd. Ed., Springer.

[L] Lackey, Brad (to appear) *A Bochner Vanishing Theorem for Elliptic Complices, (this volume).*

[N] Nakahara, M. (1990) *Geometry, Topology, and Physics*, IOP Publishing.

A GEOMETRIC INEQUALITY AND A WEITZENBÖCK FORMULA FOR FINSLER SURFACES

DAVID BAO and BRAD LACKEY *

1 Preliminaries, and the Laplace-Beltrami Operator on M

Throughout this paper, let M be a compact orientable manifold without boundary. A Finsler structure is a most natural generalization of a Riemannian metric. Rather than endow each tangent space T_xM with an inner product, the geometry is induced by a Minkowski norm $F(x,y)$ on T_xM. We will assume that the function F on TM is positive homogeneous of degree one in y, and smooth at $y \neq 0$. The fundamental tensor is defined as $g_{ij} := (\frac{1}{2}F^2)_{y^iy^j} = F\,F_{y^iy^j} + F_{y^i}\,F_{y^j}$, where F_{y^i} means $\frac{\partial F}{\partial y^i}$, etc. It is presumed to be *positive definite*. Since the g_{ij} 's are homogeneous of degree zero in y, they are well-defined quantities on the sphere bundle SM, provided that we regard the y^i 's as homogeneous coordinates.

Before analyzing SM, let us digress briefly to consider $TM \setminus 0$. The canonical projection $\pi : TM \to M$ gives rise to a covector bundle π^*T^*M which has the Hilbert form $\omega := F_{y^i}\,dx^i$ as a distinguished global section. Each fibre on $TM \setminus 0$ also has an inner product induced by the g^{ij}. The concept of a positively oriented orthonormal coframe $\{\omega^a\}$ (for π^*T^*M) with $\omega^n := \omega$ now makes sense. Here, n is the dimension of M. Expand ω^a as $v^a{}_i\,dx^i$, whereby the stipulated orientation implies that $\det(v^a{}_i) = \sqrt{\det(g_{ij})}$. Through an abuse of notation, we shall for later convenience regard each section ω^a as a differential form on $TM \setminus 0$.

Next, recall the Cartan tensor $A_{ijk} := \frac{F}{4}\,(F^2)_{y^iy^jy^k}$, the formal Christoffel symbols $\gamma^i{}_{jk}$ of the fundamental tensor, and the object $\ell := \ell^i\,\frac{\partial}{\partial x^i}$ dual to the Hilbert form. We introduce the quantity $N^i{}_j := \gamma^i{}_{jk}\,\ell^k -$

*Research supported by the Izaak W. Killam Trust.

$A^i{}_{jk}\,\gamma^k{}_{rs}\,\ell^r\,\ell^s$, then define $\delta y^i := \frac{1}{F}(dy^i + F\,N^i{}_j\,dx^j)$ and $\omega^{n+a} := v^a{}_i\,\delta y^i$. Note that $\omega^{n+n} = \ell_i\,\delta y^i = d(\log F)$; it is dual to the 'radial' vector $y^i\,\frac{\partial}{\partial y^i}$. The collection $\{\omega^a, \omega^{n+a}\}$ (where $a = 1, \dots, n$) forms an orthonormal basis for $T^*(TM \setminus 0)$ with respect to the Sasaki metric $g_{ij}\,dx^i \otimes dx^j + g_{ij}\,\delta y^i \otimes \delta y^j$ on $TM \setminus 0$.

The above constructions are invariant under the rescaling $y \mapsto \lambda\,y$, $\lambda > 0$. Hence all the objects (except F itself) make sense on SM. One also sees that, on SM, the form $d(\log F)$ vanishes.

We now list the apparatus on SM that will be relevant for our purposes. There is a Riemannian metric $g_{ij}\,dx^i \otimes dx^j + \delta_{\alpha\beta}\,\omega^{n+\alpha} \otimes \omega^{n+\beta}$, equivalently $\delta_{ab}\,\omega^a \otimes \omega^b + \delta_{\alpha\beta}\,\omega^{n+\alpha} \otimes \omega^{n+\beta}$, for which the collection $\{\omega^a, \omega^{n+\alpha}\}$ (where $a = 1, \dots, n$ & $\alpha = 1, \dots, n-1$) is an *ordered* orthonormal coframe. This metric is simply the pull-back of the Sasaki metric from $TM \setminus 0$ to SM. It defines a Hodge star with respect to the top degree form $\Pi := \omega^1 \wedge \cdots \wedge \omega^n \wedge \omega^{n+1} \wedge \cdots \wedge \omega^{n+(n-1)}$, and thus an $\mathcal{L}_2$ inner product

$$\langle\, \theta\,,\, \psi\, \rangle_k \;\; := \;\; \int_{SM} \frac{(-1)^{n-1}}{\mathrm{Vol}(S_x M)}\;\theta \wedge *\psi$$

on the differential k-forms of SM, whenever M is compact and n-dimensional. Here, $\mathrm{Vol}(S_x M)$ is the volume of the fibre $S_x M$ (treated as isometric to the indicatrix in $T_x M$); it is a function on M, hence can be regarded as one on SM. This volume function is constant for Berwald and Landsberg spaces, and reduces in the Riemannian setting to $\mathrm{Vol}(S^{n-1})$, the volume of the unit sphere in Euclidean $\mathbb{R}^n$. We use it to normalize $\langle\,,\,\rangle$ so that the objects we construct agree with the usual ones in case our Finsler structure is Riemannian. For this purpose, we could just as well have used the universal constant $\mathrm{Vol}(S^{n-1})$ in the normalization. The reason for the sign $(-1)^{n-1}$ will be made clear near the end of the next paragraph.

Let $\theta := \frac{1}{k!}\,\theta_I(x)\,dx^I$, $\psi := \frac{1}{k!}\,\psi_J(x)\,dx^J$ be smooth k-forms on M. We pull them up to SM, re-express in terms of the orthonormal ω^a 's, and then apply the normalized inner product described above. In that regard, $\theta = \frac{1}{k!}\,\theta_A\,\omega^A$ and, by limiting the multi-indices A to those of increasing order $|A|$, we have $\theta = \theta_{|A|}\,\omega^{|A|}$. Do the same for ψ. A moment's thought then gives $\theta \wedge *\psi = \sum_{|A|}\,\theta_{|A|}\,\psi_{|A|}\,\Pi$, which is $\frac{1}{k!}\,\sum_A\,\theta_A\,\psi_A\,\Pi$. Now note that $\omega^1 \wedge \cdots \wedge \omega^n = \sqrt{g}\,dx$, where $\sqrt{g}$ means $\sqrt{\det(g_{ij})}$ and dx abbreviates $dx^1 \wedge \cdots \wedge dx^n$. So $\theta \wedge *\psi = \frac{1}{k!}\,(g^{IJ}\theta_I\psi_J)\,(\sqrt{g}\,dx) \wedge \omega^{n+1} \wedge \cdots \wedge \omega^{n+(n-1)}$. This, together with the fact that θ_I and ψ_J have no y dependence of any sort, allows us to see that

$$\langle\, \theta\,,\, \psi\, \rangle_k \;\; = \;\; \frac{1}{k!}\,\int_M \theta_{i_1 \cdots i_k}\,\psi_{j_1 \cdots j_k}\,\left(\int_{S_x M} \sqrt{g}\;g^{i_1 j_1} \cdots g^{i_k j_k}\;\zeta \right)\,dx\,.$$

Here, ζ denotes $\frac{(-1)^{n-1}}{\mathrm{Vol}(S_xM)}$ times the pure dy part of $\omega^{n+1} \wedge \cdots \wedge \omega^{n+(n-1)}$, which is $\frac{1}{\mathrm{Vol}(S_xM)}$ times the volume form of S_xM. This formula says that when the $\mathcal{L}_2$ inner product on SM is applied to pulled-back forms, the answer is actually an $\mathcal{L}_2$ inner product on $\Omega^k(M)$, defined using the tensor *density* enclosed in parentheses.

Let us introduce the objects

$$\sqrt{G}\,(x) \ := \ \int_{S_xM} \sqrt{g}\;\zeta$$

and

$$G^{IJ}(x) \ := \ G^{i_1 j_1 \cdots i_k j_k}(x) \ := \ \frac{1}{\sqrt{G}} \int_{S_xM} \sqrt{g}\; g^{i_1 j_1} \cdots g^{i_k j_k}\; \zeta\ .$$

Then

$$\langle\, \theta\, ,\, \psi\, \rangle_k \ = \ \frac{1}{k!} \int_M (\, \theta_I\, G^{IJ}\, \psi_J\,)\, \sqrt{G}\, dx\ .$$

A local change of coordinates on M induces one on SM. Using the resulting transformation laws on the fundamental tensor and the invariance of the $\omega^{n+\alpha}$'s, one can check that $\sqrt{G}$ is a tensor density of weight 1, and G^{IJ} is a contravariant tensor of rank $2k$, on M. Observe that in the Riemannian setting, $\mathrm{Vol}(S_xM)$ becomes the universal constant $\mathrm{Vol}(S^{n-1})$, whence $\sqrt{G}$ and G^{IJ} reduce respectively to $\sqrt{g}$ and $g^{i_1 j_1} \cdots g^{i_k j_k}$, thereby explaining our use of a normalization.

The symmetry properties of G^{IJ} can be read off from its defining formula. It is invertible as a map from k-covectors to k-vectors; denote the inverse by $G_{IJ} := G_{i_1 j_1 \cdots i_k j_k}$. It is also clear that for each degree k, $G^{IJ}(x)$ actually define an inner product on the exterior algebra Λ^k_x. Unless the Finsler structure is Riemannian, these inner products (for $k \geq 2$) cannot be interpreted as coming from any Riemannian metric on M. Even in the case $k = 1$, the $G^{ij}(x)$ obtained is, in view of the weight factor $\sqrt{g}$, *not* exactly the S_xM average of the fundamental tensor. In later sections, we will:

- use $k!\, G_{IJ}$ to lower multi-indices of length k,
- use $\frac{G^{IJ}}{k!}$ to raise multi-indices of length k.

Note the constant multiples in front of the bundle metric and its inverse.

The codifferential d^* on M is characterized by

$$\langle\, \theta\, ,\, d^*\xi\, \rangle_k \ = \ \langle\, d\theta\, ,\, \xi\, \rangle_{k+1}\ .$$

Having d^*, the Laplace-Beltrami operator on M is then given by

$$\Delta \ := \ d^*d + d\,d^* \ .$$

One can check that in the Riemannian setting, the adjoint operator d^* reduces to the usual codifferential, and hence Δ reproduces the familiar Laplacian.

2 Covariantizing the Exterior Derivative d

Consider now the Laplace-Beltrami operator acting on k-forms. Its definition is an anti-commutator and involves two exterior differentials:

- One acts on $(k-1)$-forms

$$\phi \ = \ \phi_A \ \frac{dx^A}{(k-1)!} \ ,$$

  with output

$$d\phi \ = \ (\ \mathcal{A}^{iA}{}_I \ \partial_i \phi_A \ ) \ \frac{dx^I}{k!} \ .$$

- The other acts on k-forms

$$\theta \ = \ \theta_I \ \frac{dx^I}{k!} \ ,$$

  with output

$$d\theta \ = \ (\ \mathcal{B}^{pI}{}_P \ \partial_p \theta_I \ ) \ \frac{dx^P}{(k+1)!} \ .$$

Here, ∂_i and ∂_p refer to $\frac{\partial}{\partial x^i}$ and $\frac{\partial}{\partial x^p}$. Also, the abbreviations

$$\mathcal{A}^{iA}{}_I \ := \ \frac{\epsilon^{iA}{}_I}{(k-1)!}$$

$$\mathcal{B}^{pI}{}_P \ := \ \frac{\epsilon^{pI}{}_P}{k!}$$

are the *constant* Kronecker-epsilon tensors (albeit normalized) on M.

Presently, we have $(d\phi)_I = \mathcal{A}^{iA}{}_I \ \partial_i \phi_A$. For reasons that will become clear in a later section, we would like to re-express it as

$$(d\phi)_I \ = \ \nabla_i(\ \mathcal{A}^{iA}{}_I \ \phi_A \ ) \ + \ \mathfrak{t}_{-}{}^{A}{}_{I} \ \phi_A \ .$$

The right-hand side refers to the expression

$$\partial_i(\mathcal{A}^{iA}{}_I \, \phi_A) \; + \; (\mathcal{A}^{lA}{}_I \, \phi_A) \, \Gamma^i{}_{li} \; - \; (\mathcal{A}^{iA}{}_L \, \phi_A) \, \omega_I{}^L{}_i \; + \; \mathfrak{t}_-{}^A{}_I \, \phi_A \, .$$

As we can see, a manifold connection Γ and a bundle connection ω have been introduced. Our re-write of $(d\phi)_I$ will be successful if we choose the 'torsion' term $\mathfrak{t}_-$ to be

$$\mathfrak{t}_-{}^A{}_I \;\; = \;\; \mathcal{A}^{iA}{}_L \, \omega_I{}^L{}_i \; - \; \mathcal{A}^{lA}{}_I \, \Gamma^i{}_{li} \, .$$

In contradistinction to the above, we would like to re-write the equation $(d\theta)_P = \mathcal{B}^{pI}{}_P \, \partial_p \theta_I$ as

$$(d\theta)_P \;\; = \;\; \mathcal{B}^{pI}{}_P \, \nabla_p \theta_I \; + \; \mathfrak{t}_+{}^I{}_P \, \theta_I \, .$$

This can be accomplished if we choose the 'torsion' term $\mathfrak{t}_+$ as follows:

$$\mathfrak{t}_+{}^I{}_P \;\; = \;\; \mathcal{B}^{pL}{}_P \, \omega_L{}^I{}_p \, .$$

These formulas constitute the desired covariantization of the exterior derivative.

Let us conclude this section with a brief review of our index conventions:

- Manifold indices on M are always denoted by lower case Latin indices such as i, j, k, l and p, q, r, s. Typically, there is no canonical Riemannian metric on any given Finsler space M. Therefore these manifold indices *cannot* be raised and lowered in general.

- Multi-indices on differential forms are denoted by upper case Latin indices. We use A, B, C, D for $(k-1)$-forms, I, J, K, L for k-forms, and P, Q, R, S for $(k+1)$-forms. These are raised and lowered by a *weighted* version of the G 's, as described in §1.

3 The Divergence Lemma

We now require that our manifold connection satisfies the following trace condition:

$$\Gamma^j{}_{kj} = \partial_k \log \sqrt{G} \, .$$

This can be realized, for example, by letting $\Gamma^i{}_{jk}$ be the Christoffel symbols (of the second kind) of

$$\left[\frac{\sqrt{G}}{\sqrt{\det(h_{ij})}} \right]^{2/n} h \;\; ,$$

where h is an arbitrary Riemannian metric on M.

Given this trace condition, one has:

$$\int_M (\nabla_i V^i)\,\sqrt{G}\,dx \;=\; \int_M \partial_i(V^i\,\sqrt{G}\,)\,dx\,.$$

Indeed:

$$\begin{aligned}
&\int_M (\partial_i V^i \,+\, V^k\,\Gamma^i{}_{ki}\,)\,\sqrt{G}\,dx \\
=\;& \int_M \Big[\,(\partial_i V^i)\,\sqrt{G} \,+\, V^k \Big(\partial_k \,\log\sqrt{G}\Big)\,\sqrt{G}\,\Big]\,dx \\
=\;& \int_M \Big[\,(\partial_i V^i)\,\sqrt{G} \,+\, V^i\,\partial_i\sqrt{G}\,\Big]\,dx \\
=\;& \int_M \partial_i(V^i\,\sqrt{G}\,)\,dx\,.
\end{aligned}$$

As a corollary, we note that

$$\int_M (\nabla_i V^i)\,\sqrt{G}\,dx \;=\; 0$$

whenever

1. M is a compact orientable manifold without boundary (which is the case here), and

2. V is a *globally* defined vector field on M.

This is the divergence lemma we seek.

4 Covariantizing the Adjoint of d

Recall from §2 that our bundle connection has 'torsion'. In order to keep things simple, we now require it to be compatible with the bundle metric. Namely:

$$\nabla_k\,G^{IJ} \;=\; 0\,.$$

A straightforward calculation shows that such is equivalent to

$$(G^{-1}\partial_k G)_J^{\;I} \;:=\; G^{IL}\,\partial_k G_{LJ} \;=\; \omega_J{}^I{}_k \,+\, \omega^I{}_{Jk}\,.$$

This second characterization of metric compatibility will be useful to us in a later section.

There are two co-differentials in the formula of our Laplace-Beltrami operator Δ. One acts on k-forms θ, while the other acts on $(k+1)$-forms ξ. We would like to covariantize them.

For $d^*\theta$, we use the defining relation

$$\langle \theta , d\phi \rangle_k \quad = \quad \langle d^*\theta , \phi \rangle_{k-1} .$$

Into this we substitute the covariantized formula for $d\phi$, as described in §2. Then we use the divergence lemma of §3, and the metric compatibility stated above. The actual computation reads:

$$\begin{aligned}
& \frac{1}{k!} \int_M G^{IJ}\, \theta_I\, (d\phi)_J \sqrt{G}\; dx \\
= \; & \frac{1}{k!} \int_M G^{IJ}\, \theta_I \left[\nabla_i (\mathcal{A}^{iA}{}_J\, \phi_A) + \mathfrak{t}_{-}{}^{A}{}_{J}\, \phi_A \right] \sqrt{G}\; dx \\
= \; & \frac{1}{k!} \int_M \left[- \nabla_i (G^{IJ}\, \theta_I)\, \mathcal{A}^{iA}{}_J\, \phi_A + G^{IJ}\, \mathfrak{t}_{-}{}^{A}{}_{J}\, \theta_I\, \phi_A \right] \sqrt{G}\; dx \\
= \; & \frac{1}{(k-1)!} \int_M (k-1)!\, G_{BC} \left[- (\nabla_i \theta_I)\, \mathcal{A}^{iC}{}_J + \theta_I\, \mathfrak{t}_{-}{}^{C}{}_{J} \right] \frac{G^{IJ}}{k!} \\
& \qquad\qquad G^{AB}\, \phi_A \sqrt{G}\; dx \\
= \; & \frac{1}{(k-1)!} \int_M \left[- (\nabla_i \theta_I)\, \mathcal{A}^{i}{}_{B}{}^{I} + \theta_I\, \mathfrak{t}_{-B}{}^{I} \right] G^{AB}\, \phi_A \sqrt{G}\; dx \\
= \; & \frac{1}{(k-1)!} \int_M \left[- \mathcal{A}^{i}{}_{A}{}^{I}\, (\nabla_i \theta_I) + \mathfrak{t}_{-A}{}^{I}\, \theta_I \right] G^{AB}\, \phi_B \sqrt{G}\; dx .
\end{aligned}$$

Thus, given any k-form $\theta := \frac{1}{k!}\, \theta_I\, dx^I$, we have:

$$(d^*\theta)_A \quad = \quad - \mathcal{A}^{i}{}_{A}{}^{I}\, (\nabla_i \theta_I) + \mathfrak{t}_{-A}{}^{I}\, \theta_I ,$$

where

$$\begin{aligned}
\mathcal{A}^{i}{}_{A}{}^{I} & := \left[(k-1)!\, G_{AB} \right] \mathcal{A}^{iB}{}_J \left[\frac{G^{JI}}{k!} \right] , \\
\mathfrak{t}_{-A}{}^{I} & := \left[(k-1)!\, G_{AB} \right] \mathfrak{t}_{-}{}^{B}{}_{J} \left[\frac{G^{JI}}{k!} \right] .
\end{aligned}$$

Likewise, given any $(k+1)$-form $\xi := \frac{1}{(k+1)!}\, \xi_P\, dx^P$, we start with the relation

$$\langle \xi , d\theta \rangle_{k+1} \quad = \quad \langle d^*\xi , \theta \rangle_k$$

and perform a similar calculation as the above. The conclusion is:

$$(d^*\xi)_I \quad = \quad - \nabla_p (\mathcal{B}^{p}{}_{I}{}^{P}\, \xi_P) + \mathfrak{t}_{+I}{}^{P}\, \xi_P ,$$

where

$$\mathcal{B}^{p}{}_{I}{}^{P} \;:=\; [k!\, G_{IJ}]\; \mathcal{B}^{pJ}{}_{Q} \left[\frac{G^{QP}}{(k+1)!}\right],$$

$$t_{+I}{}^{P} \;:=\; [k!\, G_{IJ}]\; t_{+}{}^{J}{}_{Q} \left[\frac{G^{QP}}{(k+1)!}\right].$$

5 The Laplace-Beltrami Operator in Covariantized Form

Our Laplace-Beltrami operator is

$$\Delta \;:=\; d^{*}d \;+\; d\,d^{*} \;.$$

We let it act on a k-form $\theta := \frac{1}{k!}\,\theta_I\, dx^I$ to result in $\Delta\theta := \frac{1}{k!}\,(\Delta\theta)_I\, dx^I$:

- For $d^{*}d\theta$, we first substitute the covariantized formula of $\xi := d\theta$, followed by that of $d^{*}\xi$.
- For $d\,d^{*}\theta$, we first substitute the covariantized formula of $\phi := d^{*}\theta$, followed by that of $d\phi$.

These formulas are given in §2 and §4. After all is said and done, we obtain a covariantized formula for $(\Delta\theta)_I$. It reads:

$$\begin{aligned}
&-\nabla_i\Big([\, \mathcal{B}^{i}{}_{I}{}^{S}\, \mathcal{B}^{jJ}{}_{S} \;+\; \mathcal{A}^{iC}{}_{I}\, \mathcal{A}^{j}{}_{C}{}^{J}\,]\, \nabla_j\theta_J \Big) \\
&+\Big(\mathcal{A}^{jC}{}_{I}\, t_{-C}{}^{J} \;-\; t_{-}{}^{C}{}_{I}\, \mathcal{A}^{j}{}_{C}{}^{J} \;+\; t_{+I}{}^{S}\, \mathcal{B}^{jJ}{}_{S} \;-\; \mathcal{B}^{j}{}_{I}{}^{S}\, t_{+}{}^{J}{}_{S} \Big)\, \nabla_j\theta_J \\
&+\Big(\nabla_j[\, \mathcal{A}^{jC}{}_{I}\, t_{-C}{}^{J}\,] \;-\; \nabla_j[\, \mathcal{B}^{j}{}_{I}{}^{S}\, t_{+}{}^{J}{}_{S}\,] \;+\; t_{-}{}^{C}{}_{I}\, t_{-C}{}^{J} \;+\; t_{+I}{}^{S}\, t_{+}{}^{J}{}_{S} \Big)\, \theta_J\,.
\end{aligned}$$

The quantity

$$\mathcal{B}^{i}{}_{I}{}^{S}\, \mathcal{B}^{jJ}{}_{S} \;+\; \mathcal{A}^{iC}{}_{I}\, \mathcal{A}^{j}{}_{C}{}^{J}$$

we now decompose into two pieces: the symmetric and skew-symmetric parts with respect to the indices i, j.

$$\sigma^{ij}{}_{I}{}^{J} \;:=\; \tfrac{1}{2}\,(\, \mathcal{B}^{i}{}_{I}{}^{S}\, \mathcal{B}^{jJ}{}_{S} \;+\; \mathcal{A}^{iC}{}_{I}\, \mathcal{A}^{j}{}_{C}{}^{J} \;+\; \mathcal{B}^{j}{}_{I}{}^{S}\, \mathcal{B}^{iJ}{}_{S} \;+\; \mathcal{A}^{jC}{}_{I}\, \mathcal{A}^{i}{}_{C}{}^{J}\,)\,,$$

$$\alpha^{ij}{}_{I}{}^{J} \;:=\; \tfrac{1}{2}\,(\, \mathcal{B}^{i}{}_{I}{}^{S}\, \mathcal{B}^{jJ}{}_{S} \;+\; \mathcal{A}^{iC}{}_{I}\, \mathcal{A}^{j}{}_{C}{}^{J} \;-\; \mathcal{B}^{j}{}_{I}{}^{S}\, \mathcal{B}^{iJ}{}_{S} \;-\; \mathcal{A}^{jC}{}_{I}\, \mathcal{A}^{i}{}_{C}{}^{J}\,)\,.$$

We will see in §7 that $\sigma^{ij}{}_{I}{}^{J}$ is *essentially* the symbol map of Δ.

Using this decomposition, the second order term in $(\Delta\theta)_I$ becomes

$$-\,\nabla_i(\, \sigma^{ij}{}_{I}{}^{J}\, \nabla_j\theta_J\,) \;-\; (\, \nabla_i\, \alpha^{ij}{}_{I}{}^{J}\,)\, \nabla_j\theta_J \;-\; \alpha^{ij}{}_{I}{}^{J}\, \nabla_i\nabla_j\theta_J\,.$$

Now:

- The indices i and j are coordinate indices on M, so ∇_i and ∇_j refer to covariant differentiation along coordinate vector fields. Also, in the expression $\nabla_i \nabla_j \theta_J$, the operator ∇_i corrects on *both* the j and J indices.

- We would like to assume that our manifold connection is torsion-free. That is,
$$\Gamma^i{}_{kj} \;=\; \Gamma^i{}_{jk} \, .$$

- The bundle curvature $R_I{}^J{}_{ij} := \partial_i \omega_I{}^J{}_j - \omega_I{}^L{}_i \, \omega_L{}^J{}_j - \binom{i\to j}{j\to i}$ can be shown to satisfy the Ricci identity
$$(\nabla_i \nabla_j \;-\; \nabla_j \nabla_i)\, \theta_I \;=\; -\, R_I{}^J{}_{ij}\, \theta_J \, .$$

These considerations tell us that
$$-\,\alpha^{ij}{}_I{}^J\, \nabla_i \nabla_j \theta_J \;=\; \tfrac{1}{2}\, \alpha^{ij}{}_I{}^L\, R_L{}^J{}_{ij}\, \theta_J \, .$$

Therefore, having made the following structural reductions:

- our manifold connection Γ is **torsion-free and satisfies the trace condition**, namely $\Gamma^l{}_{kl} = \partial_k \log \sqrt{G}$,

- the bundle connection ω is **compatible with the bundle metric** G^{IJ} on k-forms,

we see that the Laplacian of any k-form $\theta = \frac{1}{k!}\,\theta_I\, dx^I$ has the following covariantized formula:

$$\begin{aligned}
(\Delta\theta)_I \;=\; & -\, \nabla_i\Big(\, \sigma^{ij}{}_I{}^J\, \nabla_j \theta_J \,\Big) \\
& + \Big(\, -\, \nabla_i \alpha^{ij}{}_I{}^J \;+\; \mathcal{A}^{jC}{}_I\, \mathfrak{t}_{-C}{}^J \;-\; \mathcal{A}^{jCJ}\, \mathfrak{t}_{-CI} \\
& \qquad +\; \mathfrak{t}_{+I}{}^S\, \mathcal{B}^{jJ}{}_S \;-\; \mathfrak{t}_+{}^{JS}\, \mathcal{B}^j{}_{IS} \,\Big)\, \nabla_j \theta_J \\
& + \Big(\, \tfrac{1}{2}\, \alpha^{ij}{}_I{}^L\, R_L{}^J{}_{ij} \;+\; \nabla_j[\, \mathcal{A}^{jC}{}_I\, \mathfrak{t}_{-C}{}^J \;-\; \mathcal{B}^j{}_I{}^S\, \mathfrak{t}_+{}^J{}_S \,] \\
& \qquad +\; \mathfrak{t}_-{}^C{}_I\, \mathfrak{t}_{-C}{}^J \;+\; \mathfrak{t}_{+I}{}^S\, \mathfrak{t}_+{}^J{}_S \,\Big)\, \theta_J \, .
\end{aligned}$$

6 The First Order Term

Let us find the criterion (on the bundle connection ω) under which the first order term in $\Delta\theta$ vanishes. To this end, we first carry out a tedious but straightforward calculation to re-express the coefficient, ♣, of $\nabla_j \theta_J$. It is found that

$$♣ \;=\; -\,(\sigma\omega)^{j\;\;J}_{\;I} \;+\; (\sigma\omega)^{jJ}_{\quad I} \;+\; (\mathcal{BAG})^{j\;\;J}_{\;I} \,,$$

where

$$(\sigma\omega)^{j\;\;J}_{\;I} \;:=\; \sigma^{ij\;\;L}_{\quad I}\,\omega_{L\;\;i}^{\;\;J} \;=\; \sigma^{ji\;\;L}_{\quad I}\,\omega_{L\;\;i}^{\;\;J}$$

and

$$\begin{aligned} 2\,(\mathcal{BAG})^{j\;\;J}_{\;I} \;:=\; & \Big[\big(-\,\mathcal{B}^{i\;\;S}_{\;I}\,\mathcal{B}^{jJ}_{\quad S} \;+\; \mathcal{B}^{iJS}\,\mathcal{B}^{j}_{\;IS} \big) \;+ \\ & \quad \big(\mathcal{A}^{iC}_{\quad I}\,\mathcal{A}^{j\;\;J}_{\;C} \;-\; \mathcal{A}^{iCJ}\,\mathcal{A}^{j}_{\;CI} \big) \Big]\,\partial_i \log\sqrt{G} \\ & + \mathcal{B}^{i\;\;R}_{\;I}\,(G^{-1}\partial_i G)^{\;S}_{R}\,\mathcal{B}^{jJ}_{\quad S} \;-\; \mathcal{B}^{iJR}\,(G^{-1}\partial_i G)^{\;S}_{R}\,\mathcal{B}^{j}_{\;IS} \\ & - \mathcal{A}^{iB}_{\quad I}\,(G^{-1}\partial_i G)^{\;C}_{B}\,\mathcal{A}^{j\;\;J}_{\;C} \;+\; \mathcal{A}^{iBJ}\,(G^{-1}\partial_i G)^{\;C}_{B}\,\mathcal{A}^{j}_{\;CI} \\ & + \mathcal{A}^{iC}_{\quad I}\,\mathcal{A}^{j\;\;L}_{\;C}\,(G^{-1}\partial_i G)^{\;J}_{L} \;-\; \mathcal{A}^{iCJ}\,\mathcal{A}^{j\;\;L}_{\;C}\,(G^{-1}\partial_i G)_{LI} \\ & + \mathcal{A}^{i\;\;L}_{\;C}\,\mathcal{A}^{jC}_{\quad I}\,(G^{-1}\partial_i G)^{\;J}_{L} \;-\; \mathcal{A}^{i\;\;L}_{\;C}\,\mathcal{A}^{jCJ}\,(G^{-1}\partial_i G)_{LI} \,. \end{aligned}$$

Note that the quantity $\mathcal{BAG}$ is determined as soon as we have fixed the manifold M and its Finsler structure F, and is independent of our choice of the manifold connection Γ.

The criterion for the vanishing of ♣ reads:

$$(\sigma\omega)^{jIJ} \;-\; (\sigma\omega)^{jJI} \;=\; (\mathcal{BAG})^{jIJ} \,.$$

As a matter of consistency, one should have $(\mathcal{BAG})^{jJI} = -(\mathcal{BAG})^{jIJ}$. This is indeed the case. To reduce that criterion further, we decompose the bundle connection ω into its symmetric and anti-symmetric parts:

$$\begin{aligned} {}^{s}\omega_{JIk} \;&:=\; \tfrac{1}{2}\,(\omega_{JIk} + \omega_{IJk}) \;=\; \tfrac{1}{2}\,(G^{-1}\partial_k G)_{JI} \,, \\ {}^{a}\omega_{JIk} \;&:=\; \tfrac{1}{2}\,(\omega_{JIk} - \omega_{IJk}) \,. \end{aligned}$$

It is the compatibility of ω with the bundle metric which determines its symmetric part, once the Finsler structure F is given. This was discussed in §4. Let us substitute this decomposition into our criterion. We get

$$(\sigma\,{}^{a}\omega)^{jIJ} \;-\; (\sigma\,{}^{a}\omega)^{jJI} \;=\; (\mathcal{BAG})^{jIJ} \;+\; (\sigma\,{}^{s}\omega)^{jJI} \;-\; (\sigma\,{}^{s}\omega)^{jIJ} \,,$$

where the right-hand side is completely known.

Consider the linear transformation:

$${}^{a}\omega \;\overset{\mathcal{L}}{\longmapsto}\; (\sigma\,{}^{a}\omega)^{jIJ} \;-\; (\sigma\,{}^{a}\omega)^{jJI} \,.$$

A moment's thought reveals that its domain and target spaces have the same dimension. Hence it is surjective if and only if it is injective. In other words, the above equation can be uniquely solved for the unknown ${}^{a}\omega$ if and only if the null space of $\mathcal{L}$ is trivial. We thus close this section by posing the following question:

$$\text{Does } (\sigma\, {}^{a}\omega)^{jIJ} = (\sigma\, {}^{a}\omega)^{jJI} \text{ imply that } {}^{a}\omega = 0\ ?$$

Having an affirmative answer will mean that a bundle connection can be *uniquely* chosen so that the covariantized formula for $\Delta\theta$ has no first order term.

7 Complete Positivity of the Symbol

The symbol map of our Laplace-Beltrami operator Δ is

$$\theta_J \quad \mapsto \quad -\, \sigma^{ij}{}_I{}^J\, (\sqrt{-1}\,\lambda_i)\, (\sqrt{-1}\,\lambda_j)\, \theta_J\ ,$$

where the λ_i 's are the components of a covector λ on M. Since Δ is elliptic, the above symbol map is positive. That is, for each fixed covector λ, the bilinear form

$$(\sigma^{ijIJ}\, \lambda_i\, \lambda_j\,)\, \theta_I\, \theta_J$$

is positive definite in θ. To see how that comes about, we use the identities

$$\epsilon^{jJ}{}_S\, \frac{\epsilon^S{}_R}{(k+1)!} \quad = \quad \epsilon^{jJ}{}_R\ ,$$

$$\epsilon^{iI}{}_R\; G^{RS} \quad = \quad \epsilon^S{}_R\; G^{R(iI)}$$

to re-write the bilinear form as

$$\left(\frac{1}{k!}\right)^2 [\,\epsilon^{iI}{}_R\, \lambda_i\, \theta_I\,]\, \frac{G^{RS}}{(k+1)!}\, [\,\epsilon^{jJ}{}_S\, \lambda_j\, \theta_J\,]$$
$$+ \quad \frac{1}{(k-1)!}\, [\, G^{I(iA)}\, \lambda_i\, \theta_I\,]\, G_{AB}\, [\, G^{J(jB)}\, \lambda_j\, \theta_J\,]\,.$$

In this expression:

- The second term is manifestly non-negative. It equals zero if and only if $G^{I(iA)}\,\lambda_i\,\theta_I = 0$. By writing θ_I as $\frac{1}{k!}\,\epsilon^J{}_I\,\theta_J$, one can check that the said vanishing occurs precisely when θ is G-orthogonal to the kernel of the map $\lambda \wedge \cdot\ : \Lambda^k \to \Lambda^{k+1}$.

- The first term is simply $(\frac{1}{k!})^2 \frac{G^{RS}}{(k+1)!} (\lambda \wedge \theta)_R (\lambda \wedge \theta)_S$, which is again manifestly non-negative. It is positive when the second term vanishes, because in that case $\lambda \wedge \theta \neq 0$.

This gives the asserted positivity of $\sigma^{ijIJ} \lambda_i \lambda_j \theta_I \theta_J$ on θ.

In light of the above discussion, one is tempted to ask:

$$\text{Is } \sigma^{ijIJ} \psi_{iI} \psi_{jJ} \text{ positive definite on } \psi \text{ ?}$$

If the answer is yes, the symbol of the Laplace-Beltrami operator is said to be *completely positive*. One can check that such is the case when the Finsler structure F is Riemannian.

Let us enumerate two important consequences *whenever* the symbol of Δ is completely positive:

1. In the formula for $\Delta\theta$, the second order term is $-\nabla_i(\sigma^{ij}{}_I{}^J \nabla_j\theta_J)$. Contract with θ^I and use the fact that the bundle metric is covariantly constant. We get $-\theta_I \nabla_i(\sigma^{ijIJ} \nabla_j\theta_J)$ which, after a re-write, reads:

   $$- \nabla_i(\sigma^{ijIJ} \theta_I \nabla_j\theta_J) + \sigma^{ijIJ} (\nabla_i\theta_I) (\nabla_j\theta_J) .$$

   By the divergence lemma, the first term will give zero when multiplied by $\sqrt{G}\, dx$ and integrated over M. On the other hand, complete positivity assures us that the second term is positive definite. Thus:

   $$\langle \text{ the } 2^{\text{nd}} \text{ order term in } \Delta\theta \ , \ \theta \ \rangle \quad \text{is positive unless} \quad \nabla\theta = 0 .$$

2. Recall the linear transformation $\mathcal{L}$ introduced at the end of §6. There, we wondered whether it is injective. It turns out that this issue is intimately tied to the concept of complete positivity. To unravel the relationship, *fix* any ${}^a\omega$ in the null space of $\mathcal{L}$. It satisfies the equation

   $$(\sigma\, {}^a\omega)^{jIJ} = (\sigma\, {}^a\omega)^{jJI} .$$

   Since ${}^a\omega_{IJi}$ is, by definition, skew in I, J, we have $(\sigma\, {}^a\omega)^{iIJ}\, {}^a\omega_{IJi} = 0$. Spelling out $\sigma\, {}^a\omega$ and relabeling, we get a trace-freeness criterion

   $$G^{KL}\, \mathcal{Q}_{KL} = 0 .$$

   Here,

   $$\mathcal{Q}_{KL} := \sigma^{ijIJ}\, {}^a\omega_{KIi}\, {}^a\omega_{LJj}$$

   are the matrix elements of a real, symmetric, non-negative bilinear form $\mathcal{Q}$. Diagonalize $\mathcal{Q}$ such that each diagonal entry is either -1, 0, or $+1$. The -1 entries are ruled out because $\mathcal{Q}$ is non-negative. In this eigen-basis, the trace-freeness criterion simplifies to

$$\sum_{\{J:\, \mathcal{Q}_{JJ}\neq 0\}} G^{JJ} = 0 .$$

Since (G_{IJ}) is a positive definite matrix, it follows from the formula for matrix inverses that each G^{JJ} must be positive. The above criterion then rules out the $+1$'s as well. So, for all θ,

$$\mathcal{Q}_{KL}\, \theta^K \theta^L = 0$$

or, equivalently,

$$\sigma^{ijIJ} \, ({}^a\!\omega_{KIi}\, \theta^K) \, ({}^a\!\omega_{LJj}\, \theta^L) = 0 .$$

By the hypothesized complete positivity, we must have ${}^a\!\omega_{KIi}\, \theta^K = 0$ for all θ. Thus ${}^a\!\omega = 0$, and the null space of $\mathcal{L}$ is trivial. Consequently, a metric compatible bundle connection (with 'torsion') can be uniquely specified such that

the 1st *order term in* $\Delta\theta$ *vanishes.*

8 Finsler Surfaces and a Weitzenböck Formula

The symbol of Δ is always completely positive at the function level. This is so because the quantity in question is

$$\sigma^{ij\cdot\cdot}\, \psi_{i\cdot}\, \psi_{j\cdot} \, ,$$

and one can easily check that

$$\sigma^{ij\cdot\cdot} = G^{ij} .$$

Now restrict our attention to surfaces. The complete positivity of Δ at the 2-form level turns out to be a consequence of ellipticity. We shall, however, give a direct verification. Here, I, J, K, L are 2-indices; A, B, C are simply 1-indices a, b, c. The 3-indices P, Q, R, S are all vacuous. We have

$$\sigma^{ijIJ} = \tfrac{1}{2} \left[G^{K(ia)}\, G^{L(jb)} + G^{K(ja)}\, G^{L(ib)} \right] G_{ab} \, \frac{\epsilon^I{}_K}{2!} \, \frac{\epsilon^J{}_L}{2!} \, ,$$

and

$$\psi_{i21} \;=\; -\,\psi_{i12} \;=:\; -\,\psi_i \;,$$

where the abbreviation ψ_i is introduced to reduce clutter. Using these, we find that

$$\sigma^{ijIJ}\;\psi_{iI}\,\psi_{jJ} \;=\; Q^{ij}\;\psi_i\,\psi_j \;,$$

with

$$\begin{aligned} Q^{ij} \;:=\; G_{ab}\,[\;\; & G^{(12)(ia)}\,G^{(12)(jb)} \;+\; G^{(21)(ia)}\,G^{(21)(jb)} \;- \\ & G^{(12)(ia)}\,G^{(21)(jb)} \;-\; G^{(21)(ia)}\,G^{(12)(jb)} \;\;]\,. \end{aligned}$$

Substituting the definition of the G 's, we get:

$$\begin{pmatrix} Q^{11} & Q^{12} \\ Q^{21} & Q^{22} \end{pmatrix} \;=\; (\textstyle\int \det g^{\sharp}\,)^2\; \begin{pmatrix} G_{22} & -G_{21} \\ -G_{12} & G_{11} \end{pmatrix} \;,$$

with

$$\textstyle\int \det g^{\sharp} \;:=\; \dfrac{1}{\sqrt{G}} \displaystyle\int_{S_xM} \det(g^{ij})\;\sqrt{g}\;\zeta\,.$$

Complete positivity at the 2-form level is now apparent.

It remains to address the issue for 1-forms. Here,

- A, B, C are 0-indices; so we leave them blank.
- I, J, K are 1-indices, and we shall leave each of them as is.
- P, Q, R are 2-indices.

We find that

$$\sigma^{ijIJ} \;=\; \tfrac{1}{2}\left[\; G^{R(iI)}\;\epsilon^{jJ}{}_{R} \;+\; G^{iI}\,G^{jJ} \;+\; \begin{pmatrix} i\to j \\ j\to i \end{pmatrix} \right].$$

It is more convenient to write

$$\sigma^{ijIJ}\,\psi_{iI}\,\psi_{jJ} \quad \text{as} \quad \psi_{iI}\;Q^{[iI][jJ]}\;\psi_{jJ}\;,$$

with

$$Q^{[iI][jJ]} \;:=\; \tfrac{1}{2}\left\{\;(\;G^{12iI} \;-\; G^{12Ii}\;)\,\epsilon^{jJ}{}_{12} \;+\; G^{iI}\,G^{jJ} \;+\; \begin{pmatrix} i\to j \\ j\to i \end{pmatrix} \right\}.$$

Explicitly, the coefficient matrix

$$\begin{pmatrix} Q^{[11][11]} & Q^{[11][12]} & Q^{[11][21]} & Q^{[11][22]} \\ Q^{[12][11]} & Q^{[12][12]} & Q^{[12][21]} & Q^{[12][22]} \\ Q^{[21][11]} & Q^{[21][12]} & Q^{[21][21]} & Q^{[21][22]} \\ Q^{[22][11]} & Q^{[22][12]} & Q^{[22][21]} & Q^{[22][22]} \end{pmatrix}$$

of our quadratic form is

$$\begin{pmatrix} G^{11}\,G^{11} & G^{11}\,G^{12} & G^{11}\,G^{12} & G^{12}\,G^{12} + \epsilon \\ G^{11}\,G^{12} & G^{11}\,G^{22} - 2\epsilon & G^{12}\,G^{12} + \epsilon & G^{22}\,G^{12} \\ G^{11}\,G^{12} & G^{12}\,G^{12} + \epsilon & G^{11}\,G^{22} - 2\epsilon & G^{22}\,G^{12} \\ G^{12}\,G^{12} + \epsilon & G^{22}\,G^{12} & G^{22}\,G^{12} & G^{22}\,G^{22} \end{pmatrix},$$

with

$$\epsilon := \tfrac{1}{2}\left(\det \textstyle\int g^{\sharp} - \int \det g^{\sharp} \right),$$

$$\det \textstyle\int g^{\sharp} := \det(G^{ij}) = G^{11}\,G^{22} - (G^{12})^2 .$$

Recall that

$$G^{ij} := \frac{1}{\sqrt{G}} \int_{S_xM} g^{ij}\,\sqrt{g}\,\zeta ,$$

thus the rationale behind our notation.

The four principal sub-determinants of the above coefficient matrix can be computed. Their formulas are as follows:

size	principal sub-determinant
1×1	$(G^{11})^2$
2×2	$(G^{11})^2\ \int\det g^{\sharp}$
3×3	$\frac{1}{4}\,(G^{11})^2\ [\int\det g^{\sharp} + \det\int g^{\sharp}]\ [3\int\det g^{\sharp} - \det\int g^{\sharp}]$
4×4	$\frac{1}{16}\,[\int\det g^{\sharp} + \det\int g^{\sharp}]^2\ [3\int\det g^{\sharp} - \det\int g^{\sharp}]\ [3\det\int g^{\sharp} - \int\det g^{\sharp}]$

Note that the first two principal sub-determinants are always positive.

We see that for a Finsler surface (M, F), the symbol of Δ is completely positive at the 1-form level if and only if F satisfies the inequality

$$\frac{1}{3} < \frac{\int\det g^{\sharp}}{\det\int g^{\sharp}} < 3 .$$

Or equivalently:

$$\frac{1}{3} < \frac{\det\int g^{\sharp}}{\int\det g^{\sharp}} < 3 .$$

Incidentally, when F is Riemannian, the ratio in question is 1. So the above inequality is definitely satisfied by Finsler structures which are 'small' perturbations of Riemannian structures.

Let us summarize:

Theorem 1. *Let M be a compact, oriented, boundaryless surface. Let F be a Finsler structure which satisfies the geometric inequality*

$$\frac{1}{3} \quad < \quad \frac{\int \det g^\sharp}{\det \int g^\sharp} \quad < \quad 3 \, .$$

Then there exists a unique metric-compatible bundle connection ω which, together with any torsion-free manifold connection Γ satisfying ${\Gamma^j}_{kj} = \partial_k \log \sqrt{G}$, effects the Weitzenböck formula

$$\begin{aligned}(\Delta\theta)_I \quad = \quad & - \nabla_i\Big({\sigma^{ij}}_I{}^J \, \nabla_j \theta_J \Big) \\ & + \Big(\tfrac{1}{2} \, {\alpha^{ij}}_I{}^L \, {R_L}^J{}_{ij} \, + \, \nabla_j [\, {\mathcal{A}^{jC}}_I \, {\mathfrak{t}_{-C}}^J \, - \, {\mathcal{B}^j}_I{}^S \, {\mathfrak{t}_{+S}}^J \,] \\ & \qquad + \, {\mathfrak{t}_-}^C{}_I \, {\mathfrak{t}_{-C}}^J \, + \, {\mathfrak{t}_{+I}}^S \, {\mathfrak{t}_{+S}}^J \Big) \, \theta_J \, .\end{aligned}$$

For example, this formula is definitely valid on any Randers surface (generated by a Riemannian metric $\tilde{a}$ and a 1-form $\tilde{b}$) provided that $\tilde{b}$ has uniformly small point-wise Riemannian norm.

9 Some Inequalities

We mentioned in §8 that the geometric ratio

$$\frac{\int \det g^\sharp}{\det \int g^\sharp}$$

has the value 1 for Riemannian structures. Here, we would like to say more about its numerator and denominator. But before we continue, let us state two basic inequalities that will be needed.

Lemma 2 (Jensen's inequality). *Let $(X, d\mu)$ be a regular probability space, and let $f : \mathbb{R} \to \mathbb{R}$ be a strictly convex function. Then for any integrable function g, we have*

$$f \left(\int_X g(x) \, d\mu_x \right) \quad \leq \quad \int_X f\,(g(x)) \, d\mu_x \; ,$$

and equality holds if and only if g is constant μ-almost everywhere.

Jensen's inequality is standard in analysis texts. Nevertheless, we shall sketch a proof. This exercise will put us in good stead when we come to the integral Minkowski's inequality.

Let $(X, d\mu)$, f and g be given as stated above. We approximate g by simple functions

$$g_m \quad := \quad \sum_{j=1}^{m} c_j^{(m)} \, \chi_{\Delta_j^{(m)}} \, .$$

Note that:

- $\int_X g_m(x)\, d\mu_x = \sum_{j=1}^{m} \; c_j^{(m)} \, \mu\left(\Delta_j^{(m)}\right)$.
- Since μ is a probability measure, $\sum_{j=1}^{m} \; \mu\left(\Delta_j^{(m)}\right) = 1$.

Using the convexity of f, we have:

$$\begin{aligned} f\left(\int_X g_m(x)\, d\mu_x\right) &= f\left(\sum_{j=1}^{m} c_j^{(m)} \, \mu\left(\Delta_j^{(m)}\right)\right) \\ &\leq \sum_{j=1}^{m} f\left(c_j^{(m)}\right) \mu\left(\Delta_j^{(m)}\right) \\ &= \int_X f\left(g_m(x)\right) \, d\mu_x \, . \end{aligned}$$

Taking limits gives the desired inequality.

If g is constant μ-almost everywhere, then equality in the above relation is clear. Now suppose g is not constant μ-almost everywhere. We claim that there exists two constants, $c_1 < c_2$, and two sets of positive measure, Δ_1 and Δ_2, such that

$$\int_X g(x)\, d\mu_x \quad = \quad c_1 \, \mu(\Delta_1) \; + \; c_2 \, \mu(\Delta_2) \, .$$

To see this, consider the following construction. For any $c \in \mathbb{R}$, consider the measurable sets $\Delta_1 = g^{-1}(-\infty, c]$ and $\Delta_2 = g^{-1}(c, +\infty)$. Since g is not constant μ-almost everywhere, we are able to choose c to ensure that Δ_1 and Δ_2 have positive measure. Then, one immediately sees that $c_j = \frac{1}{\mu(\Delta_j)} \int_{\Delta_j} g(x)\, d\mu_x$, $j = 1, 2$, are constants which satisfy all that is required of them. With these, we have

$$\begin{aligned} f\left(\int_X g(x)\, d\mu_x\right) &= f\left(c_1 \, \mu(\Delta_1) \; + \; c_2 \, \mu(\Delta_2)\right) \\ &< f(c_1)\, \mu(\Delta_1) \; + \; f(c_2)\, \mu(\Delta_2) \, . \end{aligned}$$

Yet, for $j = 1, 2$, we also have

$$f(c_j) \quad = \quad f\left(\int_{\Delta_j} g(x) \, \frac{d\mu_x}{\mu(\Delta_j)}\right) \quad \leq \quad \int_{\Delta_j} f(g(x)) \, \frac{d\mu_x}{\mu(\Delta_j)}$$

by the inequality we have already proven, as $\frac{d\mu_x}{\mu(\Delta_j)}$ is a probability measure on Δ_j. Thus

$$f\left(\int_X g(x)\,d\mu_x\right) \;<\; f(c_1)\,\mu(\Delta_1) \;+\; f(c_2)\,\mu(\Delta_2) \;\leq\; \int_X f(g(x))\,d\mu_x\,,$$

which finishes the proof of the lemma.

Lemma 3 (The integral Minkowski's inequality). *Suppose:*

- *$(X, d\mu)$ is a regular measure space.*
- *$A : X \to M_{n,n}$ is an integrable matrix-valued function.*
- *The matrix $A(x)$ is positive definite for every $x \in X$.*

Then

$$\left[\det\left(\int_X A(x)\,d\mu_x\right)\right]^{\frac{1}{n}} \;\geq\; \int_X \left[\det A(x)\right]^{\frac{1}{n}}\,d\mu_x\,,$$

and equality holds if and only if $A(x)$ is a scalar $c(x)$ multiple of a constant matrix A_o μ-almost everywhere.

Proof. Recall the classical version of Minkowski's inequality [HJ, 7.8.8]. It says that if A, B are positive definite real $n \times n$ matrices, then

$$\left[\det(A + B)\right]^{\frac{1}{n}} \;\geq\; (\det A)^{\frac{1}{n}} + (\det B)^{\frac{1}{n}}\,,$$

with equality if and only if $A = cB$ for some $c \in \mathbb{R}$. This extends inductively to any finite sum of positive definite matrices:

$$\left[\det\left(\sum_{j=1}^{m} A_j\right)\right]^{\frac{1}{n}} \;\geq\; \sum_{j=1}^{m} (\det A_j)^{\frac{1}{n}}\,,$$

with equality if and only if there is a fixed matrix A_o, together with constants $c_1, \ldots, c_m$, such that $A_j = c_j A_o$ for each $j = 1, \ldots, m$.

Now, let us proceed exactly as we did for Jensen's inequality. Approximate the function $x \mapsto A(x)$ by simple functions:

$$A^{(m)} \;=\; \sum_{j=1}^{m} C_j^{(m)}\,\chi_{\Delta_j^{(m)}}\,,$$

where each $C_j^{(m)}$ is a positive definite constant matrix. Using the classical Minkowski's inequality, we have:

$$\begin{aligned}
\int_X \left[\det A^{(m)}(x)\right]^{\frac{1}{n}} d\mu_x &= \sum_{j=1}^{m} \left[\det C_j^{(m)}\right]^{\frac{1}{n}} \mu\left(\Delta_j^{(m)}\right) \\
&= \sum_{j=1}^{m} \left[\det\left\{C_j^{(m)} \mu\left(\Delta_j^{(m)}\right)\right\}\right]^{\frac{1}{n}} \\
&\leq \left[\det\left(\sum_{j=1}^{m} C_j^{(m)} \mu\left(\Delta_j^{(m)}\right)\right)\right]^{\frac{1}{n}} \\
&= \left[\det\left(\int_X A^{(m)}(x)\, d\mu_x\right)\right]^{\frac{1}{n}}
\end{aligned}$$

Again, taking limits gives the desired inequality.

If $A(x) = c(x)A_o$, it is clear that equality holds in the integral Minkowski's inequality. So now suppose that $A(x)$ is not a μ-almost everywhere constant multiple of any fixed positive definite matrix. As the positive definite matrices are not totally ordered, we cannot proceed as in Jensen's inequality, Yet, we can suitably modify the argument.

- Define $A_o := \int_X A(x) d\mu_x$. We claim that there exists a set of positive measure Δ_1 such that $A_1 := \int_{\Delta_1} A(x)\, d\mu_x$ is not a multiple of A_o. Otherwise, $\int_\Delta A(x) d\mu_x = c(\Delta) A_o$ for every set Δ of positive measure. Tracing both sides and dividing by tr $\mathrm{A_o}$, we get

$$c(\Delta) = \int_X \left(\frac{\operatorname{tr} A(x)}{\operatorname{tr} A_o}\right) d\mu_x .$$

As the integrand above is positive and integrable, dc is a measure and is absolutely continuous with respect to $d\mu$. Let $c(x) = \frac{dc}{d\mu}$ and $\bar{A}(x) = c(x)A_o$. Note that

$$\int_\Delta A(x)\, d\mu_x = c(\Delta)\, A_o = \int_\Delta c(x)\, A_o\, d\mu_x = \int_\Delta \bar{A}(x)\, d\mu_x$$

for every measurable set Δ. That is, $\int_\Delta [A(x) - \bar{A}(x)]\, d\mu_x = 0$. Thus, $A(x) = \bar{A}(x)$ μ-almost everywhere, a contradiction. Therefore the asserted Δ_1 does exist after all.

- Let $\Delta_2 := X \setminus \Delta_1$, and $A_2 := \int_{\Delta_2} A(x)\, d\mu_x$. Then $A_o = A_1 + A_2$, and A_2 cannot be a positive multiple of A_1 precisely because A_1 is not a constant multiple of A_o. By the classical Minkowski's inequality,

$$\left[\det\left(\int_X A(x)\, d\mu_x\right)\right]^{\frac{1}{n}} = [\det(A_1 + A_2)]^{\frac{1}{n}}$$
$$> (\det A_1)^{\frac{1}{n}} + (\det A_2)^{\frac{1}{n}} .$$

On the other hand, using the inequality proven above,

$$(\det A_j)^{\frac{1}{n}} = \left[\det\left(\int_{\Delta_j} A(x)\, d\mu_x\right)\right] \geq \int_{\Delta_j} [\det A(x)]^{\frac{1}{n}}\, d\mu_x .$$

Combining these inequalities, we get

$$\left[\det\left(\int_X A(x)\, d\mu_x\right)\right]^{\frac{1}{n}} > \int_X [\det A(x)]^{\frac{1}{n}}\, d\mu_x ,$$

which is precisely what we need to show.

□

Theorem 4. *At every point x of a Finsler surface M, we have the following inequalities:*

$$\begin{aligned} \textstyle\int \det g^{\sharp} &\geq \frac{1}{G} \\ \det \textstyle\int g^{\sharp} &\geq \frac{1}{G} \end{aligned} .$$

In either case, equality holds if and only if the Finsler structure F is Riemannian at the point x.

Proof.
For $\int \det g^{\sharp} \geq \frac{1}{G}$:

- Let us use Jensen's inequality. By construction, ζ is a probability measure on $S_x M$. Also, $f(x) := \frac{1}{x}$ is strictly convex on the positive real numbers. But the positive reals is the range of $\sqrt{g}$, hence Jensen's inequality holds. We have:

$$\textstyle\int \det g^{\sharp} = \frac{1}{\sqrt{G}} \int_{S_x M} \frac{1}{\sqrt{g}}\, \zeta \geq \frac{1}{\sqrt{G}} \left(\int_{S_x M} \sqrt{g}\, \zeta\right)^{-1} = \frac{1}{G} .$$

- According to Lemma 2, equality holds if and only if $\sqrt{g}$ is constant on S_xM. Now, the Cartan scalar I can be expressed as

$$I \; := \; g^{ij} \left(A_{ij1} \frac{F_{y^2}}{\sqrt{g}} \; - \; A_{ij2} \frac{F_{y^1}}{\sqrt{g}} \right) ,$$

and

$$b_3 \; := \; \frac{1}{\sqrt{g}} \left(F_{y^2} \frac{\delta}{\delta y^1} \; - \; F_{y_1} \frac{\delta}{\delta y^2} \right)$$

is the unit vector (with respect to the Sasaki metric) tangent to S_xM. It is straightforward to check that

$$(d\log\sqrt{g})(b_3) \; = \; \tfrac{1}{2}\, g^{ij}\, (dg_{ij})(b_3) \; = \; I \, .$$

Hence the constancy of $\sqrt{g}$ on S_xM is synonymous with $I = 0$, which is the conclusion we seek.

For $\det \int g^{\sharp} \;\geq\; \frac{1}{G}$:

- Here, let us use the integral Minkowski's inequality with $n = 2$. We have:

$$\begin{aligned}
(\det \textstyle\int g^{\sharp})^{\frac{1}{2}} \; &= \; \left[\det(G^{ij})\right]^{\frac{1}{2}} \\
&= \; \left[\det\left(\int_{S_xM} g^{ij} \frac{\sqrt{g}}{\sqrt{G}} \zeta\right)\right]^{\frac{1}{2}} \\
&\geq \; \frac{1}{\sqrt{G}} \int_{S_xM} \left[\det(g^{ij})\right]^{\frac{1}{2}} \sqrt{g}\, \zeta \\
&= \; \frac{1}{\sqrt{G}} \int_{S_xM} \zeta \; = \; \frac{1}{\sqrt{G}} ,
\end{aligned}$$

which upon squaring gives the desired inequality.

- Lemma 3 tells us that equality holds if and only if $g_{ij}(x,y) = c(y) h_{ij}$ for some function $c = c(y)$ and a fixed metric h. Now,

$$\begin{aligned}
0 \; &= \; A_{ijk}\, \ell^j \; = \; \tfrac{1}{2} \frac{\partial g_{ij}}{\partial y^k} y^j \\
&= \; \frac{F}{2} g_{ij}\, \ell^j \frac{\partial \mathrm{logc}}{\partial y^k} \\
&= \; \frac{F}{2} \ell_i \frac{\partial \mathrm{logc}}{\partial y^k} \, .
\end{aligned}$$

This must hold for every i and k. But, at any given point of S_xM, we know that $\ell_i \neq 0$ for some i. Therefore c must be constant.

□

10 The Geometric Ratio & the Cartan Scalar

Whenever the geometric ratio

$$\frac{\int \det g^{\sharp}}{\det \int g^{\sharp}}$$

has a value different from 1, it quantifies an amount by which the Finsler function F deviates from a Riemannian structure. In this section we would like to realize this quantity as a function of the well known objects of a Finsler surface: the Cartan scalar I and the Landsberg angle θ. The quantity $d\theta$ has the significance of being the element of arc along the projective circle S_xM. Its integral over S_xM gives a certain Riemannian arc length of S_xM, say L. For Finsler surfaces, L typically varies with x, and has values different from 2π. See [M] and [BS] for more information.

The theory of Finsler surfaces is well-represented in the literature. We will merely cite [AIM, §3.5], [BCS], [M], as all these are modern presentations. For a Finsler surface (M, F), the ordered orthonormal coframe $\{\omega^1, \omega^2, \omega^3\}$ discussed in §1 is globally defined on SM. Explicitly,

$$\omega^1 \;=\; \frac{\sqrt{g}}{F}\,(y^2\,dx^1 \;-\; y^1\,dx^2\,) \;=:\; m_1\,dx^1 \;+\; m_2\,dx^2\;,$$

$\omega^2 = F_{y^1}\,dx^1 + F_{y^2}\,dx^2 = \ell_1\,dx^1 + \ell_2\,dx^2$ and $\omega^3 = \frac{\sqrt{g}}{F}(y^2\,\delta y^1 - y^1\,\delta y^2)$. Then the Cartan scalar and the Landsberg angle are given respectively by

$$A_j \;=\; I\,m_j$$
$$F\,\frac{\partial\theta}{\partial y^j} \;=\; m_j\,.$$

These are just (3.5.1.5) and (3.5.1.9) of [AIM], but for us, the Cartan tensor is defined as $A_{ijk} := \frac{F}{4}(F^2)_{y^iy^jy^k}$. We also have the interesting equation [AIM, 3.5.1.8]

$$A_j \;=\; F\,\frac{\partial \log\sqrt{g}}{\partial y^j} \;=\; \frac{\partial \log\sqrt{g}}{\partial\theta}\,m_j\,,$$

where the second equality follows from the chain rule. Putting all this together, and using $m_jm^j = 1$, yields

$$\frac{\partial}{\partial\theta}\sqrt{g} \;=\; I\,\sqrt{g}\,.$$

Or equivalently,

$$\frac{\partial}{\partial\theta}\det g \;=\; 2I\,\det g\,. \qquad (*)$$

Now, if we consider a single projective circle $S_x M$, this implies that

$$d \log \det g = 2I \, d\theta \, .$$

Thus, the Cartan scalar is not a function, but rather a pseudo-scalar on $S_x M$. This explains the statement in [AIM] that I changes sign under a change in orientation of the projective circle.

Spelling out the constituents of the geometric ratio:

$$\begin{aligned} \frac{\int \det g^\sharp}{\det \int g^\sharp} &= \left[\det(G^{jk})\right]^{-1} \frac{1}{\sqrt{G}} \int_{S_x M} \det(g^{jk}) \sqrt{g}\, \zeta \\ &= \frac{1}{\sqrt{G}} \int_{S_x M} \frac{\det(g^{jk})}{\det(G^{jk})} \sqrt{g}\, \zeta \\ &= \frac{1}{\sqrt{G}} \int_{S_x M} \left(\det(g_{js} G^{sk})\right)^{-1} \sqrt{g}\, \zeta \, . \end{aligned}$$

Now, multiply (*) by $\det(G^{jk})$ (which has no θ dependence whatsoever), and we get

$$\frac{\partial}{\partial \theta} \det(g_{js} G^{sk}) = 2I \det(g_{js} G^{sk}) \, .$$

Integrating this gives

$$\left[\det(g^\flat \cdot G^\sharp)\right](\theta) := \left[\det(g_{js} G^{sk})\right](\theta) = \left[\det(g^\flat \cdot G^\sharp)\right](\theta_0)\, e^{2 \int_{\theta_0}^{\theta} I(s)\, ds} \, ,$$

where the dot between $g^\flat$ and $G^\sharp$ signifies matrix multiplication. We have thus derived the following formula.

Proposition 5.

$$\frac{\int \det g^\sharp}{\det \int g^\sharp} = \frac{1}{\sqrt{G}} \int_{S_x M} \frac{1}{\left[\det(g^\flat \cdot G^\sharp)\right](\theta_0)}\, e^{-2 \int_{\theta_0}^{\theta} I(s)\, ds} \sqrt{g}(\theta)\, \zeta(\theta) \, .$$

This proposition is useful because one can compute the Cartan scalar explicitly in many examples. We will report on these findings elsewhere.

A Comparison of Laplacians

A.1 Introduction

In this section, we generalize the material in [BL] to the case of manifolds with parameterized metrics. Let M^n and $\mathcal{F}^m$ be compact and orientable manifolds without boundary. Let $\pi : E \to M$ be a fibre bundle with fibre $\mathcal{F}$, and let us fix a nonlinear connection $TE = HE \oplus VE$, represented in coordinates as

$$HE = \text{span}\left\{ \frac{\delta}{\delta x^j} = \frac{\partial}{\partial x^j} - \mathcal{N}^\mu_{\ j} \frac{\partial}{\partial z^\mu} \right\} .$$

Then the covectors $\delta z^\mu = dz^\mu + \mathcal{N}^\mu_{\ j} dx^j$ complete the natural coframe $\{dx, \delta z\}$ (dual to $\{\frac{\delta}{\delta x}, \frac{\partial}{\partial z}\}$) on T^*E. Here, z^μ $(\mu = 1, \ldots, m)$ are fibre coordinates on $\mathcal{F}$ and, as usual, x^j $(j = 1, \ldots, n)$ are local coordinates on M. If E should happen to be SM, then the δz^μ 's would be the 'intrinsic-coordinate' analogues of the orthonormal $\omega^{n+\mu}$'s on SM (the latter as given in §1). Likewise the δy^i 's (again, see §1), though initially defined on $TM \setminus 0$, are the 'homogeneous-coordinate' analogues of the δz^μ 's.

The curvatures of the nonlinear connection $\mathcal{N}$ are *defined* by the equation

$$d(\delta z^\mu) = \frac{1}{2} \mathcal{R}^\mu_{\ jk}\, dx^j \wedge dx^k + \mathcal{P}^\mu_{\ j\nu}\, dx^j \wedge \delta z^\nu .$$

Hence $\mathcal{R}^\mu_{\ jk} = \frac{\delta}{\delta x^j} \mathcal{N}^\mu_{\ k} - \frac{\delta}{\delta x^k} \mathcal{N}^\mu_{\ j}$ and $\mathcal{P}^\mu_{\ j\nu} = -\frac{\partial}{\partial z^\nu} \mathcal{N}^\mu_{\ j}$. These play an important role in the theory.

We take a metric on E of the form $\hat{g} = g_{jk}\, dx^j \otimes dx^k + h_{\mu\nu}\, \delta z^\mu \otimes \delta z^\nu$. The 'naive' volume form with respect to this metric is

$$d\,Vol_{\hat{g}} = \sqrt{g}\, \sqrt{h}\, dx^1 \wedge \cdots \wedge dx^n \wedge \delta z^1 \wedge \cdots \wedge \delta z^m .$$

Observe that the right-hand side is unaffected by $\delta z \mapsto dz$. The $\mathcal{L}_2$ inner product generated by this volume form will not coincide with that of a Riemannian manifold, as in general the volume of the fibre above each point is not unity. Thus we normalize this volume by defining the function

$$Vol(x) := \int_{E_x} (-1)^m \sqrt{h}\, dz^1 \wedge \cdots \wedge dz^m$$

and constructing the 'normalized fibre volume form'

$$\zeta_x := \frac{(-1)^m}{Vol(x)} \sqrt{h}\, dz^1 \wedge \cdots \wedge dz^m .$$

The $(-1)^m$ is used in order to reproduce the $(-1)^{n-1}$ in Finsler geometry (see §1). We declare the (normalized) volume form of the bundle E to be $d\,Vol_E := \sqrt{g}\, dx^1 \wedge \cdots \wedge dx^n \wedge \zeta_x$. Note that $(-1)^m\, d\,Vol_{\hat{g}} = Vol(x)\, d\,Vol_E$.

The volume form $d\,Vol_E$ induces an $\mathcal{L}_2$ structure on $\Omega^*(E)$ by

$$\langle \theta, \psi \rangle_E := \frac{1}{p!} \int_E \hat{g}^{J_1 K_1} \cdots \hat{g}^{J_p K_p}\, \theta_{J_1 \cdots J_p}\, \psi_{K_1 \cdots K_p}\, d\,Vol_E ,$$

where $\theta, \psi \in \Omega^p(E)$. In particular, suppose $\theta, \psi \in \pi^*\Omega^p(M)$; namely, suppose these forms contain only dx terms and have only x-dependent coefficients. Then an argument which mimics that in §1 will give

$$\langle \theta, \psi \rangle_E \;=\; \frac{1}{p!} \int_M \theta_{j_1 \cdots j_p}\, \psi_{k_1 \cdots k_p}\, G^{j_1 k_1 \cdots j_p k_p} \sqrt{G}\, dx \;,$$

where $\sqrt{G} = \int_{E_x} \sqrt{g}\; \zeta_x$ and $G^{j_1 k_1 \cdots j_p k_p} = \frac{1}{\sqrt{G}} \int_{E_x} g^{j_1 k_1} \cdots g^{j_p k_p} \sqrt{g}\; \zeta_x$. Thus the $\mathcal{L}_2$ structure on $\Omega^*(M)$ given by $\langle \theta, \psi \rangle_M = \langle \pi^*\theta, \pi^*\psi \rangle_E$ is realized by integration over M using the pointwise defined metric G, and volume form $\sqrt{G}\, dx$. If M is Riemannian (that is, g is a function of x alone), then $\sqrt{G} = \sqrt{g}$ and $G^{j_1 k_1 \cdots j_p k_p} = g^{j_1 k_1} \cdots g^{j_p k_p}$. Therefore this inner product reduces to the usual one in the Riemannian case; such is the importance of normalizing the fibre volume to be unity.

A.2 Two Horizontal Laplacians

Given a function $f \in \Omega^0(E)$, its exterior derivative is $df = \frac{\delta f}{\delta x^j} dx^j + \frac{\partial f}{\partial z^\mu} \delta z^\mu$. We decompose this into two derivative operators $d = d_H + d_V$ where $d_H f = \frac{\delta f}{\delta x^j} dx^j$ and $d_V f = \frac{\partial f}{\partial z^\mu} \delta z^\mu$. Note that, $f \in \pi^*\Omega^0(M)$ if and only if $d_V f = 0$, so we will primarily be interested in the horizontal derivative d_H.

On the total space E, there are two inner products induced by the two volumes $(-1)^m\, d\,Vol_{\hat{g}}$ and $d\,Vol_E$. Thus we may compute two adjoints for d_H which will produce two Laplacians. To perform both calculations simultaneously, we will work with a volume form $K(x) d\,Vol_{\hat{g}}$. The Riemannian adjoint is recovered by setting $K(x) \equiv (-1)^m$, and the normalized adjoint comes from setting $K(x) = \frac{(-1)^m}{Vol(x)}$.

Consider the following. By Stokes' theorem, we have

$$0 \;=\; \int_E d\left[\hat{g}^{IJ} \theta_I\; f\; i_{\hat{e}_J} K(x) d\,Vol_{\hat{g}} \right]$$

where $i_{\hat{e}_J}$ is the interior product with respect to $\hat{e}_J$. Here, $\hat{e}_J$ represents either $\frac{\delta}{\delta x^j}$ or $\frac{\partial}{\partial z^\mu}$, depending on the range of J. Now,

$$\begin{aligned} d\left[\hat{g}^{IJ} \theta_I\; f\; i_{\hat{e}_J} K(x) d\,Vol_{\hat{g}} \right] \;&=\; d(\hat{g}^{IJ}\theta_I) \wedge f\; i_{\hat{e}_J} K(x) d\,Vol_{\hat{g}} \\ &+ \hat{g}^{IJ}\theta_I\; (df) \wedge i_{\hat{e}_J} K(x) d\,Vol_{\hat{g}} \;+\; \hat{g}^{IJ}\theta_I\; f\; d\left[i_{\hat{e}_J} K(x) d\,Vol_{\hat{g}} \right] . \end{aligned}$$

Focusing only on the term involving df for the moment, we have two cases. The first is when $J = j$ for which

$$i_{\frac{\delta}{\delta x^j}} d\,Vol_{\hat{g}} = (-1)^{j-1} \sqrt{g}\sqrt{h}\; dx^1 \wedge \cdots \wedge \widehat{dx^j} \wedge \cdots \wedge dx^n \wedge \delta z^1 \wedge \cdots \wedge \delta z^m \,.$$

Thus the second term on the right can only contribute $df = \frac{\delta f}{\delta x^j} dx^j$ (no sum on j). In this case that term is

$$g^{ij}\theta_i \, \frac{\delta f}{\delta x^j} \, K(x) \, d\,Vol_{\hat{g}} \quad (\text{no sum on } j) \, .$$

Identically, in the case that $J = \nu$, the said term is

$$h^{\mu\nu}\theta_\mu \, \frac{\partial f}{\partial z^\nu} \, K(x) \, d\,Vol_{\hat{g}} \quad (\text{no sum on } \nu) \, .$$

Now summing over the cases, we see that the integral over E of the term involving df is just $\langle \theta, df \rangle$. By definition, this is $\langle d^*\theta, f \rangle$. So we have that

$$\begin{aligned} \langle d^*\theta, f \rangle \quad = \quad & - \int_E d(\hat{g}^{IJ}\theta_I) \wedge f \; i_{\hat{e}_J} K(x) d\,Vol_{\hat{g}} \\ & - \int_E \hat{g}^{IJ}\theta_I \; f \; d\left[i_{\hat{e}_J} K(x) d\,Vol_{\hat{g}} \right] \, . \end{aligned}$$

The argument above applies again for the first of these integrals. One discovers that it reduces to

$$- \int_E \left(\frac{\delta}{\delta x^j} (g^{ij}\theta_i) \; + \; \frac{\partial}{\partial z^\nu} (h^{\mu\nu}\theta_\mu) \right) f \; K(x) d\,Vol_{\hat{g}} \, .$$

The second integral, however, we must compute. Again we have two cases. First consider $J = j$.

$$\begin{aligned} & d\left[i_{\frac{\delta}{\delta x^j}} K(x) d\,Vol_{\hat{g}} \right] \\ & \quad = d\left[(-1)^{j-1} K(x) \sqrt{g}\sqrt{h} \; dx^1 \wedge \cdots \wedge \widehat{dx^j} \wedge \cdots \wedge \delta z^1 \wedge \cdots \wedge \delta z^m \right] \\ & \quad = \frac{\delta}{\delta x^j} [K(x)\sqrt{g}\sqrt{h}] \; dx^1 \wedge \cdots \wedge dx^n \wedge \delta z^1 \cdots \wedge \delta z^m \\ & \qquad + K(x)\sqrt{g}\sqrt{h} \; \sum_\alpha (-1)^{n+j+\alpha-1} dx^1 \wedge \cdots \wedge \widehat{dx^j} \wedge \cdots \wedge dx^n \\ & \qquad \qquad \wedge \delta z^1 \wedge \cdots \wedge d(\delta z^\alpha) \wedge \cdots \wedge \delta z^m \, . \end{aligned}$$

In this last term, the $d(\delta z^\alpha)$ can only contribute one nonvanishing term: $\mathcal{P}^\alpha{}_{j\alpha} dx^j \wedge \delta z^\alpha$ (no sum on j, α). So, if we introduce the abbreviation

$$\dot{A}_j \;\; := \;\; - \, \mathcal{P}^\mu{}_{j\mu} \; - \; \frac{\delta}{\delta x^j} \log \sqrt{h} \qquad (\mu \text{ is summed}) \, ,$$

this terms becomes

$$d\left[i_{\frac{\delta}{\delta x^j}} K(x) d\,Vol_{\hat{g}} \right] \;\; = \;\; \left[\frac{\delta}{\delta x^j} \log \sqrt{g} + \frac{\delta}{\delta x^j} \log |K(x)| - \dot{A}_j \right] \; K(x) d\,Vol_{\hat{g}} \, .$$

One can repeat the argument for the case $J = \nu$, and discovers that

$$d\left[i_{\frac{\partial}{\partial z^\nu}} K(x) d\,Vol_{\hat g}\right] = \left[\frac{\partial}{\partial z^\nu}\log\sqrt{g} + \frac{\partial}{\partial z^\nu}\log\sqrt{h}\right] K(x) d\,Vol_{\hat g}\,.$$

Putting everything together, we find

$$\begin{aligned}\langle d^*\theta, f\rangle &= -\int_E \Big[\frac{\delta}{\delta x^j}(g^{ij}\theta_i) + \frac{\partial}{\partial z^\nu}(h^{\mu\nu}\theta_\mu) \\ &\qquad + g^{ij}\theta_i\,\frac{\delta}{\delta x^j}\log\sqrt{g} + g^{ij}\theta_i\,\frac{\delta}{\delta x^j}\log|K(x)| - \theta_i\dot{A}^i \\ &\qquad + h^{\mu\nu}\theta_\mu\,\frac{\partial}{\partial z^\nu}\log\sqrt{g} + h^{\mu\nu}\theta_\mu\,\frac{\partial}{\partial z^\nu}\log\sqrt{h}\Big]\, f\, K(x) d\,Vol_{\hat g}\,.\end{aligned}$$

To identify the horizontal codifferentials, let us set $\theta = \theta_i dx^i$. If we use d^*_H to denote the horizontal Riemannian codifferential and $d^*_{\tilde H}$ for the horizontal normalized codifferential, then

$$\begin{aligned}d^*_H\theta &= -\left(\frac{1}{\sqrt{g}}\frac{\delta}{\delta x^j}(\sqrt{g}\,g^{ij}\theta_i) - \theta_i\dot{A}^i\right)\,; \\ d^*_{\tilde H}\theta &= -\left(\frac{1}{\sqrt{g}}\frac{\delta}{\delta x^j}(\sqrt{g}\,g^{ij}\theta_i) - \theta_i\dot{A}^i\right) + g^{ij}\theta_i\,\frac{\delta}{\delta x^j}\log Vol(x)\,.\end{aligned}$$

Consequently, the horizontal Laplacians, where we use the sign convention $\Delta f = d^* df$, induced by these codifferentials are

$$\begin{aligned}\Delta_H f &= -\left[\frac{1}{\sqrt{g}}\frac{\delta}{\delta x^j}\left(\sqrt{g}\,g^{ij}\frac{\delta f}{\delta x^i}\right) - \dot{A}^i\frac{\delta f}{\delta x^i}\right]\,; \\ \Delta_{\tilde H} f &= -\left[\frac{1}{\sqrt{g}}\frac{\delta}{\delta x^j}\left(\sqrt{g}\,g^{ij}\frac{\delta f}{\delta x^i}\right) - \dot{A}^i\frac{\delta f}{\delta x^i}\right] + g^{ij}\,\frac{\delta f}{\delta x^i}\,\frac{\delta \log Vol(x)}{\delta x^j}\,.\end{aligned}$$

Compare with [AK].

A.3 The Laplacian on M

In contradistinction to the complicated situation on the total space, there is only one inner product on M, and the codifferential is more easily computed.

$$\begin{aligned}\langle d^*_M\theta, f\rangle &= \langle\theta, df\rangle = \int_M G^{ij}\theta_i\,\frac{\partial f}{\partial x^j}\,\sqrt{G}\,dx^1\wedge\cdots\wedge dx^n \\ &= -\int_M \frac{1}{\sqrt{G}}\,\frac{\partial}{\partial x^j}\left(\sqrt{G}\,G^{ij}\theta_i\right)\, f\,\sqrt{G}\,dx^1\wedge\cdots\wedge dx^n\,.\end{aligned}$$

The total derivative does not contribute, in view of the divergence lemma of §3.

Thus we have,

$$d_M^*\theta \;=\; -\,\frac{1}{\sqrt{G}}\,\frac{\partial}{\partial x^j}\left(\sqrt{G}\,G^{ij}\theta_i\right)$$

and, by taking $\theta = df$, we get

$$\Delta_M f \;=\; -\,\frac{1}{\sqrt{G}}\,\frac{\partial}{\partial x^j}\left(\sqrt{G}\,G^{ij}\frac{\partial f}{\partial x^i}\right).$$

This is not yet in the precise form we wish. Let us carry out the differentiation to get the friendlier expression

$$\Delta_M f \;=\; -\,G^{ij}\frac{\partial^2 f}{\partial x^i \partial x^j} \;-\; \frac{1}{\sqrt{G}}\,\frac{\partial}{\partial x^j}\left(\sqrt{G}\,G^{ij}\right)\frac{\partial f}{\partial x^i}\,.$$

Now we may begin to re-express this Laplacian in terms of those on the total space E. In particular, the first term in the previous formula is simply

$$-\,G^{ij}\frac{\partial^2 f}{\partial x^i \partial x^j} \;=\; -\,\frac{1}{\sqrt{G}}\int_{E_x} g^{ij}\,\frac{\partial^2 f}{\partial x^i \partial x^j}\,\sqrt{g}\,\zeta_x\,.$$

It is also the highest order terms in the horizontal Laplacians, integrated over the fibre in the proper way. For the second term, we are faced with computing $\frac{\partial}{\partial x^j}\int_{E_x} g^{ij}\sqrt{g}\,\zeta_x$, which is equal to

$$\frac{\partial}{\partial x^j}\int_{\mathcal{F}} \text{the pure } dz\text{-part of } \; g^{ij}\sqrt{g}\,\frac{(-1)^m}{Vol(x)}\sqrt{h}\,\delta z^1 \wedge \cdots \wedge \delta z^m\,.$$

The asserted equality follows from a local trivialization of E into $U \times \mathcal{F}$, where U is an open subset of M. Note that the domain of integration no longer depends on x. This sort of computation was precisely the concern in [BS]. The proof in that paper tells us that

Lemma 6. *For any* $\omega \in \Omega^m(E)$,

$$\frac{\partial}{\partial x^j}\int_{\mathcal{F}} \textit{the pure dz-part of } \omega \;=\; \int_{\mathcal{F}} \textit{the pure dz-part of } \; i_{\frac{\partial}{\partial x^j}}\, d\omega\,.$$

We hasten to point out a subtlety in this lemma. Namely, the integrands on the two sides *can* differ by exact divergences *along the fibre* $\mathcal{F}$. These have the typical form $\frac{\partial}{\partial z^\mu}(\cdots)^\mu$. They integrate to zero because the fibre is, by assumption, compact and boundaryless.

For us, $\omega := g^{ij}\sqrt{g}\,\frac{(-1)^m}{Vol(x)}\sqrt{h}\,\delta z^1\wedge\cdots\wedge\delta z^m$ and, in view of the lemma, we shall calculate $i_{\frac{\partial}{\partial x^j}}\,d\omega$. Since we only need this term modulo dx, let us first compute $d\omega$ modulo $dx\wedge dx$.

$$\begin{aligned}
d\omega &= d\left(g^{ij}\sqrt{g}\,\frac{(-1)^m}{Vol(x)}\sqrt{h}\,\delta z^1\wedge\cdots\wedge\delta z^m\right)\\
&= \frac{\delta}{\delta x^k}\left(g^{ij}\sqrt{g}\,\frac{(-1)^m}{Vol(x)}\sqrt{h}\right)dx^k\wedge\delta z^1\wedge\cdots\wedge\delta z^m\\
&\qquad + \mathcal{P}^{\mu}{}_{k\mu}\;dx^k\wedge\delta z^1\wedge\cdots\wedge\delta z^m \quad \text{(all summed in } k,\mu)\\
&\qquad\qquad\qquad\qquad\qquad\qquad\qquad\qquad \pmod{dx\wedge dx}\\
&= \left[\frac{\delta}{\delta x^k}(g^{ij}\sqrt{g}) - g^{ij}\sqrt{g}\,\dot{A}_k - g^{ij}\sqrt{g}\,\frac{\delta\log Vol(x)}{\delta x^k}\right]dx^k\wedge\zeta_x\\
&\qquad\qquad\qquad\qquad\qquad\qquad\qquad\qquad \pmod{dx\wedge dx}\,.
\end{aligned}$$

Next, we carry out the interior contraction with $\frac{\partial}{\partial x^j}$. Incidentally, it can be checked that $i_{\frac{\partial}{\partial x^j}}\,d\omega$ and $i_{\frac{\delta}{\delta x^j}}\,d\omega$ share the same pure dz-parts. Thus the interior contraction could have been carried out with $\frac{\delta}{\delta x^j}$ as well. In any case, after all is said and done, we get

$$\begin{aligned}
&-\frac{1}{\sqrt{G}}\,\frac{\partial}{\partial x^j}\left(\sqrt{G}\,G^{ij}\right)\frac{\partial f}{\partial x^i}\\
&\quad = -\frac{1}{\sqrt{G}}\int_{E_x}\left[\frac{1}{\sqrt{g}}\frac{\delta}{\delta x^j}(g^{ij}\sqrt{g}) - \dot{A}^i - g^{ij}\frac{\delta\log Vol(x)}{\delta x^j}\right]\frac{\partial f}{\partial x^i}\sqrt{g}\,\zeta_x\,.
\end{aligned}$$

Now, since $f\in\pi^*\Omega^0(M)$, $\frac{\partial f}{\partial x^i} = \frac{\delta f}{\delta x^i}$. Using this, our calculations result in

$$\begin{aligned}
\Delta_M f &= \frac{1}{\sqrt{G}}\int_{E_x}\left[-\frac{1}{\sqrt{g}}\,\frac{\delta}{\delta x^j}\left(\sqrt{g}\,g^{ij}\frac{\delta f}{\delta x^i}\right) + \dot{A}^i\,\frac{\delta f}{\delta x^i}\right.\\
&\qquad\qquad\qquad\left. + g^{ij}\,\frac{\delta\log Vol(x)}{\delta x^j}\,\frac{\delta f}{\delta x^i}\right]\sqrt{g}\,\zeta_x\\
&= \frac{1}{\sqrt{G}}\int_{E_x}\left[\Delta_H f \;+\; g^{ij}\,\frac{\delta\log Vol(x)}{\delta x^j}\,\frac{\delta f}{\delta x^i}\right]\sqrt{g}\,\zeta_x\\
&= \frac{1}{\sqrt{G}}\int_{E_x}\Delta_{\hat{H}} f\;\sqrt{g}\,\zeta_x\,.
\end{aligned}$$

A.4 Connections with Finsler Geometry

Now suppose M^n is equipped with a Finsler function F, and g is the fundamental tensor on $E = SM$. So, $\mathcal{F}$ has dimension $m = n-1$. In this case it is far more convenient to work in the *homogeneous* coordinates (y^i)

of SM rather than those intrinsic coordinates (z^μ). This will also enable us to make closer contact with the literature. For Finsler geometry, the nonlinear connection $\mathcal{N}$ is equal to F times the $N^i{}_j$ encountered in §1, and $HE = \text{span}\left\{\frac{\partial}{\partial x^j} - F\,N^i{}_j\,\frac{\partial}{\partial y^i}\right\}$.

The forms $\delta y^i := \frac{1}{F}(dy^i + F\,N^i{}_j\,dx^j)$ are the homogeneous-coordinate analogues of the δz^μ 's. . There are only $n-1$ linearly independent δy 's because $\ell_i\,\delta y^i = d\log F$ vanishes on SM. This fact will be pivotal after one takes the exterior derivative of δy^i and is ready to *interpret* the calculation as having occurred on SM. In particular, one finds that

$$\mathcal{P}^i{}_{jk} \;=\; \ell^r\,P_r{}^i{}_{jk} \;-\; \Gamma^i{}_{jk}\,,$$

where $\Gamma^i{}_{jk}$ are the connection coefficients of the Chern connection, and P is its hv-curvature tensor.

We define $\dot{A}_{ijk} = \nabla_{\hat{\ell}} A_{ijk}$, where $A_{ijk} = \frac{F}{2}\frac{\partial g_{ij}}{\partial y^k}$ is the Cartan tensor and $\hat{\ell} = \frac{y^i}{F}\frac{\delta}{\delta x^i}$ is the canonical line. The covariant derivative may be any one of Cartan type (such as the Chern connection), as they all coincide along $\hat{\ell}$. For all such covariant derivatives, a remarkable Bianchi identity says that $\ell^r P_r^i{}_{jk} \;=\; -\dot{A}^i{}_{jk}$.

Next, the bundle metric $\hat{g}$ is the Sasaki metric, pulled back onto SM. The resulting non-normalized fibre volume form is given by the expression $\sqrt{g}\sum_j \frac{(-1)^{j-1}}{F^n}\,y^j\,dy^1\wedge\cdots\wedge dy^{j-1}\wedge dy^{j+1}\wedge\cdots\wedge dy^n$. That sum multiplying the $\sqrt{g}$ is the homogeneous-coordinate version of $(-1)^m dz^1\wedge\cdots\wedge dz^m$. This observation allows us to identify the density factor $\sqrt{h}$ as $\sqrt{g}$.

Explicitly, the Chern connection is $\Gamma^i{}_{jk} = \frac{g^{si}}{2}\left(\frac{\delta g_{ij}}{\delta x^k} - \frac{\delta g_{jk}}{\delta x^i} + \frac{\delta g_{ki}}{\delta x^j}\right)$, from which it can be shown that $\Gamma^s{}_{ks} = \Gamma^s{}_{sk} = \frac{\delta}{\delta x^k}\log\sqrt{g}$. Also, recall the *abbreviation* $\dot{A}_j := -\mathcal{P}^s{}_{js} - \frac{\delta}{\delta x^j}\log\sqrt{h}$ that we introduced. We now see that it equals $(\dot{A}^s{}_{js} + \frac{\delta}{\delta x^j}\log\sqrt{g}) - \frac{\delta}{\delta x^j}\log\sqrt{g} = \dot{A}^s{}_{js}$, thereby justifying our previous notation.

Among Finsler spaces, some classes are identified for their nice structure. One of the most general classes of Finsler spaces which admit a Gauss-Bonnet theorem consists of those whose volume function, $Vol(x)$, is constant over M, [BC]. Such spaces are again being singled out here as precisely those in which the Riemannian horizontal Laplacian of SM coincides with the normalized horizontal Laplacian.

Within this class of spaces are those for which $\dot{A}_j = 0$. For these spaces, the horizontal Laplacian reduces to the familiar formula,

$$\Delta_H f \;=\; -\,\frac{1}{\sqrt{g}}\,\frac{\delta}{\delta x^j}\left(\sqrt{g}\,g^{ij}\,\frac{\delta f}{\delta x^i}\right).$$

One can expand this and write the derivatives in terms of the Cartan connection, and finds that this yields the generator for the h-stochastic diffusion described in [AZ].

References

[AIM] Antonelli, P.L., Ingarden, R.S. and Matsumoto, M. (1993) *The Theory of Sprays and Finsler Spaces with Applications in Physics and Biology*, D. Reidel and Kluwer Academic Press, 1993.

[AK] Anastasiei, M. and Kawaguchi, H. (1993) Absolute energy of a Finsler space can't be harmonique, *Tensor* (NS), **53**, 108-114.

[AZ] Antonelli, P.L. and Zastawniak, T.A. (1994) Stochastic Calculus on Finsler Manifolds and An Application in Biology, *Nonlinear World*, **1**, 149-171.

[BC] Bao, D. and Chern, S.S. (1996) A Note on the Gauss-Bonnet Theorem for Finsler Spaces, *Ann. Math.*, **143**, 233-252.

[BCS] Bao,D., Chern, S.S. and Shen, Z. (to appear) Rigidity Issues on Finsler Surfaces, *Revue Roumaine de Mathématiques Pures et Appliquées.*

[BL] Bao, D. and Lackey, B. (1996) A Hodge Decomposition Theorem for Finsler Spaces, *C.R. Acad. Sc. Paris*, **323**, 51-56.

[BS] Bao, D. and Shen, Z. (1994) On the Volume of Unit Tangent Spheres in a Finsler Manifold, *Results in Math.*, **26**, 1-17.

[HJ] Horn, R.A. and Johnson, C.R. (1985) *Matrix Analysis*, Cambridge.

[L] Lackey, B. (to appear) A Bochner Vanishing Theorem for Elliptic Complices, *(this volume).*

[M] Matsumoto, M. (1992) The Main Scalar of Two-Dimensional Finsler Spaces with Special Metric, *J. Math. Kyoto*, **32**, 889-898.

[R] Rund, H. (1959) *The Differential Geometry of Finsler Spaces*, Springer-Verlag.

SPINORS ON FINSLER SPACES

FRANCIS J. FLAHERTY

1 Introduction

In an earlier article, I considered the problem of constructing a Dirac operator on a Finsler space from the functional analytic viewpoint. The resulting theory is unwieldy since the Lichnerowicz formula involves curvature terms from a superconnection. The resulting index formula uses the Atiyah-Singer theorem for families, further complicating matters. Because this approach centers on the fibers, I have changed my focus to the pullback tangent bundle sitting over the projectivized tangent bundle or tangent sphere bundle of the Finsler space, depending on the precise type of homogeneity of the Finsler metric.

Indeed, the horizontal bundle and the vertical bundle split the tangent bundle of the slit tangent bundle of the Finsler space (tangent bundle with the zero section removed). Since the Cartan connection is a metric connection, a natural Dirac operator, exists on a Finsler space. Although other metric connections may be used, what the operator acts on are spinors constructed from the pullback tangent bundle.

The square of the Dirac operator (Lichnerowicz formula) contains torsion and curvature terms from the Cartan connection. In the future, I plan to relate this formula to the classical Bochner formula and the harmonic integral theory of Bao and Lackey (Cf. these proceedings).

2 Connections and Splittings

Suppose that $\mathbf{p}$ is the projection mapping of the slit tangent bundle $TM_o = TM - (0)$ and that π is the projection mapping of the tangent bundle. The pull-back bundle of the tangent bundle over $\mathbf{p}$ is denoted by

$$p^*(TM)$$

If you have a splitting of $T(TM_o)$

$$T(TM_o) = (\ker \mathbf{p}_*) \bigoplus \mathcal{H}$$

the space $\mathcal{H}_z$ is called the horizontal space at $z \in TM$ and $(\ker \mathbf{p}_*)_z$ is called the vertical space at z denoted by $\mathcal{V}_z$. This type of splitting can be mediated by a connection mapping, which is defined via a connection in the bundle $\mathbf{p}^*(TM)$:

$$\nabla_{\tilde{X}} V$$

for V a section of the pull-back and $\tilde{X}$ a vector field on TM_o. The connection mapping is then defined by

$$\mathcal{K}_z(\tilde{X}) = \nabla_{\tilde{X}} V$$

in which $\mathbf{p}(z) = v$ and V extends v. The horizontal space may then be defined as the kernel of the connection mapping. Refined versions of this construction use non-linear connections and will not be discussed further in this article.

3 Finsler and Riemannian Structures

Suppose that M is a Finsler manifold with Finsler metric, F, that is, a smooth function on TM such that

$$F(x, \lambda v) = \lambda F(x, v), \quad \lambda \in \mathbb{R}, \quad \lambda > 0.$$

The slit tangent bundle TM_o then has a Riemannian structure if you assume that the bilinear form

$$g_{ij}(x, v) = \frac{1}{2}(F^2(x, v))_{v_i v_j}$$

is positive definite. Here (x, v) are local fiber coordinates in TM_o or TM.

In fact, if dx^i denote the pullbacks of the corresponding forms on M and if

$$\vartheta^i := \delta^i v = dv^i + \gamma^i_{jk} v^k dx^j$$

the coframe fields $dx^i, \delta^i v$ form a local basis for the cotangent spaces to TM_o. The γ^i_{jk} are the formal Christoffel symbols associated to g_{ij}

$$\gamma^i_{jk} = \frac{1}{2}(g^{ir}(\partial_j g_{rk} + \partial_k g_{jr} - \partial_r g_{jk})$$

and the partial derivatives ∂_j indicate differentiation with respect to the pullback coordinates in TM_o.

With respect to this local coframe, the Riemannian structure on TM_o is defined by

$$ds^2 = g_{ij}dx^i \otimes dx^j + g_{ij}\vartheta^i \otimes \vartheta^j.$$

4 Spinors

To produce spinors on the pull-back bundle, you have to consider the bundle of Clifford algebras on TM_o. Now the Riemannian structure on TM_o, can be represented by the direct sum

$$g_{ij} \oplus g_{ij}.$$

More precisely, with respect to the local coframe defined in the previous section

$$ds^2 = g_{ij}dx^i \otimes dx^j + g_{ij}\vartheta^i \otimes \vartheta^j.$$

The direct sum depends on the notion of horizontal subspace or connection. Given a splitting of TM_o you get

$$Cl(TM_o) = Cl(\mathbf{p}^*(TM))\widehat{\bigotimes}Cl(\mathcal{H})$$

where Cl denotes the Clifford bundle and the tensor product is the graded tensor product over $\mathbb{Z}/2$. The spinors on TM_o can then be considered as the tensor product of the spinors on the pull-back with the vertical spinors. Note that the Riemannian structure on TM_o induces a Riemannian structure on both vertical and horizontal spaces so this procedure makes sense. I am indebted to Peter Antonelli for telling me about the work of S. Vacaru, which follows a somewhat similar path, although Vacaru concentrates on splittings of vector bundles. In earlier work, I ignored the metric on $\mathbf{p}^*(TM)$, considering the totally degenerate metric on $\mathbf{p}^*(TM)$. In that case,

$$Cl(\mathbf{p}^*(TM) = \wedge\mathbf{p}^*(TM)$$

the pull-back of the exterior algebra on M.

5 Differential Operators

Let me digress to develop a formula for the Dirac operator, with respect to a metric connection with torsion, on any spin manifold X. So suppose that $\tilde{\nabla}$ is the Levi-Civita connection on X and suppose that ∇ is any metric connection on X with torsion τ. It is essential that the connections involved with spinors be metric for then they extend to the Clifford bundle. Further the assumption that X be spin simply implies that

$$\tilde{\nabla}\mathbf{c} = 0$$

where $\mathbf{c}$ denotes Clifford multiplication. Returning to the two connections define the difference tensor B by the equation

$$\nabla = \tilde{\nabla} + B$$

Using the connection form matrices $(\tilde{\omega})$ and (ω) for the covariant derivatives $\tilde{\nabla}$ and ∇ you can easily see that

$$\tau^i = d\omega^i + \omega^i_j \wedge \omega^j = d\omega^i + \tilde{\omega}^i_j \wedge \omega^j + b^i_j \wedge \omega^j = b^i_j \wedge \omega^j$$

in which (ω^j) is the dual coframe to the orthonormal frame field (e_i). Recall that the Dirac operator $\mathcal{D}$ applied to a spinor field s is defined by

$$\mathcal{D}s = e_i \tilde{\nabla}_{e_i} s.$$

Proposition 1.

$$\mathcal{D} = e_i \nabla_{e_i} + \frac{1}{2}\mathbf{c}(B(e_i)e_i)$$

using the notation introduced previously.

If you express the torsion tensor τ in a local coframe

$$\tau^i = \tau^i_{jk}\omega^j \wedge \omega^k.$$

From the expression for τ define a differential form T via lowering indices, so

$$T = t_{ijk}\omega^i \wedge \omega^j \wedge \omega^k = g_{il}\tau^l_{jk}\omega^i \wedge \omega^j \wedge \omega^k.$$

Again all this amounts to doing is lowering indices and anti-symmetrizing.

Theorem 2.

$$\mathcal{D}^2 = \sum(\tilde{\nabla}_{e_i} + \frac{1}{4}\mathbf{c}(\iota_{e_i}T))^2 + \frac{1}{4}\kappa + \frac{1}{4}\mathbf{c}(dT) - \frac{1}{8}\langle T, T\rangle.$$

where κ is the scalar curvature of M.

More general formulae such as these occur in the work of J.-M. Bismut.

Returning to the case at hand, the space TM_o supports several covariant derivatives.

First, the $(0,2)$-tensor field

$$g_{ij}dx^i \otimes dx^j + g_{ij}dv^i \otimes dv^j$$

defines a Riemannian structure on TM_o by the Finslerian condition. Consequently this Riemannian structure determines a Levi-Civita connection, denoted by, $\tilde{\nabla}$.

Next, there is the Cartan connection on $T(TM_o)$, which can be described by its effect on horizontal and vertical vectors. The development I use goes back to Akbar-Zadeh. First, some preparation on torsion, again.

The torsion of a connection (ω^i_j) on the pull-back bundle $\mathbf{p}^*(TM)$ is defined by

$$\tau^i = d\vartheta^i + \omega^i_j \wedge \vartheta^j$$

where ϑ^i is the pull back of a coframe field on M. The torsion can then be decomposed

$$\tau^i = \frac{1}{2}S^i_{jk}\vartheta^i \wedge \vartheta^k + U^i_{jk}\nu^k \wedge \vartheta^j$$

where $\nu^k = dv^k + \omega^k_j v^j$. The v^j are the fiber coordinates in TM_o.

Theorem 3. *For any Finsler manifold M there is a unique metric connection, ∇, whose torsion satisfies*

$$S_{ijk} = 0, \quad U_{ijk} = U_{jik}.$$

As usual the indices are lowered with the aid of the metric tensor (g_{ij}).

A similar fact holds for the vertical bundle, that is, there is a connection, ∇, on $\mathcal{V}$ such that $\nabla g = 0$ and the $v - v$ torsion of ∇ vanishes. Matsumoto characterizes such pairs of Finsler connections by an additional concept of null deflection which does not enter here. The resulting connection is called the Cartan connection of the Finsler manifold, denoted by, ∇.

Now the difference between the Cartan and Levi-Civita connections can be written as

$$\nabla = \tilde{\nabla} + B.$$

Consider now a local orthonormal frame field (e_a) on TM_o, adapted to the splitting, where a goes from 1 to $2n$.

The torsion of ∇ satisfies

$$\tau^a = \tau^a_{i,n+j}\vartheta^i \wedge \nu^j.$$

and the Dirac operator, $\mathcal{D}$ satisfies

$$\mathcal{D}(\tilde{\nabla}) = \mathcal{D}(\nabla) + \mathbf{c}(B(e_i)e_{n+i})$$

where $\mathcal{D}(A)$ is the Dirac operator of the metric connection A. To apply the theorem, T must be redefined as

$$T = g_{ab}\tau^{b}_{i,n+j}\omega^{a} \wedge \vartheta^{i} \wedge \vartheta^{j}$$

where

$$\omega^{a} = (\vartheta^{i}, \nu^{i}).$$

The main theorem on $\mathcal{D}^2$ yields a formula involving the scalar curvature of TM_o and since the Cartan scalar curvature is minorized by it, no theorem is readily available.

Finally, a horizontal Dirac operator does occur. In fact, define the operator $\mathcal{D}^h$ on $S(\mathcal{H})$ via

$$\mathcal{D}^h = e_i \tilde{\nabla}_{e_i}$$

using the Cartan connection. It may happen that this horizontal Dirac operator provides the link with the Bao-Lackey-deRham complex.

References

[1] Akbar-Zadeh, H. (1963) Les Espaces de Finsler et Certaines de Leurs Généralisations, *Ann. de l'É.N.S.*, **80**, 1–79.

[2] Bismut, J.-M. (1989) A Local Index Theorem for Non-Kähler Manifolds, *Math. Ann.*, **284**, 681–699.

[3] Flaherty, F.J. (1996) Dirac Operators on Finsler Spaces, *Contemporary Mathematics*, **196**, 59-66.

[4] Matsumoto, M. (1986) *Foundations of Finsler Geometry and Special Finsler Spaces*, Kaiseisha Press, 108–115.

[5] Vacaru, S. (1996) Spinor Structures and Nonlinear Connections in Vector Bundles, Generalized Lagrange and Finsler Spaces, *J. Math. Phys.*, **37**, 508–523.

www.ingramcontent.com/pod-product-compliance
Ingram Content Group UK Ltd.
Pitfield, Milton Keynes, MK11 3LW, UK
UKHW020134290926
15471UKWH00003B/487
* 9 7 8 9 4 0 1 1 5 2 8 3 9 *